Body Contouring Following Bariatric Surgery And Massive Weight Loss

Post-Bariatric Body Contouring

Edited by

Bishara S. Atiyeh, MD, FACS

Professor of Plastic & Reconstructive Surgery
American University of Beirut, Lebanon
General Secretary
Euro- Mediterranean Council for Burns and Fire Disasters – MBC
Beirut, Lebanon

Co-Editor

M. Costagliola, MD

Professor of Plastic & Reconstructive Surgery
Scientific Coordinator
Euro- Mediterranean Council for Burns and Fire Disasters – MBC
Toulouse, France

CONTENTS

FOREWORD

Individuals who lose weight either through diet and exercise or surgical intervention physically feel better. However, rather than having their efforts rewarded and looking better, most have to deal with excessive folds of lax skin all over their body.These individuals are now seeking surgical removal of excess folds in order to restore their self confidence and body image. Our specialty has responded well to this growing demand. The need is not limited to one country or one continent. Obesity is a worldwide epidemic!

Body contouring following bariatric surgery and massive weight loss has become one of the most rapidly growing fields within plastic surgery. The techniques which have rapidly evolved over the past few years are now established and have a record of safety and efficacy.

In this timely publication Dr. Bishara Atiyeh and Michel Costagliola, who are themselves leaders and contributors in this field, bring a truly international group of authors together from all across the globe. Each and every one is an innovator, who has contributed to rapid advances in this most exciting field.

Because traditional or classical body contouring procedures are not suitable for the massive weight loss patient's, procedures described in this e-Book were developed. Procedures which maximize the removal of excess and unwanted skin whilst preserving volume and restoring contour. Patient safety with minimization of surgical risk is paramount for any operative procedures and is particularly so in this group of patients who may well be under nourished and otherwise at high risk. The issue of patient safety is addressed well by these experts and an entire chapter is devoted to this subject.

I commend the two editors and all the contributors for this comprehensive, beautifully illustrated, well written volume which will further advance the art and science of recontouring following massive weight loss.

Foad Nahai
ISAPS President

PREFACE

Morbid obesity is a chronic condition that is extremely difficult to treat. Unfortunately it is growing at an alarming rate in all age groups and is becoming a real epidemic in Europe and North America. Bariatric surgery is becoming the current standard treatment for severe obesity and its overall safety and effectiveness continues to improve. Even though success of bariatric surgery can be expressed by net reduction of BMI (Body Mass Index) global patient satisfaction does not usually exceed 6.2 on a visual analogue scale (VAS) of 10. Hand in hand with massive weight loss obtained with bariatric procedures comes the need to effectively and aesthetically manage the excess skin that remains as more patients are at present seeking the help of plastic surgeons to address the resulting aesthetic concerns.

Massive weight loss (MWL) patients present a very different profile than those who have not been obese. Their deformities are more severe and present with bi-dimensional skin excess which is usually poor in tone. Patients who experience massive weight loss are left looking "deflated" with disfiguring skin laxity circumferentially around the torso including the breasts in addition to redundant tissues on the upper arms, buttocks, and thighs leading to poor social acceptance and quality of life. To answer these problems, body contouring is mandatory. When expertly and effectively performed global satisfaction of patients increases to 8.5 VAS.

MWL patients present extremely diverse, disordered, and often complex and unpredictable contour deformities requiring lifts in areas previously rarely seen. Body contouring for these patients is unlike routine cosmetic surgery. It is evolving from traditional body contouring procedures, in particular from abdominoplasties, to a series of very specialized surgical procedures designed to aesthetically improve specific areas of the body. The subspecialty of body contouring surgery after massive weight loss is indeed a rapidly evolving field of plastic surgery and as an individual plastic surgeon continue to gain more experience with the massive weight loss patient and continues to refine their techniques, it is important that new learning and insights are shared within aesthetic surgical subspecialty.

Bishara S. Atiyeh, MD, FACS
Professor of Plastic & Reconstructive Surgery
American University of Beirut, Lebanon
General Secretary
Euro- Mediterranean Council for Burns and Fire Disasters – MBC
Beirut, Lebanon

List of Contributors

Agullo, Francisco J.

Department of Surgery, Texas Tech University Health Sciences Center
1800 Alberta Avenue
El Paso, Texas, 79912, USA

Atiyeh, Bishara S.

Clinical Professor
Plastic and Reconstructive Surgery
American University of Beirut
Beirut, Lebanon

Ayoub, Chakib

Associate Professor of Anesthesiology
American University of Beirut Medical Center
Beirut, Lebanon

Dibo, Saad

Resident
Plastic and Reconstructive Surgery
American University of Beirut
Beirut, Lebanon

Gheita, Alaa

Emeritus Professor Plastic and Reconstructive Surgery
Faculty of Medicine Cairo University Egypt
Cairo, Egypt

Habal, Mutaz B.

Director, Tampa Bay Craniofacial Center
Research Professor, University of South Florida
Adjunct Professor of Material Science, University of Florida
205 W. Dr. Martin L. King, Jr. Blvd Suite 103
Tampa, FL 33603, USA

Hamdi, Moustapha.

Professor and Chairman of Plastic Surgery Department
Brussels University Hospital
Laarbeeklaan 101
1090 Brussels
Belgium

Hayek, Shady N.

Assistant Professor
Plastic and Reconstructive Surgery
American University of Beirut
Beirut, Lebanon

Ibrahim, Amir E.
Chief Resident

Plastic and Reconstructive Surgery
American University of Beirut
Beirut, Lebanon

Karam, Cynthia J.

Resident
Department of Anesthesiology
American University of Beirut
Beirut, Lebanon

le Roux, Carel W.

Department of Investigative Medicine,
Imperial College Faculty of Medicine, Hammersmith Campus,
Du Cane Road, London W12 ONN, UK

Magdaleno, Ronis Júnior

Department of Medical Psychology and Psychiatry,
Faculty of Medical Sciences,
University of Campinas—UNICAMP,
Campinas, Brazil

Masri, Sami

Research Fellow
Surgery Department
American University of Beirut
Beirut, Lebanon

Morales, Gracia Héctor Javier

Belisario Dominguez 2501, Col.
Obispado, C.P. 64060,
Monterrey, N.L., Mexico

Pournaras, Dimitrios J.

Department of Bariatric Surgery, Musgrove Park Hospital, Taunton, UK
Department of Investigative Medicine,
Hammersmith Hospital, Imperial College
London, UK

Rebane, Mari

Fellow Division of Plastic and Reconstructive Surgery University of Miami School of Medicine
1475 NW 12 Ave.
Miami, Florida 33136, USA

Rieger, Ulrich M.

Department of Plastic, Aesthetic & Reconstructive Surgery
Innsbruck Medical University
Anichstrasse 35
6020 Innsbruck
Austria

Rizk, Marwan S.

Instructor

Department of Anesthesiology
American University of Beirut
Beirut, Lebanon

Rueda, Steven

Third year medical student
University of Miami School of Medicine
Miami, Florida, 33131, USA

Safadi, Bassem.

Associate Professor of Clinical Surgery
American University of Beirut Medical Center
Beirut, Lebanon

Sarhane, Karim A.

Master of Science student
Department of Biochemistry and Molecular Genetics
American University of Beirut
Beirut, Lebanon

Sozer, Sadri O.

El Paso Cosmetic Plastic Surgery Center
1600 Medical Center Suite 400
El Paso, Texas, 79902, USA

Thaller, Seth R.

Professor and Chief, Division of Plastic Surgery
The DeWitt Daughtry Family Department of Surgery
University of Miami School of Medicine
Miami, Florida, USA

Tuncer, Serhan

Department of Plastic Surgery - Gent University Hospital
Gent, Belgium
Medical Institute Edith Cavell
Brussels, Belgium

von Finckenstein, Joackim

Department of Plastic and Aesthetic Surgery
Wittelsbacherstrasse 2a, 82319 Starnberg
Klinikum Starnberg, Germany

2

CHAPTER 1

Overweight and Obesity: A True Global Epidemic

Bishara S. Atiyeh* and Karim A. Sarhane

American University of Beirut Medical Center, Beirut, Lebanon

Abstract: Overweight and obesity have reached epidemic proportions, with more than 1 billion adults overweight globally. Childhood obesity is already epidemic in some areas and on a rise in others. An estimated 22 million children under age five are considered to be overweight worldwide. The rising obesity epidemic reflects profound changes in society and in the behavioral patterns of communities over the recent decades. There is a variety of factors that play a role in obesity which makes it a complex health issue to address; nevertheless, overweight and obesity are primarily diet-induced. The principal causes of the epidemic are first a sustained excess in ingestion of energy-dense foods and second an increasingly sedentary lifestyle. Although obesity should be considered an avoidable chronic disease in its own right, it is also one of the most substantial key risk factors for other chronic diseases. The morbidities associated with obesity reduce a patient's quality of life and contribute to escalating medical costs.

Keywords: Obesity, overweight, body mass index.

1. INTRODUCTION

Although the cause of obesity is multifaceted, it is clear that chronic overconsumption of food plays a fundamental role [1]. The foods we eat every day contribute to our well-being. Foods provide us with the nutrients we need for healthy bodies and the calories we need for energy. If we eat too much, however, the extra food turns to fat and is stored in our bodies. Initially fat cells increase in size. When they can no longer expand, they increase in number. If we overeat regularly, excess fat causes weight gain and we may become obese. When this type of overeating becomes compulsive and out of control, it is often classified as a "food addiction" [1, 2].

2. OVERWEIGHT AND OBESITY MEASUREMENT

Human Body Shape (HBS; body mass relative to standing height) is of interest in various scientific disciplines. Exact determination of body composition to define the quantity and distribution of muscle and fat requires complex measurements unfortunately not available in the clinical setting. HBS can be quantified, however, using several metrics. Customarily used methods are based on geometric similarity models that use readily obtainable data: body mass (M) and body height (H). For adults, overweight and obesity ranges are usually determined by using weight and height to calculate a number called the "Body Mass Index" (BMI) [3] defined as $BMI = M/H^2$. Overweight and obesity are both labels for ranges of weight that are greater than what is generally considered healthy for a given height (Tables **1** and **2**). The terms also identify ranges of weight that have been shown to increase the likelihood of certain diseases and other health problems [3, 4].

Table 1: BMI Calculation Formula.

Measurement Units	BMI Formula
Kilograms and meters (or centimeters)	$BMI = \text{weight (kg)} / [\text{height (m)}]^2$
Pounds and inches	$BMI = \text{weight (lb)} / [\text{height (in)}]^2 \times 703$

*****Address correspondence to Bishara S. Atiyeh:** American University of Beirut Medical Center, Beirut, Lebanon; Tel: 961-3-340032; E-mail: batiyeh@terra.net.lb

Table 2: Classification of Overweight and Obesity.

BMI	
Below 18.5	Underweight
18.5 to 24.9	Healthy weight
25.0 to 29.9	Overweight
30 to 34.9	Obese: Class I Obesity
35to 39.9	Class II Obesity
40 to 49.9	Morbidly Obese: Class III Obesity
> 50	Super Obese

BMI is an indicator that measures body fat indirectly and correlates well to direct measures of body fat, such as skinfold thickness measurements (with calipers), underwater weighing, bioelectrical impedance, dual-energy x-ray absorptiometry (DXA), and computerized tomography. BMI calculation is an inexpensive and easy-to-perform method of screening for weight categories that may lead to health problems. It is a screening tool, not a diagnostic tool. The correlation between the BMI number and body fatness is fairly strong. Despite the fact that this correlation varies by sex, race, and age, for adults 20 years old and older, BMI is interpreted using standard weight status categories that are the same for all ages and for both men and women. However, at the same BMI, women tend to have more body fat than men, older people, on average, tend to have more body fat than younger adults, and highly trained athletes may have a high BMI because of increased muscular mass rather than increased body fat [5].

For children and teens, the interpretation of BMI is both age- and sex-specific and is often referred to as BMI-for-age [5]. The BMI number is plotted on the CDC (center for disease control) BMI-for-age growth charts (for either girls or boys) to obtain a percentile ranking. Percentiles are the most commonly used indicator to assess the size and growth patterns of individual children. A percentile indicates the relative position of the child's BMI number among children of the same sex and age [5, 6]. (Table **3**, Fig. **1**).

Table 3: BMI-for-Age Weight Status Categories and the Corresponding Percentiles.

Weight Status Category	Percentile Range
Underweight	Less than the 5th percentile
Healthy weight	5th percentile to less than the 85th percentile
Overweight	85th to less than the 95th percentile
Obese	Equal to or greater than the 95th percentile

Although the World Health Organization has adopted BMI to quantify obesity, BMI remains a misunderstood empiric 19th-century observation that may be an illogical parameter for this task [7]. It appears also that the ability of BMI growth charts to detect obesity in children is limited because of the confounding influence of growth on HBS. The use of BMI may be justified in adults, because an increase in body dimensions can occur only in 2 dimensions, whereas growth occurs in 3 dimensions in children [4]. It may not be apparent also that, for objects of identical shape and density, BMI is directly and exactly proportional to height. Differences in height may overwhelm differences in thickness: Short overweight patients may have a lower BMI than tall thin patients. Because volume and mass are cubic functions of the linear dimension, the ponderal index ($PI = kg/m^3$) may be a more appropriate index being a statistic proportional to the cube of the height instead of the square of the height. It depends on shape but is insensitive to height [7]. For children, development of sex-specific HBS indices ($HBSI_G$ for girls and $HBSI_B$ for boys), based on powers other than 2 or 3 (of body height as a function of body mass) and on derived models of human growth may produce substantial reductions in index variability and sensitivity to age-related factors and could be a better method to determine obesity in children [4].

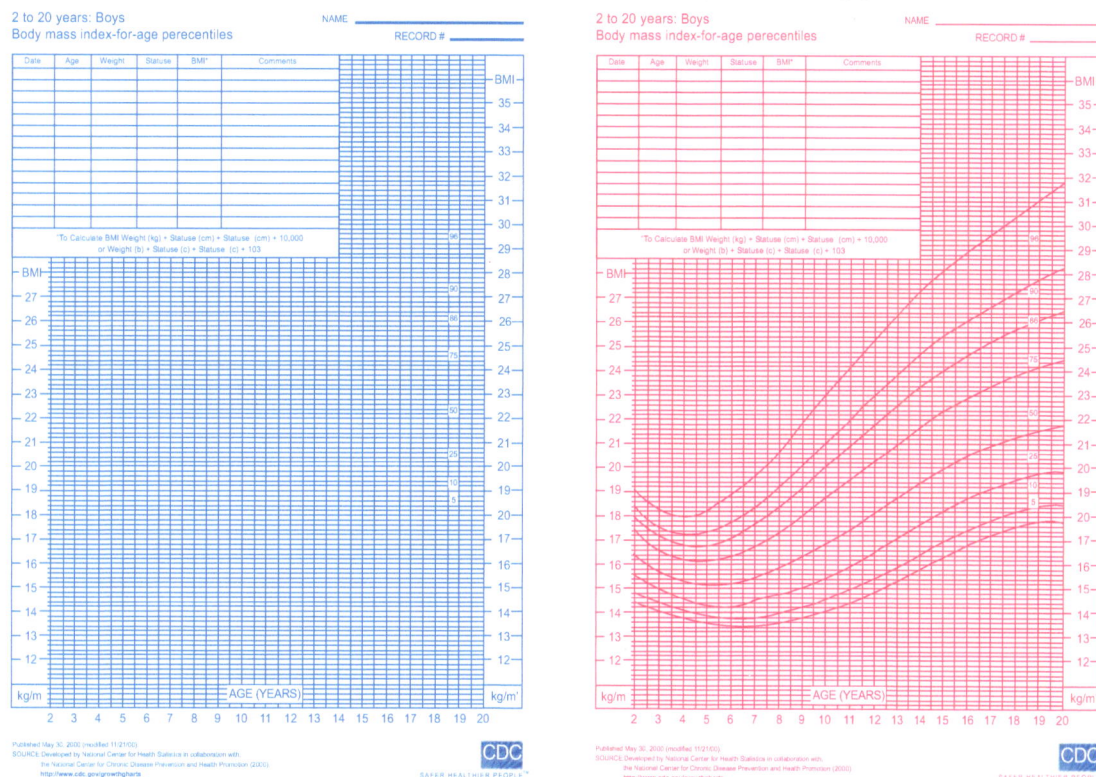

Fig. 1. Body Mass Index-For-Age Percentile Charts for Boys and Girls in the United States Available at: CDC Growth Charts: United States http://www.cdc.gov/growthcharts/.

Irrespective, BMI has probably become the most common indicator used to assess overweight and obesity in a wide variety of settings, including clinical, public health, and community-based programs. Although it is certainly not a perfect surrogate for total body fatness and not without its technical limitations, BMI has been recommended as the most appropriate single indicator of overweight and obesity for adults and children alike. There is little evidence that other indexes or measures of body fat such as skinfolds, waist circumference, or bioelectrical impedance are sufficiently practicable or provide appreciable added information. These alternative measures of fatness remain, however, important for research and perhaps in some specialized screening situations [8].

3. THE OBESITY EPIDEMIC

Obesity has reached epidemic proportions globally, with more than 1 billion adults overweight - at least 300 million of them clinically obese. In some countries rates have doubled over just a few years. Obesity is rising at an alarming rate throughout Europe constituting a pan-European epidemic. Rates have risen threefold or more since 1980 in some areas of North America, the United Kingdom, Eastern Europe, the Middle East, the Pacific Islands, Australasia and China. During the past 20 years there has been a dramatic increase in obesity in the United States where it is considered a chronic (long-term) disease, like high blood pressure or diabetes. In 2007, only one state had a prevalence of obesity less than 20%. Thirty states had a prevalence equal to or greater than 25%; three of these states had a prevalence of obesity equal to or greater than 30%. In many countries now significantly more than half the adult population is overweight and up to 30% of adults are clinically obese. The obesity epidemic is not restricted to industrialized societies and the increase in obesity rates is often faster in developing countries than in the developed world. In countries undergoing nutritional transition, over-nutrition and obesity often co-exists with under-nutrition. Current obesity levels range from below 5% in China, Japan and certain African nations, to over 75% in urban Samoa. But even in relatively low prevalence countries like China, rates are almost 20% in some cities [2, 3, 9].

Childhood obesity is already epidemic in some areas and on the rise in others. An estimated 22 million children under five are considered to be overweight worldwide. In the USA, the prevalence of obese children aged 6-to-11 years has increased considerably since the 1960s. According to the US Surgeon General, the number of overweight children has doubled and the number of overweight adolescents has trebled since 1980. Obesity prevalence in youths aged 12-17 has increased dramatically from 5% to 13% in boys and from 5% to 9% in girls between 1966-70 and 1988-91. The problem is global and increasingly extends into the developing world; for example, in Thailand the prevalence of obesity in 5-to-12 year olds children rose from12.2% to 15.6% in just two years [10].

4. ETIOLOGY OF OBESITY AND THE CONCEPT OF "FOOD ADDICTION"

Overweight and obesity are primarily diet-induced and result from an energy imbalance. Caloric balance is like a scale. To remain in balance and maintain body weight, caloric intake (from foods) must be balanced by the calories consumed (in normal body functions, daily activities, and exercise) [3].

Changing Lifestyle

Reasons for caloric imbalance vary by individual. However, beyond this equation, obesity is a complex mixture of psychological, environmental, social, cultural, economic, geographic, and genetic influences which makes it a difficult health issue to address [2, 3, 11-13]. The rising obesity epidemic in recent decades reflects profound changes in society and in behavioral patterns of communities in terms of economic growth, modernization, urbanization and globalization of food markets [10]. The principal causes of the epidemic are first a sustained excess in ingestion of energy-dense foods, and second an increasingly sedentary lifestyle and changing environments which curtail opportunities for physical activity [9, 10]. As incomes rise and populations become more urban, diets high in complex carbohydrates give way to more varied diets with a higher proportion of fats and sugars with insufficient consumption of fruits and vegetables. At the same time, large shifts towards less physically demanding work have been observed worldwide. Moves towards less physical activity are also found in the increasing use of automated transport, technology in the home, and more passive leisure pursuits [10]. Observations demonstrate, however, that the modern way of living has favored a change in human activities whose impact goes well beyond what has traditionally been attributed to a lack of physical exercise [14]. An increase in knowledge-based work characteristic of modern lifestyle appears as a stimulus favoring a significant enhancing effect on food intake. Television viewing is another example of sedentary activity that has been shown to increase the intake of high-density foods [14]. Even more intriguing are findings that changing the circadian rhythm in experimental animals by creating artificially shorter days induces weight gain, changes in body temperature rhythms, and levels of metabolic hormones (including leptin and insulin). These findings may be relevant to humans, given the widespread occurrence of both shift work and jet lag [15].

Neurologic Control of Energy Homeostasis

Obesity as a biological problem rather than an issue of overindulgence has been largely supported by recent experimental studies [15]. There is increasing evidence that obesity is under the control of multiple mechanisms in the brain often working in a "parallel" manner to regulate appetite and weight [16]. The hypothalamus is the specialized brain area that integrates the control of energy homeostasis, regulating both food intake and energy expenditure. Clinical data suggest that, once provided with sufficient calories to gain weight, some individuals have a predisposition to develop and maintain an elevated body weight set-point mediated by an integrated neural network. Several orexigenic and anorexigenic neuropeptides in the brain are involved, although their relative contributions are different. Hypothalamic neurons respond to peripheral signals, such as hormones and nutrients, by modifying the synthesis of neuropeptides acting as neurotransmitters and neuromodulators. Moreover, obese subjects have increased metabolism in the somatosensory cortex, which suggests an enhanced sensitivity to the sensory properties of food. Genetic, environmental, as well as pharmacologic factors all alter neural pathways involved in energy homeostasis by modulating lipid metabolism in the hypothalamus. Furthermore, neuronal histamine and its receptors widely distributed throughout the brain are also involved in the regulation of obesity [11, 12, 17, 18].

Genetic Influence

Science shows that genetics plays a role in obesity. A majority of human obesity is inherited as a polygenic trait, however, genes and behavior may both be needed for a person to become overweight [3, 12]. Obesity tends to run in families. This may be caused both by genes and by shared diet and lifestyle habits [2]. Possible mechanisms for overeating and for the preference of high-sugar and high-fat foods are innate biological factors that interact with the environment to predispose individuals to overeat. Studies have shown that humans who were heterozygous for the gene CD36, the expression of which is related to a receptor in the tongue found to play a role in fat detection and preference, could not discriminate fat and were more likely to eat more high-fat meats, sweets, and spreads. It was recently suggested also that persons with a gene variant associated with fewer dopamine receptors in the caudate nucleus, part of the brain's reward system and an important area for habit learning, predicts an individual's susceptibility for future weight gain [15]. Epigenetic changes (heritable changes in gene expression that act independently of changes in DNA sequence) may also play a role in obesity. Similarly, studies suggest that a high-fat diet in pregnant and lactating mice can affect their pups' brain development, increasing their vulnerability to becoming obese and to engaging in addiction like behaviors in adulthood [15].

Gut Bacteria

Recent studies have shown that the intestinal bacteria in obese individuals differ from those in lean persons. This could trigger a low-grade systemic inflammation. Moreover, gut bacteria that participate in food digestion play a critical role in the onset of obesity probably by inducing failure of homeostatic mechanisms that regulate appetite, food consumption, and energy balance. Incretins, cholecystokinin, brain-derived neurotrophic factor, leptin, long-chain fatty acid coenzyme A, endocannabinoids and vagal neurotransmitter acetylcholine all play a role in the regulation of energy intake, glucose homeostasis, insulin secretion, and pathobiology of obesity and type 2 diabetes mellitus indicating a cross-talk among the gut, liver, pancreas, adipose tissue, and hypothalamus [19].

Associated Morbid Conditions

Certain medical conditions and medications, such as hypothyroidism, Crushing syndrome, steroids, antidepressants, and birth control pills, can also cause or promote obesity, although these are much less common causes of obesity than overeating and inactivity [3].

"Food Addiction"

The concept "food addiction", or more accurately addiction to specific components of food, can be described in much the same way as other addictive behaviors. Studies suggest that multiple but similar brain circuits are disrupted in obesity and drug addiction. Gastric stimulation in obese subjects activates cortical and limbic regions involved with self-control, motivation, and memory. These brain regions are also activated during drug craving in drug-addicted subjects [18]. Highly palatable foods produce pleasure and reduce pain in a manner similar to other addictive substances. Both food and drugs induce tolerance over time, and withdrawal symptoms, such as distress and dysphoria, often occur upon discontinuation of the drug or during dieting [1]. From an evolutionary perspective, it seems that the mesolimbic reward pathway has evolved to reinforce the motivation to approach and engage in naturally rewarding behaviors such as eating highly palatable foods rich in fat and sugar that can be rapidly converted into energy, thereby promoting survival in times of famine. Specific areas of the mesolimbic system, such as the caudate nucleus, the hippocampus and the insula, are activated similarly by drugs and by food both of which are considered pleasurable behavioral activities. Both stimulate also the release of striatal dopamine, a fundamental neurotransmitter in the reward system. Endogenous opiates, another group of players in the reward pathways, are also activated by drugs and by foods, especially sweet foods, whereas naltrexone, an opioid blocker, has been shown to reduce cravings for both [1]. Obese individuals display more impulsive behaviors and tend to be more sensitive to reward and punishment. For these individuals, the forces driving food consumption are likely to go beyond physiologic hunger [1].

Over the last few generations our food environment has changed radically. Food is available more abundantly and certain foods have been artificially modified to enhance their rewarding properties. However, not everyone who is exposed to drugs becomes an addict, and, similarly, not everyone exposed to high-fat, high calorie foods becomes a compulsive overeater. Vulnerability may stem from various personality traits but may also be attributed, in part, to a genetic predisposition and/or to brain adaptations and specifically to downregulation of the dopamine (DA) D2 receptors linked to addictive behavior [1]. DA is a neurotransmitter known to play a major role in motivation that is involved with reward and prediction of reward. Brain imaging studies show that obese individuals have significantly lower D2/D3 receptor levels, similar to drug-addicted subjects, which make them less sensitive to reward stimuli. In turn this may predispose obese subjects to seek food as a means to temporarily compensate for understimulated reward circuits and would make them more vulnerable to food intake. The decreased D2/D3 receptor levels are also associated with decreased metabolism in prefrontal regions of the brain involved in inhibitory control and processing of food palatability, which may underlie inability of obese individuals to control food intake. The reduction in DA D2 receptors in obese subjects coupled with the enhanced sensitivity to food palatability could make food their most salient reinforcer putting them at risk for compulsive eating and obesity [17]. Although considering compulsive overeating as an addiction may not yet be accepted, evidence highlighting the addictive component of obesity and the role played by biologic vulnerability and environmental triggers cannot continue to be overlooked [1].

5. IMPACT OF OBESITY ON HEALTH

Obesity is a global health problem [1]. It is the second leading cause of preventable deaths in the United States [2]. Morbidities associated with obesity reduce a patient's quality of life and contribute to escalating medical costs [13]. Although it should be considered an avoidable chronic disease in its own right, obesity is also one of the substantial key risk factors for other chronic diseases together with smoking, high blood pressure and high blood cholesterol [6]. The non-fatal, but debilitating health problems associated with obesity include respiratory difficulties, chronic musculoskeletal problems, skin problems (Fig. **2**), and infertility. The more life-threatening problems fall into four main areas: cardiovascular diseases; conditions associated with insulin resistance such as type 2 diabetes; gallbladder disease and certain types of cancers [9, 10].

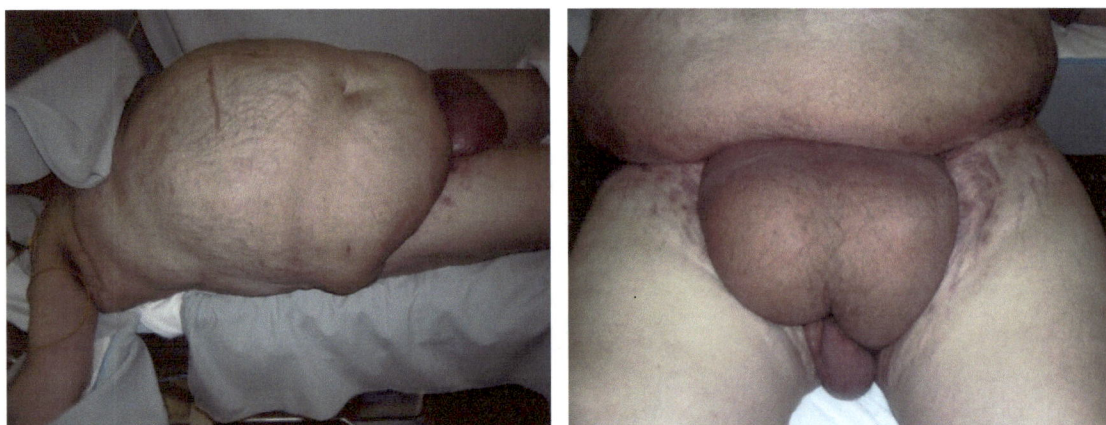

Fig. 2. Severe Cellulitis with Impending Skin Necrosis in Suprapubic Area in a Male Patient with Morbid Obesity.

Overweight and obesity lead to adverse metabolic effects on blood pressure, cholesterol, triglycerides and insulin resistance. A weight gain of 11 to 18 pounds increases a person's risk of developing type 2 diabetes to twice that of individuals who have not gained weight. More than 80 percent of people with diabetes are overweight or obese and approximately 85% of people with diabetes are type 2, and of these, 90% are obese or overweight [20]. While the link between obesity and type 2 diabetes is clear on an epidemiological level, the underlying mechanism linking these two common disorders is not as clearly understood. One hypothesis linking positive energy balance and type 2 diabetes is the adipose tissue expandability hypothesis which postulates that individuals possess a maximum capacity for adipose expansion determined by both genetic and environmental factors. Once the adipose tissue expansion limit is reached, adipose tissue ceases to store energy efficiently and lipids begin to accumulate in other tissues. Ectopic lipid accumulation in non-adipocyte cells causes lipotoxic insults including insulin resistance, apoptosis and inflammation [21]. Visceral White Adipose Tissue (WAT) is currently believed to be the key depot linked with obesity-related systemic metabolic disturbances while subcutaneous abdominal adipose tissue (SAAT) may be a 'protective fat depot' [22]. Another hypothesis linking obesity to insulin resistance is inhibition of receptor activation by reactive lipid species [22].

One reason for the major impact of obesity on the development of cardiovascular disease is that it is often accompanied by the Metabolic Syndrome (MetS) [23] a cluster of atherogenic dyslipidemia, hyperglycemia with insulin resistance, hypertension, proinflammatory and prothrombotic states, and abdominal obesity [24]. A diet high in excess calories from fat and simple sugars combined with a sedentary lifestyle fuels a proinflammatory state which leads to the development of this syndrome. However, overweight and obesity even in the absence of MetS or insulin resistance place individuals at higher risk for major cardiovascular events, as well as for total death, compared with normal-weight men without metabolic derangements [23]. High blood pressure is twice as common in adults who are obese than in those who are at a healthy weight and the incidence of heart disease (heart attack, congestive heart failure, sudden cardiac death, angina or chest pain, and abnormal heart rhythm) is increased [9, 12, 20]. Neuroadrenergic abnormalities, comprising increased resting sympathetic nervous system activity and blunted sympathetic neural responsiveness are also recognized features of metabolic syndrome obesity, which may represent both a cause and consequence of the obese state and contribute importantly to both the pathophysiology and adverse clinical prognosis of this high-risk population [25].

It has been observed also that the absolute level of Adipose Tissue Blood Flow (ATBF) decreases with increasing BMI. The ATBF is highly variable over time, responding to nutritional stimuli for a far greater range than skeletal muscle blood flow. Maximum ATBF decrease is observed in women with both increased age and BMI. Overweight postmenopausal women have a 71% decrease in maximum ATBF compared with normal-weight premenopausal women. A reduced ATBF may lead to decreased adipose tissue glucose and triglyceride extraction with subsequent postprandial hyperlipidemia and hyperglycemia and increased risk of metabolic and vascular complications [26].

Chronic overweight and obesity contribute as well significantly to osteoarthritis, a major cause of disability in adults. For every 2-pound increase in weight, the risk of developing arthritis is increased by 9 to 13 percent [9, 12, 20]. Obesity increases also the risks of cancer of the breast, colon, prostate, endometroium, kidney and gallbladder [9, 10]. Indeed an estimated 78, 000 new cancer cases in the EU each year have been attributed to overweigh [9]. Moreover, at a BMI ≥ 40 the rates of maternal comorbidities and complications such as pre-eclampsia, gestational diabetes, impending foetal hypoxaemia, foetal macrosomia, as well as neonatal infections and hyperbilirubinaemia are significantly higher [27].

Adiposity is an important risk factor for low-grade inflammation [22] and obesity is a known chronic inflammatory state affecting cytokine and proteinase profiles and characterized by elevated levels of acute-phase reactants, cytokines, and cell surface receptors. More specifically, obesity produces a proteolytic environment with circulating Matrix Metallo Protease (MMP) levels comparable to serum MMP concentrations found in cancer and burn populations, which predisposes to wound problems and delayed healing [28]. WAT becomes inflamed during adipose tissue hypertrophy due to an influx of macrophages that secrete proinflammatory cytokines, including TNF-α, virgule and data suggest that in women an increase of 0.8 kg in

Visceral Adipose Tissue (VAT) corresponds to an elevated C-Reactive Protein (CRP) concentration of 1.8 mg/l, and a corresponding increase of 0.7 mg/l in men. Several cytokines, including IL-6 and TNF-α, are produced in adipose tissue and induce hepatic production of CRP. In addition, inflammatory cells such as monocytes and macrophages are components of adipose tissue and accumulate in obese states [19]. It is hypothesized also that the state of chronic inflammation and dysmetabolism observed in visceral obese patients associated with MetS negatively influences non-bariatric post-operative outcomes and is potentially key to the pathogenesis of other comorbidities due to the prolific metabolic activity of abdominal adipocytes [23, 29].

6. CONCLUSION

Obesity certainly is one of the most important worldwide present day public health issues. It reflects an imbalance between energy intake and expenditure that is mediated by the interaction of energy homeostasis and hedonic food intake behavior. Research has clearly shown the deleterious impact of obesity on health and psychosocial functioning. It increases the risk of disability, lost productivity, and medical costs. The costs of obesity have been estimated at up to 8% of overall health budgets and represent an enormous burden both in individual illness, disability and early mortality as well as in terms of the costs to employers, tax payers and society [9]. In addition to the medical costs associated with the treatment of obesity, consumers spend in excess of $30 billion a year to lose weight or prevent weight gain by paying for a variety of things, including diet drinks and food, appetite suppressants, diet books, videos/cassettes, commercial programs, and fitness clubs [12, 18, 30, 31].

Increasing strength of the evidence establishing an association between obesity and premature mortality and a wide variety of comorbid conditions including type 2 diabetes, obstructive sleep apnea, and multiple malignancies, among many others, has become more definitive relatively recently. Medical therapy for severe obesity consists of diet and exercise and is enhanced by various forms of instruction, group therapy, individual counseling, and medication, however, on a population basis, only bariatric surgery results in sustained meaningful weight loss [28]. Three long-term prospective clinical trials have all demonstrated sustained weight loss ranging from 15% to 30% or more of total body weight for more than 10 years. Associated with this weight loss, multiple studies have demonstrated improved survival in comparison to nonsurgical subjects identified through various databases. A major component of the explanation for this improved survival is, presumably, a direct result of induction of long-term remission of various life-threatening co-morbidities, including type 2 diabetes, hypertension, obstructive sleep apnea, dyslipidemia, and related cardiovascular risk [13].

While obesity exacerbates a host of life-threatening, age-related chronic diseases, a somewhat paradoxical finding in old age is that being somewhat overweight appears to be a benefit with regard to longevity. For those with osteoarthritis, coronary heart disease, and type 2 diabetes mellitus, weight loss interventions in overweight/obese older subjects definitely lead to significant benefits, while having slightly negative effects on bone mineral density and lean body mass. In contrast, higher BMIs are associated with increased survival after age 65 years. Because of this contradictory state of the science, there is a critical need for further study of the relationship of weight and weight loss/gain to health in the various age groups [32].

REFERENCES

[1]   Taylor VH, Curtis CM, Davis C. The obesity epidemic: the role of addiction. CMAJ 2009 Dec 21. [Epub ahead of print].

[2]   Obesity. eMedicine Health, practical guide to health. Viewed 22 May 2009, <http://www.emedicinehealth.com/obesity/article_em.htm>.

[3]   Overweight and Obesity. CDC, Department of Health and Human Services. Centers for disease control and prevention. Viewed 22 May 2009, <http://www.cdc.gov/nccdphp/dnpa/obesity/defining.htm>.

[4]   Lebiedowska MK, Alter KE, Stanhope SJ: Human Body Shape Index based on an experimentally derived model of human growth. J Pediatr 2008; 152: 45-9.

[5]   Healthy Weight - it's not a diet, it's a lifestyle! CDC, Department of Health and Human Services. Centers for Disease Control and Prevention. Viewed 23 May 2009, <http://www.cdc.gov/healthyweight/assessing/bmi/adult_bmi/index.html>.

[6]   Healthy Weight - it's not a diet, it's a lifestyle! CDC, Department of Health and Human Services. Centers for Disease Control and Prevention. Viewed 23 May 2009, <http://www.cdc.gov/healthyweight/assessing/bmi/childrens_bmi/about_childrens_bmi.html>.

[7]   Riess ML, Connolly LA, Woehlck HJ. Body mass index: an illogical correlate of obesity. Anesthesiology 2009; 111: 920-921.

[8]   Himes JH. Challenges of accurately measuring and using BMI and other indicators of obesity in children. Pediatrics 2009 ; 124 Suppl 1: S3-22.

[9]   International Obesity Task Force and European Association for the Study of Obesity. Obesity in Europe. Viewed 22 May 2009, <http://www.iotf.org/media/euobesity.pdf>.

[10]   WHO, obesity and overweight. Viewed 22 May 2009, <http://www.who.int/dietphysicalactivity/publications/facts/obesity/en/>.

[11]   Diéguez C, Fruhbeck G, López M. hypothalamic lipids and the regulation of energy homeostasis. Obes Facts 2009; 2:126-35.

[12]   Levin BE. Interaction of perinatal and pre-pubertal factors with genetic predisposition in the development of neural pathways involved in the regulation of energy homeostasis. Brain Res 2010 Jan 5. [Epub ahead of print].

[13]   The Obesity Epidemic. Plast Reconstr Surg 2006; 117S: 5S-7S.

[14]   Chaput JP, Tremblay A. Obesity and physical inactivity: the relevance of reconsidering the notion of sedentariness. Obes Facts 2009; 2: 249-54

[15]   Chang HJ. Scientists probe brain's role in obesity. JAMA 2010; 303:19.

[16]   Andersson J, Sjostrom LG, Karlsson M, Wiklund U, *et al.* Dysregulation of subcutaneous adipose tissue blood flow in overweight postmenopausal women. Menopause 2009 Nov 24. [Epub ahead of print].

[17]   Masaki T, Yoshimatsu H. Molecular mechanisms of neuronal histamine and its receptors in obesity. Curr Mol Pharmacol 2009; 2: 249-52.

[18]   Wang GJ, Volkow N, Thanos PK, Fowler J. Imaging of brain dopamine pathways: implications for understanding obesity. J Addict Med 2009; 3: 8-18.

[19]   Das UN. Obesity: Genes, brain, gut, and environment. Nutrition 2009 Dec 17. [Epub ahead of print].

[20]   Mokdad AH, Ford E, Bowman BA, *et al.* Prevalence of obesity, diabetes, and obesity related health risk factors. JAMA 2003; 289: 76-9.

[21]   Virtue S, Puig AV. Adipose tissue expandability, lipotoxicity and the Metabolic Syndrome - an allostatic perspective. Biochim Biophys Acta 2010 Jan 4. [Epub ahead of print].

[22]   Hamer M, O'Donovan G. Cardiorespiratory fitness and metabolic risk factors in obesity. Curr Opin Lipidol 2009 Sept 18. [Epub ahead of print].

[23]   Arnlov J, Ingelsson E, Sundstrom J, Lind L. Impact of body mass index and the metabolic syndrome on the risk of cardiovascular disease and death in middle-aged men. Circulation 2010; 121: 230-36.

[24]   Ripsin C. The metabolic syndrome: underdiagnosed and undertreated. South Med J 2009; 102: 1194-95.

[25]   Straznicky NE, Gavin W. Lambert GW, A. Lambert EA. Neuroadrenergic dysfunction in obesity: an overview of the effects of weight loss. Curr Opin Lipidol 2009 Oct 3. [Epub ahead of print].

[26]   Andersson J, Sjostrom LG, Karlsson M, Wiklund U, *et al.* Dysregulation of subcutaneous adipose tissue blood flow in overweight postmenopausal women. Menopause 2009 Nov 24. [Epub ahead of print].

[27]   Briese V, Voigt M, Hermanussen M, Wittwer-Backofen U. Morbid obesity: Pregnancy risks, birth risks and status of the newborn. Homo 2009 Dec 28. [Epub ahead of print].

[28]   Albino FP, Koltz PF, Gusenoff JA. A comparative analysis and systematic review of the wound-healing milieu: implications for body contouring after massive weight loss. Plast Reconstr Surg 2009; 124: 1675-82.

[29]   Doyle SL, Lysaght J, Reynolds JV. Obesity and post-operative complications in patients undergoing non-bariatric surgery. Obes Rev 2009 Dec 16. [Epub ahead of print].

[30]   Spence-Jones G. Overview of Obesity. Crit Care Nurs Q 2003; 26: 83-8.

[31]   Wolfe B. Bariatric surgery: An evolving field. World J Surg 2009; 33: 1981-82.

[32]   Bales CW, Buhr GT. Body mass trajectory, energy balance, and weight loss as determinants of health and mortality in older adults. Obes facts 2009; 2: 171-8.

CHAPTER 2

Bariatric Surgery: An Overview

Bassem Safadi[*] and Sami Masri

American University of Beirut Medical Center, Beirut, Lebanon

Abstract: Bariatric surgical procedures were designed and developed to induce weight loss and hence improve or eliminate obesity related co-morbidities. The field of bariatric surgery has grown appreciably over the last decade with a rapid rise in the utilization of bariatric surgery worldwide, partly fueled by the introduction of minimally invasive surgery techniques and partly by the growing epidemic of obesity. This chapter reviews the most common bariatric operations performed with emphasis on technique, short and long-term outcomes.

Keywords: Bariatric surgery, sleeve gastrectomy, gastric band, gastric bypass, sleeve gastrectomy, biliopancreactic bypass.

1. INTRODUCTION

The word "bariatric" is derived from the Greek word "Bari" or "Baros" which denotes pressure or heavy [1]. Surgeons engaged in this field of surgery that treats obesity began organizing in the late 70's and in the year 1983 the American Society of Bariatric Surgery (ASBS) was founded [2]. This was followed by the foundation of the International Federation of the Surgery of Obesity (IFSO) in 1995 [3]. More recently the ASBS was renamed the American Society of Bariatric and Metabolic Surgery (ASMBS) because of the recognition of the importance of this field of surgery in the treatment and amelioration of metabolic problems such as diabetes mellitus [3]. Bariatric surgery was developed by a relatively small group of surgeons and now has become a widely practiced discipline with several 100 thousand surgical procedures performed worldwide.

2. HISTORICAL PERSPECTIVE

Bariatric surgical procedures were divided into two broad categories: malabsorptive and restrictive [4]. Malabsorptive procedures induce weight loss by reducing the absorptive capacity of the small bowel and by limiting the gut's ability to absorb calories. This idea was proposed by Kremen and colleagues in Minneapolis in the 1950's who observed that shortening the guts in dogs induced a predictable weight loss [4, 5]. In the same context, it was observed that patients who lost the majority of the small intestine from trauma or surgical resection for tumors lost weight and maintained weight loss. The Jejuno-Ileal (JI) bypass was one of the first types of malabsorptive operations performed, in which the proximal jejunum was disconnected and anastomosed to the terminal ileum [4] (Fig. **1**). There were several modifications that followed but the most popular form was the one proposed by Payne and DeWind in 1969 in which the proximal 35cm of the jejunum was anastomosed to the distal 10cm of the terminal ileum in an end-to-side fashion [4, 6]. This left behind a large blind small bowel loop that became susceptible to bacterial overgrowth. Patients lost weight significantly but developed diarrhea, gastrointestinal side effects, electrolyte, vitamins and mineral deficiencies. Tragically, a significant number of patients developed liver cirrhosis and ultimately died. This was attributed to the large blind loop and the resultant bacterial overgrowth that lead to the release of toxins into the portal circulation. This unacceptably high rate of mortality and morbidity led to the abandonment of this procedure [4, 7, 8]. The gastric bypass procedure was introduced in the mid 1960's as a procedure with both restrictive and mal-absorptive features (4, 9). It has remained to date one of the most successful and popular bariatric operations.

*****Address correspondence to Bassem Safadi:** American University of Beirut Medical Center, Beirut, Lebanon; Tel: +961-3-04602; E-mail: bs21@aub.edu.lb

In the 1970's bariatric surgeons began exploring methods to restrict the stomach's capacity. Procedures such as the gastroplasties were developed [4, 10]. The Vertical Banded Gastroplasty (VBG) depicted in (Fig. **2**) was once a popular procedure but was gradually abandoned because of a high failure rate and high rate of side effects such as vomiting and Gastro-Esophageal Acid Reflux (GERD) [11, 12]. Other surgeons re-explored the idea of "malasbsorption" and found means to eliminate the "long blind loop' with procedures such as the Biliopancreatic Diversion (BPD) and the duodenal switch (DS) [4, 13-16]. All these developments that took place over the past 5 decades or so led to the maturation of bariatric surgery as a discipline. In 1994, the first laparoscopic gastric bypass was performed in the United States by Whitgrove and Clark [17]. The advent of "minimally invasive" approach and the rising epidemic of obesity worked synergistically to create an unprecedented rise in the number of bariatric procedures done worldwide.

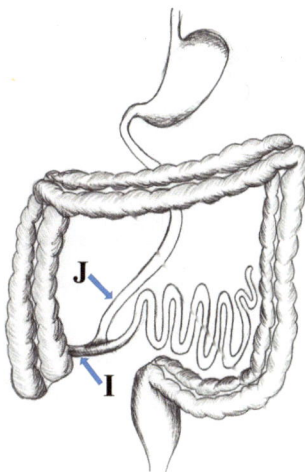

Fig. 1. The Jejuno-Ileal Bypass (JIB): The Proximal Jejunum (J) is Anastomosed to the Distal Ileum (I) in an End to Side Fashion.

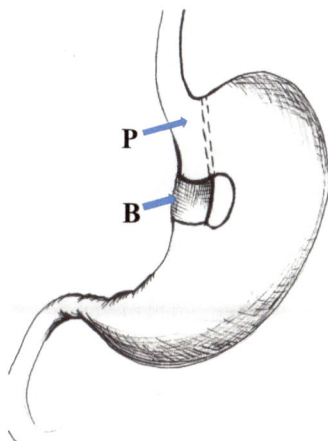

Fig. 2. The Vertical Banded Gastroplasty (VBG): A Vertical Staple Line Partitions a Small Gastric Pouch (P). A Fixed Band (B) Creates a Tight Stoma Outlet.

3. RATIONALE FOR SURGERY AND INDICATIONS

The accumulation of fat in excess amount in the body is unhealthy and potentially has lethal implications. Many prospective observational studies have linked obesity to morbidity and shortened survival; put in simple term: Obesity kills [18, 19]. In addition obesity increases the risk of developing disease states that affect practically every organ in the body as well as the mental and psychological well-being of an individual. Obesity is a chronic health problem with genetic, environmental and behavioral components. The enormous changes to human societies have made calorie dense food widely available and have made us less physically active. Therefore the solution to the epidemic of obesity can only be through major societal changes. Diet programs, diet pills, appetite suppressants may work short-term but most morbidly obese patients will regain weight and will go through a: "YO-YO" cycle of weight loss and ultimate weight regain [20].

In the absence of major societal change and with no real medical therapy surgery remains the only effective treatment of morbid obesity [21]. As we will discuss later bariatric surgery has been shown to effectively reduce weight, ameliorate obesity related co-morbidities and prolong survival in morbidly obese patients [21-24]. Because of the inherent risk of surgery there should be a cut-off for patient selection where the potential benefit of surgery exceeds the risk. The first attempt at defining this cut-off was a National Institutes of Health (NIH) consensus panel conference that convened in 1991 which suggested at the time that suitable patients for bariatric surgery are those with Body Mass Index (BMI) exceeding 40 kg/m2 or 35 kg/m2 in the presence of serious obesity related co-morbidities [25]. The panelists recommended that:

1. Patients seeking therapy for severe obesity for the first time should be considered for treatment in a nonsurgical program with integrated components of a dietary regimen, appropriate exercise, and behavioral modification and support.

2. Gastric restrictive or bypass procedures could be considered for well-informed and motivated patients with acceptable operative risks.

3. Patients who are candidates for surgical procedures should be selected carefully after evaluation by a multidisciplinary team with medical, surgical, psychiatric, and nutritional expertise.

4. The operation be performed by a surgeon substantially experienced with the appropriate procedures and working in a clinical setting with adequate support for all aspects of management and assessment, and

5. Lifelong medical surveillance after surgical therapy is a necessity.

At the time the NIH consensus panel conference was held, the two commonest procedures were open gastroplasty and open gastric bypass. There were few bariatric surgeons and a relatively small number of procedures were performed annually. A lot has changed since then: laparoscopic surgery has made the post-operative recovery shorter and less painful and potentially safer [26]. New procedures such as Laparoscopic Adjustable Gastric Banding (LAGB), BPD, DS and Sleeve Gastrectomy (SG) have emerged. Surgeons are performing an ever increasing number of procedures and are becoming more proficient. Those factors have made current bariatric procedures safer. As a result the recent trend has been to recommend bariatric surgery for patients with mild obesity (BMI 30-35 kg/m2) if they have significant obesity related co-morbidities particularly type II diabetes. As a matter of fact, some investigators are pushing the envelope and are studying the role of procedures in resolving metabolic diseases, such as gastric bypass patients with type II diabetes and a normal BMI [27].

4. PATIENT PREPARATION

The ideal set up of a bariatric surgical practice includes a multidisciplinary team comprising of health care professionals, dietitians, mental health specialists, and exercise physiologists. Most morbidly obese patients have medical and psychological co-morbidities that have to be addressed before and after surgery. The facility should also be well equipped to handle patients with severe obesity such as proper waiting areas, seats, beds, transportation means, *etc.* The worldwide trend nowadays is to concentrate bariatric surgery in high volume centers or what is referred to as "centers of excellence" where the team is versed not only in the technical part of the operation but also peri-operative care and handling of problems and complications [28].

The medical work-up will vary depending on the severity of medical problems but might involve an endocrine work-up, gastrointestinal endoscopy, pulmonary function testing, cardiac evaluation, *etc.* A period of pre-operative weight loss has been shown to be beneficial as it decreases visceral fat and the size of the left lobe of the liver hence facilitating surgery [29]. An added benefit to pre-operative weight loss is selecting patients who are motivated and who generally will do better than others. Patients who failed prior bariatric procedures and who present for revisional surgery require careful operative planning based on imaging and endoscopy to define the anatomy and also require extensive psychological assessment and intervention. Perhaps the most important part of the pre-operative evaluation is patient education and informed consent. Bariatric surgery should be explained to patients as a tool that provides them with enormous help to eat healthier and become more physically active. It should not be viewed as a cure or an easy fix to a difficult and chronic problem.

5. TYPES OF PROCEDURES

I-Roux-Y-Gastric Bypass

Introduction

The gastric bypass was developed by Mason and Ito in 1967 [9]. At that time surgery for gastric and duodenal ulcers was common and surgeons observed at that time that patients who underwent gastric resection lost weight and maintained weight loss. The idea behind the gastric bypass is to create a small gastric passage and divert the food directly into the small intestine to induce some restriction and malabsorption at the same time. It can be performed open *via* a midline incision or Left upper quadrant incision. The laparoscopic approach is done with 4-6 ports in the upper abdomen.

Technique

The Roux-y-Gastric Bypass can be done with numerous technical alterations but should have the following elements (Fig. **3**).

1. A small gastric pouch measuring less than 50 ml and preferably between 15 to 30 mls. The pouch is constructed along the lesser curvature *i.e.* should contain no fundus as fundic tissue has a tendency to distend with time. The pouch should be stapled and divided completely from the rest of the stomach.

2. The small bowel is divided a distance away from the Ligament of Treitz and the bowel distally is anastomosed to the gastric pouch. This segment of the small bowel is called the Alimentary Limb (AL). The Gastro-Jejunostomy (GJ) can be fashioned by suturing or stapling techniques.

3. The entero-enterostomy (EE) is fashioned about 100-150 cm distal to the GJ connecting the alimentary limb to the Bilio-Pancreatic Limb (BPL). The aim of the Roux-Y reconstruction is to reduce the risk of bile reflux esophagitis and gastritis. A distal limb can be extended to more than 250 cm for induction of some malabsorption.

Complications

The most serious early complications with the RYGB are GI leak, bleeding, intestinal obstruction and pulmonary embolus [26, 28, 30]. The most common site of leak is at the gastro-jejunostomy but can occur at any staple line or at the entero-enterostomy. These complications carry considerable morbidity and mortality risk if not diagnosed and treated early and aggressively. Early re-exploration for the suspicion of a leak (ex. tachycardia) is justified even without evidence of a leak on imaging. Small bowel obstruction may occur early due to kinking or intraluminal bleeding and should be treated promptly because delay may cause dilation and subsequent leak from the divided stomach. Stenosis at the gastro-jejunostomy is more common with the circular stapler but unusual with anastomoses constructed manually or with the linear stapler.

Long-term complications include inadequate weight loss, excess weight loss, anastomotic ulcers, small bowel obstruction from adhesions or internal hernias, gall stones, Vitamin B-12 deficiency, osteoporosis. Incisional hernias are very common after open RYGB and are rare after laparoscopic RYGB. Excessive and redundant skin and wrinkling are long-term sequalae to massive weight loss that take place with any bariatric operation.

Recent prospective data from the United States demonstrate that the mortality in expert centers is 0.2% for laparoscopic RYGB and 2.1% for open RYGB. The risk of major complications is around 5-8% [26].

Long-Term Results

The reports of long-term results with the RYGB show weight loss at 60-70% Excess Weight Loss (EWL) in the first two years after surgery [21-24]. After which there is some weight regain with a sustained plateau of 50-60% EWL even beyond 10 years of follow-up. At ten years more than 80% of patients have maintained a BMI less than 30kg/m2 [31]. More importantly RYGB effectively reduced the risk of diabetes, hyperlipidemia and hypertension in the long-term, improved longevity and reduced mortality from cardiovascular disease and from cancer [21-24].

Special Considerations

The mechanisms of weight reduction with RYGB are still not entirely understood. Simplistically this operation was classified as a combination restrictive and malabsorptive operation. However patients after RYGB seldom vomit in contrast to patients who undergo restrictive operations such as VBG and LAGB and do not have diarrhea or appreciable carbohydrate and protein malabsorption. Patients after RYGB eat less and report a feeling of early satiety. In 2002, Cummings published a report that for the first time demonstrated a reduction in the appetite hormone Ghrelin after successful RYGB [32]. This hormonal change in the gut might explain why the RYGB has remained a successful bariatric operation after more than 40 years of development. More on the hormonal changes with bariatric surgery will be reviewed in Chapter 3.

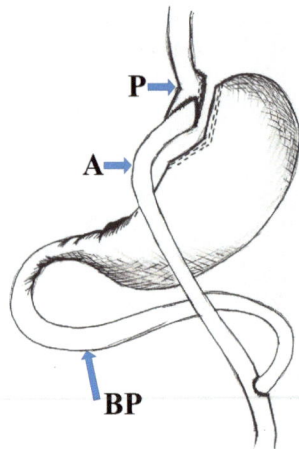

Fig. 3. The Roux-Y-Gastric Bypass (RYGB). A Small Gastric Pouch (P) is Anastomosed to the Proximal Jejunum. The Alimentary Limb (A) is Anastomosed to the Biliopancratic Limb (BP).

II-Single Anastomosis (Mini) Gastric Bypass

In the Mini-GB also referred to as the single anastomosis or loop GB the gastric pouch is fashioned along the lesser curvature but made longer than in the classical RYGB. There is only a single anastomosis to the small intestine at 200cm distal to the Ligament of Treitz. Proponents of the MGB say that it is a technically simpler operation and hence faster and safer than the RYGB. It is also easier to reverse [33]. Opponents of the MGB criticize this operation because of the risk of developing bile reflux. There is also concern over the long-term effect of gastric exposure to bile and the subsequent increased risk of cancer of the stomach [34]. More studies are needed to establish the efficacy and the long-term safety of the Mini-GB.

III-Sleeve Gastrectomy

Introduction

The Sleeve Gastrectomy (SG) originally was described as part of the DS procedure [16]. The DS procedure carried a relatively high rate of morbidity and mortality partly because it was chosen for super-obese patients. In an attempt to reduce morbidity surgeons at Mt. Sinai Hospital in New York decided to perform the operation in two stages; the SG, a purely restrictive first intervention to induce significant weight loss, followed by the intestinal component of the SG or alternatively RYGB [35]. Initial reports of the sleeve gastrectomy primarily involved patients at extremely high risk for conventional bariatric surgery. That strategy worked in reducing surgical complications and mortality. Since then an increasing body of literature supported that the SG can be a very effective standalone primary procedure [36-38].

Technique

The procedure can be done laparoscopically with relative ease through 4-6 ports. The main idea being the SG is to remove as much of the gastric reservoir while maintaining the antral pump (Fig. **4**). Starting from a point around 5-6cm proximal to the pylorus all the vessels along the greater curvature are sealed and divided all the way to the Angle of His. All posterior gastric attachments to the pancreas are released; adequate retrogastric mobilization decreases the risk of leaving a large posterior stomach. A bougie 36-42Fr is placed transorally and oriented toward the antrum. Then alongside the bougie the stomach is stapled and divided in a vertical fashion all the way to the Angle of His. Although the procedure does not involve anastomoses, the length of the staple line still renders the patient at risk for bleeding or leakage. Oversewing the staple line or use of other sealants has been suggested by authors [39, 40].

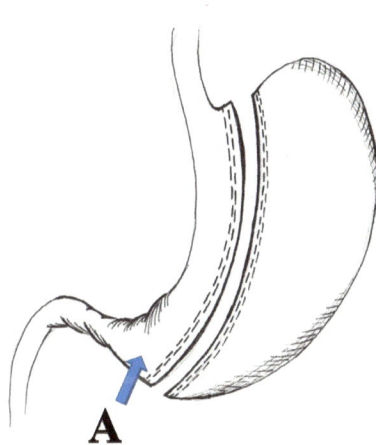

Fig. 4. The Sleeve Gastrectomy (SG): Resection of 60-70% of the Gastric Volume Leaving a Long Tubularized Stomach Preserving Some on the Antral Pump (A).

Complications

The most serious complication of the SG is staple line leak which takes place in 1-6% of patients [39, 40]. The patients who had prior gastric procedures such as gastric banding, VBG and fundoplication are at much higher risk and the SG may not be suitable for those patients. Unlike leaks in RYGB leaks in SG may present later even after weeks of operation and are much more difficult to manage. Bleeding from the staple line is another short term risk that seems to be reduced with the addition of buttress material to the staple line. Long-term complications particular to SG includes GERD and Vitamin B-12 deficiency.

Long-Term Results

The SG is a new bariatric procedure so there isn't much long-term data on efficacy. Mid-term data show weight loss comparable to the RYGB. The SG may be as effective in ameliorating and resolving diabetes than RYGB [36-38].

Special Considerations

The SG has quickly become a very popular bariatric operation because it is technically easier than operations since it there is no anastomosis involved and the work is done in one quadrant. It has attracted many patients because of the absence of a foreign object and no major diversion of the GI tract. The sleeve resection reduces the size of the stomach achieving restriction and early satiety with documented reduction in serum Ghrelin levels post-operatively [41]. However, the Achiles Heel of this operation remains the staple line leak and the seriousness of this complication should not be underestimated.

IV-Piliopancreatic Diversion – Duodenal Switch

Introduction

The Bilio-Pancreatic Diversion (BPD) was described by Scopinaro in 1979 after groundwork in the animal laboratory (4, 13). The intent in the procedure is to reduce the effective absorptive capacity of the small intestine without leaving a bling intestinal loop. The operation includes a distal gastrectomy with a long Roux-en-Y reconstruction. The Duodenal Switch (DS) was developed along similar lines with the idea that pyloric preservation lessens the possibility of dumping symptoms and decreases the risk of marginal ulcers [14-16]. These operations have not gained much popularity although they are quite effective in weight control, particularly in super obese patients where the gastric bypass operation may not yield good weight control. Most surgeons view the DS as a technically difficult procedure and are concerned regarding long-term metabolic derangements, protein malnutrition and other deficiencies.

Technique

These procedures are performed through an upper midline incision or laparoscopic. In the BPD the distal stomach is resected leaving a gastric capacity around 200-250ml. The Roux-en-Y gastro-jejunostomy constitutes an Alimentary Limb (AL) measuring 250cm and a Common Channel (CC) measuring 50-75cm (Fig. **5**).

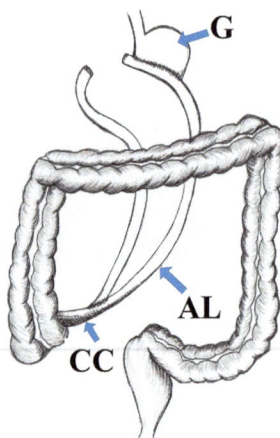

Fig. 5. The Biliopancreatic Diversion (BPD). A Distal Gastrectomy Leaves 200-250cm of Gastric Reservoir (G). The Alimentary Limb (AL) Measures 250cm and the Common Channel Measures 50-75cm.

As described earlier a sleeve gastrectomy is done in the DS. The duodenum is divided few cm distal to the pylorus. The measurements of the AL and CC are based on the total length of the small intestines with the CC being roughly 10% of the total length of the small intestines which comes out to be 50-100cm. The distance from stomach to cecum comes out to around 40% of the total length of the small intestines [41] (Fig. **6**).

Complications

The DS and BPD have not gained as much widespread use as the RYGB so they have remained limited to few centers and have been performed on super-obese patients. The reported mortality and morbidity may be higher than with the RYGB [35]. Common complications include anastomotic leaks, stricture, bleeding, small intestinal obstruction, DVT and pulmonary emboli. The gastro-jejunostomy in the BPD is technically much easier than the duodeno-ileostomy so anastomotic complications are seldom encountered in the BPD. Marginal ulcers are more common in the BPD. Long-term metabolic and nutritional derangements such as protein and vitamin deficiencies are more likely to occur after such procedures and patients should be followed and supplemented life-long.

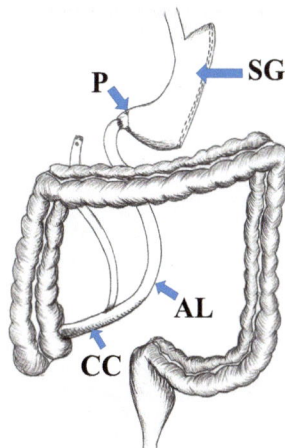

Fig. 6. The Duodenal Switch (DS) Constituting a Sleeve Gastrectomy (SG). The Pylorus (P) is Preserved and a Duodeno-Enterostomy with Roux-Y Reconstruction is Fashioned.

Long-Term Results

Weight loss maintenance and resolution of co-morbidities such as diabetes is excellent following BPD and DS and perhaps exceeding results with the RYGB with durable long-term results (16, 21).

V-Laparoscopic Adjustable Gastric Banding

Introduction

The idea of banding the proximal stomach evolved from the early fixed bands that were placed *via* laparotomy to the modern day Laparoscopic Adjustable Gastric Band (LAGB). Most bands are made of Silicon and have an inflatable inner cuff that wraps the cardia circumferentially. The inflatable cuff is connected *via* tubing to a reservoir placed subcutaneously that allows percutaneous "adjustment" of the diameter of the band. The LAGB has distinct advantages including relative ease of operation, relative low operative risk, adjustability and no major alterations in the digestive system. The impact on satiety is not dramatic as in the RYGB so patients need to make more effort in implementing dietary and lifestyle changes and avoid maladaptive eating behavior such as shifting to sweet liquids. In addition, follow-up is crucial for the success of the band since serial adjustment as well as dietary and psychological support determines the outcome with this operation [42].

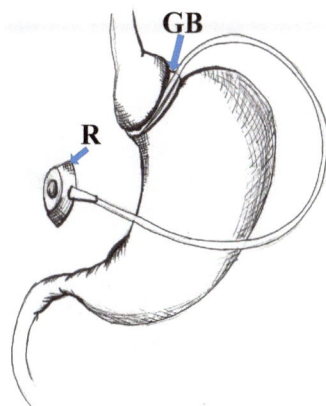

Fig. 7. The Laparoscopic Adjustable Gastric Band (LAGB). The Gastric Band (GB) Partitions a Small Gastric Pouch and is Attached to a Subcutaneous Reservoir (R) that is Used for Adjustments.

Technique

The technique of LAGB has universally been changed from a peri-gastric to a Pars flaccida approach so the band is not directly in contact with the serosa along the lesser curve [42, 43]. That has been shown to reduce complications such as slippage and erosion. The final lay of the LAGB takes on an oblique position few cm below the gastro-esophageal junction on the lesser curve to the Angle of His (Fig. 7). Once the band is secured and locked in place, the fundus is wrapped anterior to the band again to keep the band in place and reduce the risk of prolapse and slippage. The tubing of the band is exited through the abdominal wall and attached to the reservoir which is housed in a subcutaneous pocket.

Complications

The LAGB has the lowest immediate peri-operaive risk with very low risk of mortality and major complications [26, 42]. What plagues the LAGB are the long-term risks of complications and re-operations which can approach 50% on long-term follow-up. Long-term problems include failure to lose weight, esophageal dilation, gastric pouch dilation, band slippage, intra-gastric erosion in addition to mechanical problems with the tubing and device. Some of these complications may present in an acute fashion and will require prompt surgical correction. More chronic problems are best managed by converting the gastric band to another procedure such as the RYG or the BPD [43-45].

Long Term Results

Long term results and impact on weight and co-morbidities vary widely among centers and are in large part determined by patient selection and quality of follow-up. Good to excellent results can be achieved with 50% of patients or more and resolution of co-morbidities will be achieved in patients who lose and maintain weight loss [42, 46].

VI-Procedures in the Horizon

The growth in bariatric surgery seen in the past two decades has generated new ideas and concepts. Operations on the digestive tract are now being studied in the treatment of type II diabetes in patients with normal weight or slight obesity [47]. Endoscopic endoluminal and stapling procedures are being developed [48]. Electric pacing of the vagus nerve and stomach are among many new alternatives being studied for the treatment of obesity [49].

6. CONCLUSION

Bariatric surgery remains the most effective and durable treatment option for patients with morbid obesity. The long-term benefits include sustained weight reduction and resolution and improvement in obesity related complications. This translates into better quality of life and longevity. There are many types of bariatric operations that vary in mechanisms of action and in benefit/risk profile. The choice of operation should be based on the surgeon and center's experience and on thorough patient evaluation and education.

REFERENCES

[1]   Deitel M, Melissas J. The origin of the word "Bari". Obes Surg 2005; 15: 1005-8.

[2]   American Society for Metabolic & Bariatric Surgery. Mission and Purpose. Viewed 01 Nov 2010, <http://www.asmbs.org/Newsite07/aboutasbs_mission.htm>.

[3]   International Federation for the Surgery of Obesity and Metabolic Disorders (IFSO). Viewed 01 Nov 2010, <http://www.ifso.com/Index.aspx?id=About IFSO>.

[4]   Buchwald H, Buchwald JN. Evolution of operative procedures for the management of morbid obesity 1950-2000. Obes Surg 2002; 12: 705-17.

[5]   Kremen A, Linner J, Nelson C. An experimental evaluation of the nutritional importance of proximal and distal small intestine. Ann Surg 1954; 140: 439-47

[6]   Payne JH, DeWind LT. Surgical treatment of obesity. Am J Surg 1969; 118:141-7.

[7]   Griffen WO Jr, Bivins BA, Bell RM. The decline and fall of the jejunoileal bypass. Surg Gynecol Obstet 1983; 157:301-8.

[8]   Halverson JD, Wise L, Wazna MF, *et al.* Jejunoileal bypass for morbid obesity. A critical appraisal. Am J Med 1978; 64: 461-75.

[9]   Mason EE, Ito C. Gastric bypass in obesity. Surg Clin North Am 1967; 47: 1345-51.

[10] Mason EE. Vertical Banded Gastroplasty for Obesity. Arch Surg 1982; 117: 701-6.

[11] Balsiger BM, Poggio JL, Mai J, *et al.* Ten and more years after vertical banded gastroplasty as primary operation for morbid obesity. J Gastrointest Surg 2000; 4: 598-605.

[12] Ikramuddin S, Kellogg TA, Leslie DB. Laparoscopic conversion of vertical banded gastroplasty to a Roux-en-Y gastric bypass. Surg Endosc 2007; 21: 1927-30.

[13] Scopinaro N, Gianetta E, Civalleri D, *et al.* Bilio-pancreatic bypass for obesity: (I and II). Br J Surg 1979; 66: 616-20.

[14] DeMeester TR, Fuchs KH, Ball CS, *et al.* Experimental and clinical results with proximal end-to-end duodenojejunostomy for pathologic duodenogastric reflux. Ann Surg 1987; 206: 414-26.

[15] Hess DS, Hess DW. Biliopancreatic diversion with a duodenal switch. Obes Surg 1998; 8: 267-82.

[16] Anthone GJ. The duodenal switch operation for morbid obesity. Surg Clin North Am 2005; 85: 819-33.

[17] Wittgrove AC, Clark GW, Tremblay LJ. Laparoscopic Gastric Bypass, Roux-en-Y: Preliminary report of five cases. Obes Surg 1994; 4: 353-7.

[18] Pischon T, Boeing H, Hoffmann K, *et al.* General and abdominal adiposity and risk of death in Eurpoe. N Eng J Med 2008; 359: 2105-20.

[19] Ferrari P, Riboli E, Calle EE, *et al.* Overweight, obesity, and mortality from cancer in a prospectively studied cohort of U.S. adults. N Eng J Med 2003; 348: 1625-38.

[20] North American Association for the Study of Obesity and the National Heart, Lung, and Blood Institute. *The Practical Guide: Identification, Evaluation, and Treatment of Overweight and Obesity in Adults.* Bethesda, Md: National Institutes of Health; 2000. NIH publication 00-4084.
      <http://www.nhlbi.nih.gov/guidelines/obesity/prctgd_c.pdf>.

[21] Buchwald H, Avidor Y, Braunwald E, *et al.* Bariatric surgery: a systematic review and meta-analysis. JAMA 2004; 292: 1724-37.

[22] Sjöström L, Lindroos A, Peltonen M, *et al.* Lifestyle, Diabetes, and Cardiovascular Risk Factors 10 Years after Bariatric Surgery. N Eng J Med 2004; 351: 2683-93.

[23] Sjöström L, Narbro K, Sjöström D, *et al.* Effects of bariatric surgery on mortality in swedish obese subjects. N Eng J Med 2007; 357: 741-52.

[24] Christou V, Sampalis JS, Liberman M, *et al.* Surgery decreases long-term mortality, morbidity, and health care use in morbidly obese patients. Ann Surg 2004; 240: 416-23; discussion 423-4.

[25] NIH conference. Gastrointestinal surgery for severe obesity. Consensus Development Conference Panel. Ann Intern Med 1991; 115: 956-61.

[26] The Longitudinal Assessment of Bariatric Surgery (LABS) Consortium. Perioperative safety in the longitudinal assessment of bariatric surgery. N Engl J Med 2009; 361: 445-54.

[27] Pories WJ, Dohm LG, Mansfield CJ. Beyond the BMI: the search for better guidelines for bariatric surgery. Obesity 2010; 18: 865-71.

[28] Hollenbeak CS, Rogers AM, Barrus B, *et al.* Surgical volume impacts bariatric surgery mortality: a case of centers of excellence. Surgery 2008; 144: 736-43.

[29] Benotti PN, Still CD, Wood GC, *et al.* Preoperative weight loss before bariatric surgery. Arch Surg 2009; 144: 1150-5.

[30] Podnos Y, Jimenez J, Wilson S, *et al.* Complications after laparoscopic gastric bypass: a review of 3464 cases. Arch Surg 2003; 138: 957-61.

[31] Christou NV, Look D, Maclean LD. Weight gain after short-and long-limb gastric bypass in patients followed for longer than 10 years. Ann Surg 2006; 244: 734-40.

[32] Cummings DE, Weigle DS, Frayo RS, *et al.* Plasma ghrelin levels after diet-induced weight loss or gastric bypass surgery. N Eng J Med 2002; 346: 1623-30.

[33] Rutledge R. The mini-gastric bypass: experience with the first 1,274 cases. Obes Surg 2001; 11: 276-80.

[34] Johnson WH, Fernandez AZ, Farrell TM, *et al.* Surgical revision of loop ("mini") gastric bypass procedure: multicenter review of complications and conversions to Roux-en-Y gastric bypass. Surg Obes Relat Dis 2007; 3: 37-41. Epub 2006 Dec 27.

[35] Regan JP, Inabnet WB, Gagner M, *et al.* Early experience with two-stage laparoscopic Roux-en-Y gastric bypass as an alternative in the super-super obese patient. Obes Surg 2003; 13: 861-4.

[36] Lee CM, Cirangle PT, Jossart GH. Vertical gastrectomy for morbid obesity in 216 patients: report of two-year results. Surg Endosc 2007; 21: 1810-6.

[37] Himpens J, Dobbeleir J, Peeters G. Long-term results of laparoscopic sleeve gastrectomy for obesity. Ann Surg 2010; 252: 319-24.

[38] Brethauer SA, Hammel JP, Schauer PR. Systematic review of sleeve gastrectomy as staging and primary bariatric procedure. Surg Obes Relat Dis 2009; 5: 469-75.

[39] Chen B, Kiriakopoulos A, Tsakayannis D, *et al.* Reinforcement does not necessarily reduce the rate of staple line leaks after sleeve gastrectomy. A review of the literature and clinical experiences. Obes Surg 2009; 19: 166-72. Epub 2008 Sep 16.

[40] Karamanakos SN, Vagenas K, Kalfarentzos F, *et al.* Weight loss, appetite suppression, and changes in fasting and postprandial ghrelin and peptide-YY lebels after Roux-en-Y gastric bypass and sleeve gastrectomy: a prospective, double blind study. Ann Surg 2008; 247: 401-7.

[41] Hess DS. Limb measurements in duodenal switch. Obes Surg 2003; 13: 966.

[42] Fielding GA, Ren CJ. Laparoscopic adjustable gastric band. Surg Clin North Am 2005; 85: 129-40.

[43] Di Lorenzo N, Furbetta F, Favretti F, *et al.* Laparoscopic adjustable gastric banding *via* pars flaccid versus perigastric positioning: technique, complications, and results in 2,549 patients. Surg Endosc 2010; 24: 1519-23. Epub 2010 Mar 31.

[44] Mittermair RP, Obermüller S, Perathoner A, *et al.* Results and complications after Swedish adjustable gastric banding-10 years experience. Obes Surg 2009; 19: 1636-41.

[45] Mognol P, Chosidow D, Marmuse JP. Laparoscopic conversion of laparoscopic gastric banding to Roux-en-Y gastric bypass: a review of 70 patients. Obes Surg 2004; 14: 1349-53.

[46] Angrisani L, Lorenzo M, Borrelli V. Laparoscopic adjustable gastric banding versus Roux-en-Y gastric bypass: 5-year results of a prospective randomized trial. Surg Obes Relat Dis 2007; 3: 127-32; discussion 132-3. Epub 2007 Feb.

[47] Wang TT, Hu SY, Gao HD, *et al.* Ileal transposition controls diabetes as well as modified duodenal jejunal bypass with better lipid lowering in a nonobese rat model of type II diabetes by increasing GLP-1. Ann Surg 2008; 247: 968-75.

[48] Hashiba K. Endoscopic bariatric procedures and devices. Gastrointest Endosc Clin N Am 2007; 17: 545-57.

[49] Greenway F, Zheng J. Electrical stimulation as treatment for obesity and diabetes. J Diabetes Sci Technol 2007; 1: 251-9.

CHAPTER 3

Physiological and Nutritional Impact of Bariatric Surgery

Dimitrios J. Pournaras[1,2] and Carel W. le Roux[1*]

[1]Department of Bariatric Surgery, Musgrove Park Hospital, Taunton, [2]Imperial Weight Centre of Investigative Medicine, Hammersmith Hospital, Imperial College London, UK

Abstract: Bariatric surgery is the only and effective treatment for morbid obesity and leads to diabetes remission. The mechanism of action is reviewed in this chapter. The role of gut hormones as facilitators of appetite control, weight loss and an improved glycemic control after bariatric surgery is described. Furthermore nutritional considerations and their management after bariatric surgery are discussed.

Keywords: Appetite control, gut-brain axis, malabsorption, diabetes, vitamin supplementation.

1. INTRODUCTION

Bariatric surgery has been proven to be the only effective, long-term treatment for morbid obesity [1]. The effects of surgery are not simply restricted to substantial loss of weight. Improvements in co-morbidities associated with a high body mass index, including hypertension, sleep apnea, hyperlipidemia and type 2 diabetes, have been demonstrated [2]. Of particular interest is the improved glycemia, which sometimes can be indistinguishable from remission of type 2 diabetes as evidenced by the discontinuation of diabetic medications and normal glycemic control [2]. The mechanism that leads to sustained weight loss as well as diabetes remission after bariatric operations, however, remains to be elucidated.

Energy intake and expenditure are adjusted through the process of energy homeostasis leading to remarkable stability in body mass over time. It is likely from an evolutionary standpoint that the dominant rule is selection bias toward homeostatic systems that respond to reduced energy intake rather than energy excess. In fact weight loss leads to an increase in the perceived hunger and a decrease in the metabolic rate. Leptin and insulin are the key messengers of the status of energy stores from the periphery to the central nervous system [3].

2. APPETITE CONTROL AND CHANGES IN THE GUT-BRAIN AXIS AFTER BARIATRIC SURGERY

Bariatric procedures were designed to reduce gastric volume (Laparoscopic Adjustable Gastric Banding [LABG], Laparoscopic Sleeve Gastrectomy [LSG]), to cause malabsorption of nutrients (Biliopancreatic Diversion [BPD], Duodenal Switch [DS]) or to produce both (Roux-en-Y Gastric Bypass [RYGB]) with the objective to achieve weight loss. It is now known that calorie malabsorption does not occur (with the exception of the BPD and DS); however the effects of bariatric procedures cannot be fully attributed to the reduced stomach volume. Anorexigenic and orexigenic hormones, namely: glucagon-like peptide-1 (GLP-1), peptide YY (PYY) (YY being an abbreviation for tyrosine), ghrelin, Cholecystokinin (CCK), Glucose-Dependent Insulinotropic Polypeptide (GIP), Oxyntomodulin (OXM), and Pancreatic Polypeptide (PP) were all shown to be implicated to variable degrees [3]. The discovery of these appetite-signaling peptides, the gut hormones, which have hunger and satiety effects and thus have an integral role in appetite regulation, has led to the establishment of the concept of the gut-brain axis.

A crucial role in the regulation of energy homeostasis is played by the gut-brain axis. The central melanocortin system is influenced by signals of hunger and satiety mediated through peptides made in the gut and released into the circulation. Changes in gut hormone concentrations probably may be responsible,

*****Address correspondence to Carel W. le Roux:** Imperial Weight Centre of Investigative Medicine, Hammersmith Hospital, Imperial College London; Tel: +44-7-970719453; Fax: +44-2-033130673; E- mail: c.leroux@imperial.ac.uk

at least in part, for the weight loss following bariatric surgery. PYY and GLP-1 are present throughout the intestinal tract with higher concentrations in the distal segments and are released postprandially by intestinal endocrine L cells [4]. GLP-1, together with PYY and OXM act synergistically and cause satiety.

PYY is released in proportion to the calories ingested and is not altered by gastric distension. It has an inhibitory effect on gastrointestinal mobility as well as the gastric, pancreatic and intestinal secretion [5]. PYY and GLP-1 act synergistically and have been shown to induce satiety and reduce food intake in both the obese and the non-obese [3, 6-8]. Interestingly, obese individuals have a PYY deficiency that would reduce satiety and could thus reinforce obesity [9]. GLP-1 augments insulin response to nutrients acting as an incretin. Incretins enhance glucose-stimulated insulin secretion and are hormones secreted by the gastrointestinal tract in response to nutrient ingestion. GLP-1 inhibits also glucagon secretion in a glucose-dependent manner [10].

Ghrelin, a 28-amino acid peptide produced from the fundus of the stomach and the proximal intestine, is the only known orexigenic gut hormone. It leads to increased hunger and food intake [10]. Obese individuals have lower fasting ghrelin levels, and significantly reduced postprandial ghrelin suppression compared to normal weight individuals. Cummings *et al.* showed a profound suppression of ghrelin following RYGB and brought the interaction of weight loss operations and the gut-brain axis into focus [11]. Interestingly, it has been shown that an intact vagus nerve is required for ghrelin to have an appetite effect which has direct technical implications on the surgical procedure [3]. The role of ghrelin in the success of bariatric surgery, however, is only partial and remains to be further elucidated [3]. The gene that encodes ghrelin is also responsible for the encoding of another peptide named obestatin. The role of obestatin is currently controversial, although it might have a role as an anti-appetite agent [3].

CCK is secreted by I cells in the mucosa of the duodenum, jejunum, and proximal ileum in response to a meal. It is involved in the regulation of food intake by inducing satiety following a meal and has been implicated in gastric emptying and distension, gallbladder contraction, pancreatic secretion, and intestinal motility [3]. Though some studies have suggested that CCK is not a mediator of appetite control and weight loss after bariatric surgery, others have detected some changes in its peak concentration following gastric restrictive operations that could have an effect on satiety [3].

A number of studies have shown that changes in gut hormone concentrations may partially explain the weight loss following bariatric surgery [11]. Patients after RYGB, the commonest bariatric procedure worldwide, experience a behavioral change with reduced food intake and increased satiety. We have recently shown that satiety, as measured with visual analogue scores after a standard mixed meal, increases after gastric bypass and this is associated with increased levels of PYY and GLP-1 [12]. This occurs within days postoperatively, prior to any significant weight loss, and the effect is sustained for at least 24 months [13].

3. DIABETES REMISSION AND METABOLIC SURGERY

Bariatric surgery leads to improved glycemic control in type 2 diabetic patients. The reported effects are impressive with a substantial proportion of patients achieving adequate glycemic control without the aid of medication, a condition indistinguishable from remission. A meta-analysis reported improved glycemic control and remission of type 2 diabetes in 83.8% of patients following gastric bypass and 47.8% following gastric banding, many within days of the operation and independent of weight loss [2, 12]. Surgically induced decrease of caloric intake, weight loss, carbohydrate and fat malabsorption, or alterations in gut hormone release have all been suggested as possible explanations for the dramatic effect of bariatric surgery on diabetes [14]. Animal studies suggest that the underlying mechanism that is targeted by bariatric surgery includes altered β-cell secretion, and/or improved entero-insulinar responses, specifically the main incretin hormones GLP-1 and GIP [15].

Gastric restrictive surgeries restrict stomach capacity and limit the intake of solid food and calories and as a result facilitate weight loss. Improved insulin sensitivity and remission of diabetes that is primarily mediated by weight loss occur usually several months following surgery in patients with early stage type 2 diabetes mellitus. On the other hand, rapid improvement of hyperglycemia and even diabetes resolution following bariatric procedures such as RYGB and BPD which share the common feature of bypassing the proximal small intestine may be observed within days after surgery often allowing discontinuation of diabetes medications and suggesting that rapid β-cell enhancing effects related to intestinal bypass occur long before weight loss [14, 15]. Two mechanisms have been proposed to explain this rapid normalization of glucose control after RYGB. The first "foregut hypothesis" suggests that exclusion of the duodenum and proximal jejunum from the transit of nutrients may reduce insulin resistance by preventing secretion of a putative signal that promotes insulin resistance and type 2 diabetes [14, 16]. The second "hindgut hypothesis" holds that diabetes control results from the expedited delivery of nutrient chyme to the distal intestine, enhancing exaggerated physiologic responses from the distal small bowel to nutrients that improve glucose metabolism [14]. In the latter hypothesis gut hormones produced in the distal small bowel such as GLP-1 may act as an incretin stimulating the β-cells in the pancreas to restore insulin responses [17]. However, the improvement of glycemia immediately after RYGB may not be completely attributable to an incretin effect [15].

Though there may be mounting evidence that exclusion of the proximal small intestine in diabetic subjects from contact with ingested nutrients is a critical component to improve glucose tolerance very rapidly and more effectively than great restriction of food intake [14], the mechanisms that lead to sustained weight loss as well as diabetes remission after bariatric operations remain to be elucidated [10]. Several comparative clinical studies have shown that, even though duodenal exclusion results in improvement or even resolution of type2 diabetes, exclusion of the duodenum from the passage of food in non-diabetics disrupts the physiologic entero-insular axis and impairs glucose tolerance [14]. Moreover, as different types of surgery may present different stimuli to the pancreas and gut, the most effective type of surgery remains unclear as well [15].

4. NUTRITIONAL CONSIDERATIONS AFTER BARIATRIC SURGERY

The nutritional status of patients following bariatric surgery is of particular interest especially in view of the fact that it affects healing and the outcomes of further surgical procedures such as body contouring procedures.

Bariatric procedures induce weight loss by a combination of restriction and malabsorption. Even with procedures such as RYGB for which the mechanism of weight loss is not due malabsorption and are rarely associated with protein-calorie deficiency, postoperative deficiencies in iron, and other vitamins and minerals are common. Micronutrient deficiencies mostly occur from malabsorption secondary to bypassing segments of gastrointestinal tract. Micronutrient deficiency needs active surveillance and requires supplementation to maintain normal levels [18, 19]. Considering the prevalence of vitamin deficiencies among obese patients, patients should be screened for these preoperatively as part of their assessment [20].

Protein is one of the main nutrients affected by bariatric surgery. Protein malnutrition, characterized by hypoalbuminemia, anemia, edema, asthenia and alopecia represents a serious potential late complication of some malabsorptive bariatric procedures such as BPD. The pathogenesis is multifactorial, but is most commonly related to excessive malabsorption from bypassing segments of small intestine and to a lesser degree from food limitation [19]. Patients who undergo BPD with or without DS are at high risk of developing nutritional complications which will need specialized care [20].

Patients undergoing bariatric surgery, be it restrictive or malabsorptive, are at risk of developing iron deficiency. The bypass of the primary site of absorption in the duodenum and proximal jejunum may contribute to the development of anemia postoperatively. However, anemia is usually only seen in the setting of other chronic sources of bleeding, such as menstruation or stomal ulceration. There is also inconsistent data regarding iron deficiency after restrictive operations [19].

Regarding purely restrictive procedures with no bypass component a concern has been raised that decreased nutrient intake and an avoidance of certain types of food due to intolerance could lead to vitamin deficiency although no evidence to support this hypothesis is available [19]. Deficiencies have been reported to be frequent after gastric bypass surgery for patients who do not receive vitamin supplementation as well as for those who do [18, 20, 21]. Vitamin B12 and folate deficiencies are highly prevalent though purely restrictive operations may not demonstrate the same deficiency rates [19]. The duodenum and proximal jejunum are selective sites for calcium absorption, while vitamin D is absorbed preferentially in the jejunum and ileum. Calcium and vitamin D deficiencies occur when these segments are bypassed. With a relative lack of calcium, the production of Parathyroid Hormone (PTH) may be increased potentially causing bone loss and long-term risk of osteoporosis [19].

It is important to keep in mind that morbidly obese individuals often have a pre-existing degree of nutritional deficiency before surgery [19]. Identifying which patients will develop deficiencies or complications following bariatric surgery is, however, more challenging. Therefore it is well recognized that regular and lifelong follow-up of all patients after bariatric surgery is necessary [18]. There is a wide variability in practices in due to the lack of available data. However our practice is to monitor calcium, vitamin B_{12}, ferritin, folate and iron indices with blood tests at 3, 6, 12, 24 months postoperatively and yearly thereafter for all patients who undergo bariatric operations with a bypass component.

It is widely accepted that lifelong multivitamin supplementation is required after bariatric surgery. However no controlled trials are available to support the type and dosage of vitamin supplements [18]. The current recommended dietary allowances of these vitamins and nutrients are published by the United States Department of Agriculture (USDA) [19]. Of note is the fact that wide variation in the use of supplements has been reported in the past [22]. Most clinicians arguing from first principles would suggest that multivitamin supplementation is essential postoperatively and would reinforce this in the follow-up appointments [18]. Data do not exist to recommend one specific product, but licensed, commercially available multivitamin tablets which may also be available in chewable or liquid form are usually suggested.

Table **1** [18] outlines our policy for different bariatric procedures and illustrates the difference in management of patients undergoing procedures where clinical malabsorption is present (biliopancreatic diversion and duodenal switch) and those undergoing procedures where malabsorption may not be clinically evident (gastric banding, sleeve gastrectomy and gastric bypass). This difference is supported even further by a recent randomized trial in which gastric bypass and biliopancreatic diversion with duodenal switch were compared [23]. The patients who underwent the latter procedure had a higher risk of vitamin A and D deficiencies in the first year after surgery and of thiamine deficiency in the initial months after surgery [23].

Table 1: Summary of the Different Bariatric Procedures including the Nutritional Challenges Imposed by Each Type of Procedure and the Current Policy for Vitamin Supplementation and Monitoring [10].

PROCEDURE	Protein-Calorie Malabsorption	Vitamin deficiency	Vitamin supplementation	Vitamin monitoring
LAGB	No	Infrequent	Yes	No
SG	No	Infrequent	Yes	Yes
RYGB	Rare	Frequent	Yes	Yes
BPD/DS	Yes	Very frequent	Yes	Yes

LAGB: Laparoscopic Adjustable Gastric Banding.
SG: Sleeve Gastrectomy.
RYGB: Roux-en-Y Gastric Bypass.
BPD: Biliopancreatic Diversion.
DS: Duodenal Switch.

5. CONCLUSION

Nutritional deficiencies develop following bariatric surgery. These deficiencies are seen to a greater extent with the more malabsorptive procedures. Although nutrient deficiencies are common following all types of bariatric surgery, there is a lack of consensus in the form of extent of supplementation. The potential for bariatric patients to develop complications exists if the patients are not adequately followed and managed [19].

Multivitamin supplementation is essential following bariatric surgery. Screening for nutritional deficiencies prior to weight loss surgery and monitoring vitamin levels postoperatively at regular intervals are required. A well-recognized persistent problem of bariatric surgery, however, is the return of appetite and weight regain of a number of patients after the operation. It seems that patients with poor weight loss or poor weight loss maintenance have attenuated postprandial responses of PYY and GLP-1 compared with patients with good sustained weight loss. Appetite and weight loss associated with increased PYY and GLP-1 do not, however, prove causality, and neither does poor weight maintenance associated with attenuated PYY and GLP-1 [12].

Regardless of this fact, it is impossible to approach bariatric surgery outside the context of gut hormones and vice versa. The change in anatomy after gastric bypass surgery with a small stomach pouch, a bypassed stomach and jejunal segment of approximately 100 cm, and a common limb of approximately 300 to 500 cm seem to result in early postoperative changes in profiles of hormones released from the distal gastrointestinal tract [12]. Although gut hormones are affected by bariatric surgical procedures in multiple ways and the effect of bariatric surgery on weight loss and diabetes remission is well recognized, the mechanism for some of these procedures has not been fully elucidated yet. Some of these effects are independent of weight loss, are associated with changes in the gut brain-axis following the altered anatomy and are facilitated by gut hormones.

REFERENCES

[1]   Sjöström L, Narbro K, Sjöström CD, *et al.* Swedish obese subjects study. Effects of bariatric surgery on mortality in Swedish obese subjects. N Engl J Med 2007; 357: 741-52.

[2]   Buchwald H, Avidor Y, Braunwald E, *et al.* Bariatric surgery: a systematic review and meta-analysis. JAMA 2004; 292: 1724-37.

[3]   Pournaras DJ, Le Roux CW. Obesity, gut hormones, and bariatric surgery. World J Surg 2009; 33: 1983-1988.

[4]   Adrian TE, Ferri GL, Bacarese-Hamilton AJ, *et al.* Human distribution and release of a putative new gut hormone, peptide YY. Gastroenterology 1985; 89: 1070-77.

[5]   Deng X, Wood PG, Sved AF, *et al.* The area postrema lesions alter the inhibitory effects of peripherally infused pancreatic polypeptide on pancreatic secretion. Brain Res 2001; 902: 18-29.

[6]   Batterham RL, Cowley MA, Small CJ, *et al.* Gut hormone PYY(3-36) physiologically inhibits food intake. Nature 2002; 418: 650-54.

[7]   Batterham RL, Cohen MA, Ellis SM, *et al.* Inhibition of food intake in obese subjects by peptide YY3-36. N Engl J Med 2003; 349: 941-48.

[8]   Naslund E, Barkeling B, King N, *et al.* Energy intake and appetite are suppressed by glucagon-like peptide-1 (GLP-1) in obese men. Int J Obes Relat Metab Disord 1999; 23: 304-11.

[9]   le Roux CW, Batterham RL, Aylwin SJ, *et al.* Attenuated peptide YY release in obese subjects is associated with reduced satiety. Endocrinology 2006; 147: 3-8.

[10]   Wren AM, Seal LJ, Cohen MA, *et al.* Ghrelin enhances appetite and increases food intake in humans. J Clin Endocrinol Metab 2001; 86: 59-92.

[11]   Cummings DE, Weigle DS, Frayo RS, *et al.* Plasma ghrelin levels after diet-induced weight loss or gastric bypass surgery. NEJM 2002; 346: 1623-30.

[12]   le Roux CW, Welbourn R, Werling M, *et al.* Gut hormones as mediators of appetite and weight loss after Roux-en-Y gastric bypass. Ann Surg 2007; 246: 780-85.

[13]   Pournaras DJ, Osborne A, Hawkins SC, *et al.* The gut hormone response following Roux-en-Y gastric bypass. Cross-sectional and prospective study. Obes Surg 2010; 20: 56-60. Epub 2009 Oct 14.

[14]   Rubino F, Forgione A, Cummings DE, *et al.* The mechanism of diabetes control after gastrointestinal bypass surgery reveals a role of the proximal small intestine in the pathophysiology of type 2 diabetes. Ann Surg 2006; 244: 741-49.

[15]   Kashyap SR, Daud S, Kelly KR, *et al.* Acute effects of gastric bypass versus gastric restrictive surgery on beta-cell function and insulinotropic hormones in severely obese patients with type 2 diabetes. Int J Obes (Lond) 2010; 34: 462-71. Epub 2009 Dec 22.

[16]   Wickremesekera K, Miller G, Naotunne TD, *et al.* Loss of insulin resistance after Roux-en-Y gastric bypass surgery: a time course study. Obes Surg 2005; 15: 474-81.

[17] Polyzogopoulou EV, Kalfarentzos F, Vagenakis AG, *et al.* Restoration of euglycemia and normal acute insulin response to glucose in obese subjects with type 2 diabetes following bariatric surgery. Diabetes 2003; 52: 1098-03.

[18] Pournaras DJ, le Roux CW. After bariatric surgery, what vitamins should be measured and what supplements should be given? Clin Endocrinol (Oxf) 2009; 71: 322-25.

[19] Bloomberg RD, Fleishman A, Nalle JE, *et al.* Nutritional deficiencies following bariatric surgery: what have we learned? Obes Surg 2005; 15: 145-54.

[20] Kaidar-Person O, Person B, Szomstein S, *et al.* Nutritional deficiencies in morbidly obese patients: a new form of malnutrition? Part A: vitamins. Obes Surg 2008; 18: 870-76.

[21] Gasteyger C, Suter M, Gaillard RC, *et al.* Nutritional deficiencies after Roux-en-Y gastric bypass for morbid obesity often cannot be prevented by standard multivitamin supplementation. Am J Clin Nutr 2008; 87: 1128-33.

[22] Brolin RE, Leung M. Survey of vitamin and mineral supplementation after gastric bypass and biliopancreatic diversion for morbid obesity. Obes Surg 1999; 9: 150-54.

[23] Aasheim ET, Björkman S, Søvik TT, *et al.* Vitamin status after bariatric surgery: a randomized study of gastric bypass and duodenal switch. Am J Clin Nutr 2009; 90: 15-22.

CHAPTER 4

Psychosocial Aspects of Massive Weight Loss after Bariatric Surgery

Ronis Magdaleno Júnior[*]

Department of Medical Psychology and Psychiatry, Faculty of Medical Sciences, University of Campinas UNICAMP, Rua Padre Almeida 515, sala 14, Campinas, Brazil 13025-251

Abstract: The objective of the present chapter is to understand the gamut meanings for patients when undergoing bariatric surgery, the impact that this represents in their lives, the psychosocial complications and to offer some recommendations for a better psychological evolution of patients undergoing bariatric surgery.

Bariatric surgery is a procedure that results in a complex network of emotional experiences. The main emotional experiences in the post-operative procedures are: social re-insertion, personal acceptance, risk of disillusion, recovery of self esteem, improvement in quality of life and body image. Body contouring following a significant weight loss can re-establish a good psychosocial functioning, but it must be stated with well-established criteria. Some practical considerations to deal with the operated patients are proposed.

Keywords: Psychology, bariatric surgery psychology, body image, self esteem, personal acceptance.

1. INTRODUCTION

Morbid obesity is usually accompanied by profound impairment of the body image construct and by the presence of severe psychological problems concerning body weight and body shape. Psychological consequences of extreme obesity are: anxiety, depression, low self-esteem, and negative body image [1-3]. Discrimination due to obesity, social isolation and stigmatic experiences start in the earliest social contacts [3-5], and this prejudice may contribute to depression, eating disorders, body image disturbance, and other forms of suffering [6-8].

The increase in the prevalence of obesity, failures in conventional treatments, accompanied by development of new surgical techniques, above all laparoscopies, have led to an expressive increase in the number of bariatric surgeries throughout the last decade [9-13]. The decision to undergo bariatric surgery is commonly prompted by life-threatening medical risks, but psychosocial factors, including social isolation, depression, discrimination and inability to perform desired tasks, are primary reasons for deciding to have obesity surgery. The surgical treatment not only leads to substantial weight loss reduction, but also to the improvement or cure of co-morbidities, including improved quality of life [2, 14-16] and body image [1], diminished psychopathology [17] and abnormal eating behavior [18], and enhanced psychosocial functioning [1, 19, 20].

However, in spite of the evolutions of surgical techniques [21] and the pre and post operative handling of these patients, the long term maintenance of weight loss is not a guaranteed result [22] since the success of therapeutic surgery depends on significant behavioral changes, being strongly influenced by the individuals' ability to implement consistent and permanent changes in their life style [5], as well as to acquire the capacity to diminish the use of food as a means of fulfilling emotional needs [2, 23].

During the National Institutes of Health (NIH) Consensus Development Conference of 1991, bariatric surgery was recommended for well-informed and motivated patients with acceptable operative risks and class III

*__Address correspondence to Ronis Magdaleno Júnior:__ Department of Medical Psychology and Psychiatry, Faculty of Medical Sciences, University of Campinas—UNICAMP, Rua Padre Almeida 515, sala 14, Campinas, Brazil 13025-251; Tel: +55-19-32542103; Fax: +55-19-32032103; E-mail: ronism@uol.com.br

Bishara S. Atiyeh and M. Costagliola (Eds)

obesity, and for those with class II obesity and high-risk pre-morbid conditions. Careful selection of candidates for the surgery carried out by a multidisciplinary team has been also suggested as a crucial aspect [21].

Actually, the objective of bariatric surgery is, besides the reduction of weight and co-morbidity, to bring about an improvement in the psychosocial functioning and in the quality of life [1, 2, 19, 20, 24]. However, a significant number of patients do not benefit psychologically from surgery, and present premature cessation of weight loss, develop alimentary and/or psychopathological disorders, and experience deterioration of their quality of life [5, 9, 19, 24, 25]. This is still a controversial topic. Better understanding of the meanings that obesity as well as surgery has for these patients is an urgent necessity.

Factors such as sound self-esteem and good social and personal adjustment, positive changes in pre-op psychological parameters, satisfaction with the results and confidence in having the necessary skills to adopt and maintain new behavioral models are crucial to a satisfactory post-operative evolution [20]. Dziurowicz-Kozlowska *et al.* [2] suggested that the psychological evaluation of candidates for bariatric surgery should focus on patients' general psychological and mental health condition, time and circumstances of development of obesity, dieting history, lifestyle, eating behavior, functions of eating, structure and quality of the social support network, knowledge of the essence of surgery, motivations to undergo surgery and realistic expectations regarding its outcomes as well as an awareness of the need to make permanent modifications in habits and remain under physician's supervision.

However, there are still no reliable predictors for therapeutic success [26, 27] and many patients fail because of psychological difficulties [5, 28], a fact that reinforces the need for specific studies on the psychological dynamic, with a view to establishing a psychosocial approach during the evaluation of the candidates [2, 25, 29]. To achieve this, besides the objective measures relative to the weight loss and the improvement of health conditions, it is fundamental to know the psychosocial factors involved in the patient's evolution.

The impact of the dramatic weight loss brought about by the surgery on the psychological and social well-being of these patients is yet to be fully understood. Thus, the objective of the present chapter is to understand the gamut of meanings for patients when undergoing bariatric surgery, the impact that this represents in their lives and the psychosocial complications.

2. THE MAIN EXPERIENCES OF PATIENTS AFTER BARIATRIC SURGERY

Bariatric surgery is a procedure that results in a complex network of emotional experiences. These results are linked to several factors, including:

- Rapid and massive weight loss, without the patient having time to elaborate the physical, psychological and social changes.

- Imbalance in eating habits caused by anatomical impediment to ingesting copious amounts of food.

- Unreal expectations *vis-à-vis* the scope of treatment.

- Greater social re-insertion with greater visibility, which exposes them to previously inexistent demands.

Due to these factors, the pre-operative preparation of patients becomes essential, not only from the clinical point of view, but above all, emotionally. Patients must undergo surgery duly informed as to what to expect in the post-operative period. Possible psychic conflicts which may arise from the patient's increased social visibility and contact with others must be addressed and worked through (Table **1**).

Psychological evaluation of a candidate for bariatric surgery is one of the most important and difficult elements of the clinical assessment [20, 30]. The candidate understanding of surgery meanings and of obesity plays a fundamental role in determining whether he/she is suitable for undergoing surgery and for ensuring a good prognosis.

Table 1: Important Psychological Risk Factors to be Considered in Patient Selection.

Psychological risk factors
Presence of eating disorders, especially Binge Eating Disorder
Cognitive impairment or limited resources for elaborating anguish
Presence of depressive disorders
Patients who imagine the surgery will solve all their problems
Denial of difficulties inherent in the surgery
Association of weight loss with the solution of emotional problems
Personality with strong compulsive traits
Presence of psychotic structure
Defensive and protective role of obesity
Lack of social and family support
Serious conflicts related to sexuality
Lack of motivation for the surgery and post-operative

3. EMOTIONAL EXPERIENCES IN THE POST-OPERATIVE PERIOD (Table 2)

Social Re-Insertion

From the emotional point of view, the most evident consequence of a long period of obesity is the damage caused by rejection and by lack of a place in society. Offenses that obese individuals suffer, "jests", criticisms, inadequacy in public and even private places, all start little by little to undermine their self esteem, with intense feelings of worthlessness and alterations of identity.

Besides exacerbating tendency for social isolation, social structure, adjusted primarily for non-obese individuals, causes constant embarrassments. Decreasing life's options as years go by gives rise to increasing exclusion and loss of self esteem that foment a vicious circle within which the individual becomes isolated and introverted. This increases addictive behavior to counteract impoverishment in the capacity to solve problems, reinforcing feelings of worthlessness and sadness.

Surgery, after an odyssey through all the possible ways of losing weight, presents itself as an option and hope for a cure and a restart of the process of an active social life. The first source of relief, after post surgical recovery, comes from a strong sensation of acceptance and social reinsertion. Patients feel that they are part of a world which previously they felt not belonging to. They experiment a feeling of genuine happiness, albeit, continually threatened by the fear of going back to gaining weight and losing the new stage they have achieved [31].

Personal Acceptance

The obese patients' anguish is further reinforced by the progressive medical discoveries and evidence as to the devastating effects of excess weight on an individual's health. In addition to feelings of exclusion and isolation, this generates a feeling of despair and urgency to get free from the problem of obesity in a definitive and rapid manner. After surgery, patients experience an improved social acceptance [31]. The whole process is accompanied by the hope of finding a new place within the social context, a place that is marked by acceptance and admiration by others as well as by themselves, something long forgotten and replaced by discrimination, stigmatization, and social inadequacy [31].

The stigmatization and discrimination they suffer, besides the low self esteem, are directly related to depressive symptoms commonly found associated with obesity. There are several studies showing the high prevalence of depression in people with obesity [6, 9, 32-34], a number that is significantly higher than in the general population [22]. The expectation of a decrease in stigmatization and discrimination, accompanied by their consequent acceptance, is a fundamental element in the decision to undergo surgery. After weight loss, the feelings of shame, discrimination and exclusion start to be replaced by other more positive feelings related to improving quality of life and social functioning [31].

Risk of Disillusion

During the initial period of weight loss an increase of self esteem, self confidence, assertiveness and expressiveness occurs, revealing a feeling of euphoria and a denial of difficulties with the patient feeling triumphant against obesity. This behavior is strongly reinforced by the health team that overvalues the consequent visual results of surgery. This very frequent phase in the post-operative period has no definite duration; it generally begins as soon as the physical recovery from the surgery is concluded and lasts up to one to one and a half year. It is a self-limited period, which when it comes to an end, can jeopardize the patient's motivation to go on with the treatment. After this phase, the patients have to encounter new life experiences, such as jealousy, mistrust, fear and the envy of others that, a short while previously, simply did not exist.

Most obese patients show psychological and interpersonal improvements after surgery directly related to weight loss and the impression of having found a solution to their problems. Many of the patients have high expectations and hope solving all their lives' problems with surgery. Unfortunately, some become disappointed when realizing that their lives did not improve dramatically after weight loss. Moreover, reinforced by information obtained from non reliable sources [30], most bariatric surgery candidates frequently believe that they will lose more weight than anticipated and tend to underestimate the efforts and minimize the risks involved [35].

A complexity of feelings have been witnessed to evolve post-operatively, sometimes ending in a breakout of psychic symptoms present pre-operatively, or in new situations of stress that lead to new psychic or even psychiatric symptoms.

Van Hout *et al.* [1, 19] have shown that gradual weight regain, as time goes by, is related to lessening of psychosocial gains, with disappointments linked to the fact that life has not improved substantially complicated by difficulties in adapting psychologically to the consequences of surgery.

Part of the favorable emotional gains is reinforced by an illusion induced by rapid and evident weight loss during the first months, when it appears that, besides the limited physical discomforts generated by the surgery itself, the major problems of their lives appear to be definitely overcome without much effort. Loss of the illusion in solving problems through surgery may explain decreasing medium and long-term psychological gains observed in the literature [31].

The so important role of seduction, particularly for females, which was previously paralyzed and imprisoned by the excess fat and its psychic consequences, becomes above all shame a possibility available to obese patients. Nevertheless, as patients become exposed to new roles and tensions they are not always able to deal with, an imbalance in relationships with others emerges [31].

When the reactions of others, precisely of those whose "acceptance" was believed guaranteed by the fact of loosing weight and getting thin (and here is the central point of the illusion), begin to appear, patients can feel profoundly disillusioned, sometimes confused, which could lead to depressive, phobic and anxious states, as well as to lack of motivation to continue the treatment for further weight loss [31].

Another element that strongly contributes to the disillusion is the fact of weight regain either by other patients they know or by themselves, and, particularly, when the patients realize that their bodies, even though thinner following massive weight loss, still maintain the stigma of obesity, most importantly flaccidity and excess skin.

The shame, previously attributed to obesity, becomes associated with flaccidity, skin folds and scars. It is at this moment that there is a great risk of isolation that requires special attention by the psychological team because the disillusion and the end of the euphoric triumphant period over the illness can have unfortunate consequences on the whole process. The patients become disinterested in the follow up care required to maintain the surgery's objectives which constitutes a psychic and even physical threat [36].

Recovery of Self Esteem and Improvement in Quality of Life

Self esteem is a fundamental factor for the treatment's success [31] and is a strong factor linked with the best results of bariatric surgery [20]. There is evidence in fact that it is significantly improved by bariatric surgery.

A drive to recover their own identity lost or hidden by excess weight is the main stimulus for surgery [31]. The sensation of finding themselves again is experienced with great pleasure and relief. To them a new life begins when they are discharged home after surgery and the process of recovering their identity is lived as born-again. Patients must learn to eat and learn to get acquainted with reality in a different way. All this is experienced with a lot of satisfaction [31]. The possibility of reconstructing one's identity is the most important factor in the recovery of self esteem, and in completing the vicious circle that leads to a greater motivation to continue the struggle against regaining weight [31].

Among the obese, the indexes of dissatisfaction with the quality of life are very high. These are very significantly improved after bariatric surgery [14, 31, 37]. Mitchell *et al.* [32] reported notable improvements in the quality of life of obese patients after surgery. Van Hout *et al.* [1, 19] and Dixon *et al.* [38] showed, however, evidence that, despite the early significant gains in the quality of life after surgery, these gains tend to become more modest as time goes by Kolotkin *et al.* [39] argued that even though there is a long term improvement in physical fitness of patients following bariatric surgery, there is still a lot of controversy regarding improvement in their mental state.

Improvement in Body Image: Acceptance and Defenselessness

The importance of an unsatisfactory body image in causing psychological distress in obese patients has been thoroughly investigated [40]. This psychological distress encompasses lack of self-esteem, depression, and tendency to avoid social and sexual relationships.

Weight loss after surgery leads to marked improvement in body image and attractiveness, and would be one of the main reasons for post-surgical psychological improvement, better social integration, and an enhanced quality of life [31, 36]. Surgery and weight loss are experienced by patients as a valuable opportunity to recover their place in society as well as their identities long hidden by excess weight [31]. The surgically obtained weight loss is accompanied by a marked improvement in the attitudes towards body weight and shape, and by a substantial normalization of the body image and self esteem. Following stable weight loss, the post-obese subjects are highly satisfied with their new lean body shape, and their quality of life markedly improves both in a socially-accepted somatic morphology and in improved mobility. It could gradually lead to the cessation of distressing and pathological body-related behavior.

However, in addition to the visibility they generate, surgery and weight loss are also a motive of apprehension and fear due to the sense of losing the protective barrier that obesity provides [36]. As a result of weight loss, many obese persons find themselves confronted with emotional and psychosocial problems they encounter for the first time [37]. After all the efforts to find their place in the world again and to be admired, they are faced with the reappearance of their sexual bodies: a new situation with which they are unable to cope [36].

King *et al.* uncovered a relation between morbid obesity and childhood traumas and postulated that accumulation of excess fat is a mean of protection from any close contacts. Weight loss is thus experienced as a danger to the integrity of the individual who once again feels exposed to a reality he is afraid of. Obesity, thus, can be associated to secondary gains in subjects who suffered sexual abuse and would feel threatened by being physically attractive.

Obesity is not just a problem; rather it can also be used as means of protection in interpersonal relations. Weight loss may hence be experienced with intense affective ambivalence [36]. At the time when patients become attractive and start being admired, phobic symptoms appear. There is a feeling of lack of protection when the shield represented by excess body fat is lowered. The feeling of helplessness is evident primarily in women who have lost their protective barrier and have to face the world. What obesity used to justify, that is isolation in relation to others and the feeling of rejection, must now be re-expressed in a new structure, in which obesity plays no role [36].

The period following massive weight loss is that in which new life experiences emerge, ranging from feelings of genuine happiness to clear expressions of jealousy, mistrust, fear and envy felt by others. Sometimes it happens that the patient's obesity may serve certain functions which satisfy the needs of the family system; hence the patient's weight loss may be perceived by his/her most immediate entourage as an undesirable and threatening phenomenon. Partners' jealousy is a new factor, which they are not used to deal with. Weight loss may create an imbalance in the relationship between partners [36].

What is frequently confusing to massive weight loss patients is the forced mourning of the loss of a place which, however bad it may have appeared, provided protection from the world and from their own impulses that were imprisoned within and compensated by compulsive eating. A feeling of defenselessness ensues by the shattering of a certain feeling of ease in the protective shelter created by excess body fat.

Table 2: Summary of the Main Post-Operative Experiences.

Social re-insertion
Personal acceptance
The risk of disillusion
The recovery of self esteem and improvement in quality of life
Improvement in body image: Acceptance and defenselessness

4. THE PSYCHOLOGICAL IMPACT OF POST-BARIATRIC BODY CONTOURING SURGERY (Table 3)

Patients who have acquired an increased life expectancy and a new outlook through massive weight loss are often left with functional sequelae as a consequence of generalized cutaneous excess and ptosis that result. Once weight loss has been achieved, the body image of the "cured" obese patient is not infrequently cosmetically unacceptable. Deformities engendered by massive weight loss are constant reminders to the patient of physical and psychological difficulties. To complete treatment, body contouring procedures are performed to correct those functional and aesthetic impairments.

In clinical practice, in spite of satisfactory weight loss and maintenance, some individuals still disparage their body image after bariatric surgery, and seek body contouring procedures in order to improve their physical appearance. Patients are very discontent with the increasing skin folds or continue to be concerned with their body shape and size. The same issue of shame, that had previously been attributed to obesity is, now attributed to flaccidity, skin folds, scars, and an empty bag-like appearance [36]. These remnants of obesity strongly contribute to the frustrated expectations of having once again a beautiful, healthy and functional body. Skin and soft tissue redundancy of the trunk, buttocks, breasts, upper arms, and thighs following massive weight loss are extremely unsightly, and plastic surgical procedures may be necessary to restore a pleasant body image. Body contouring surgery following post-bariatric massive weight loss allows the patient to overcome functional disorders that hinder social interactions, and enables him/her to achieve a balance in psychological functioning and improved image perception.

Some formerly obese individuals may perceive their body as aesthetically worsened following massive weight loss increasing the risk of weight regain. Due to this perception, the risk of isolation is great, demanding special attention from the psychological team, as disillusion can threaten the patient both psychologically and physically, making the patients to lose interest in the care demanded to fulfill the objectives of surgery.

At this point, in their minds, plastic surgery begins to take the place previously occupied by bariatric surgery, that is, the hope of achieving a position in which they are accepted and admired. Undergoing plastic surgery to correct the "leftovers" of the bariatric surgery becomes a vital issue. The expectation of undergoing corrective plastic surgery is experienced with the same anxiety and urgency as they felt when awaiting bariatric surgery and they are caught up in the hope of finding their bodies and their sexuality. Body contouring following significant weight loss can re-establish a good psychosocial functioning because of the perception of improved body image.

Table 3: Psychosocial Benefits of Plastic Surgery for Bariatric Patients.

Increased self-esteem
Improved self-confidence
More intense social activities
Better interpersonal relationships
Relief from depression and anxiety
Improvement in emotional stability
Better relationship with colleagues and partner
Less stigmatization
Better sexual performance
Fewer physical limitations during sexual intercourse

Datta *et al.*, 2006

5. PRACTICAL CONSIDERATIONS

Classically, results of bariatric surgery have been evaluated by the degree of weight loss and the improvement in the general health condition, besides the psychosocial benefits attained. However, these parameters tend to be evaluated separately in the literature, which limits the value of the conclusions attained. For example, there are patients who present an improvement in their ability to control food impulses, even though they have not reached their goal in weight loss. Others can achieve their goal in weight loss but are incapable of adapting to their new body image, or feel a strong anxiety associated with eating experienced as a loss in the quality of life even though their weight loss goal has been attained. For this reason, it appears obvious that other forms of evaluation of surgical results are necessary taking into consideration the true significance surgical procedures have for each particular patient.

It is important to identify those aspects of a patients' psychological make-up which would be expected to improve or worsen their prognosis, and to provide the necessary pre- and post-operative psychosocial counseling. A better understanding of patients' psychosocial functioning after bariatric surgery is also needed to facilitate the identification of the psychological variables so as to improve interventions and to ensure adequate adjustment and success.

At present, professionals are asked to make decisions about bariatric candidates without genuine evidence for support. We hope to offer some markers to assist the professionals in their decision regarding surgery and to help in developing a specific assessment protocol of the psychosocial aspects of candidates for bariatric surgery. Such markers will allow an appropriate psychosocial plan to be developed and the health team to identify factors that may negatively or positively affect prognosis.

There are some areas on which the attention of the health team should be focused:

- Understanding the influences that massive weight loss has on the new life experiences, and the ambivalence with which bariatric surgery is viewed, on one hand as intensely desired and on the other apprehended being a source of profound suffering depriving patients the joy of eating, something extremely pleasing to them.

- Being attentive to the great influx of new life experiences, some of which the patients are not prepared for, such as, feelings of jealousy, envy and competition.

- Rapidly identifying and valuing the experience of reinsertion and social acceptance as the most significant gain of surgery by the health team since it directly influences self esteem that, a strong positive predictive factor.

- Considering self esteem by mental health professionals as the center of attention since it is the starting point for the vicious circle: low self esteem – anxiety – search for food.

- Recognizing shame and discrimination as strong dissimulating factors inversely related to self esteem. These can lead to physical and mental isolation, that are factors that can cause the installation of a depressive state and other psychiatric disorders. At this stage body contouring surgery plays a very important role.

- Identifying when disillusion becomes critical, specially when patients seek the services with unreal expectations. Disillusion is strongly reinforced whenever patients have the urge to believe that surgery is the solution to all their other problems. The best way to approach this issue is by psychotherapeutic preoperative intervention.

- Focusing, besides psychotherapeutic treatment, on helping patients with morbid obesity to better adjust their eating habits without neglecting the defensive function of obesity. Patient's re-encounter with his/her sexual body involves re-emergence of conflicts so far hidden by obesity. Special attention should be given to conflicts related to patient's sexuality, to feelings of helplessness and the phobic symptoms that can appear after massive weight loss.

6. CONCLUSION (Table 4)

Table 4: Recommendations for a Better Psychological Evolution of Patients Undergoing Bariatric Surgery.

Psychotherapy focusing on improving self-esteem
Identifying feelings of shame related to a greater exposure
Identifying how patients deal with feelings of competitiveness, envy and jealousy
Assessing if obesity has a defensive function
Providing realistic parameters with regards the results of surgery

Table 4: cont....

Distinguishing between realistic necessity for plastic surgery and unreal expectations
Observation of deviations to other compulsions
Acceptance of skin folds and scars
Clarifying the misunderstanding between emotional and physical hunger
Confidence in own ability to adopt new behavioral models

REFERENCES

[1]　van Hout GCM, Fortuin FAM, Pelle AJM, *et al.* Psychosocial functioning, personality, and body image following vertical banded gastroplasty. Obes Surg 2008; 18: 115-20.

[2]　Dziurowicz-Kozlowska AH, Wierzbicki Z, Lisik W, *et al.* The objective of psychological evaluation in the process of qualifying candidates for bariatric surgery. Obes Surg 2006; 16: 196-202.

[3]　Wadden T, Sarwer DB. Behavioral assessment of candidates for bariatric surgery: a patient-oriented approach. Obesity 2006; 14 (Suppl 2): 53S-62S.

[4]　Papageorgiou GM, Papakonstantinou A, Mamplekou E, *et al.* Pre- and postoperative psychological characteristics in morbidly obese patients. Obes Surg 2002; 12: 534-9.

[5]　Bocchieri LE, Meana M, Fischer BL. A review of psychosocial outcomes of surgery for morbid obesity. J Psychosom Res 2002; 52: 155-65.

[6]　Kalarchian MA, Marcus MD, Levine MD, *et al.* Psychiatric disorders among bariatric surgery candidates: relationship to obesity and functional health status. Am J Psychiatry 2007; 164: 328-34.

[7]　Puhl RM, Moss-Racusin CA, Schwartz MB. Internalization of weight bias: implications for binge eating and emotional well-being. Obesity 2007; 15: 19-23.

[8]　Organização Mundial da Saúde. Obesity and overweight. http://www.who.int/dietphysicalactivity/publications/facts/obesity/en (accessed August 28, 2009).

[9]　Bauchowitz AU, Gonder-Frederick LA, Olbrisch ME, *et al.* Psychosocial evaluation of bariatric surgery candidates: A survey of present practices. Psychosomatic Medicine 2005; 67: 825-32.

[10]　Pareja JC, Pilla VF, Callejas-Neto F, *et al.* Gastroplastia redutora com bypass gastrojejunal em Y-de-Roux: conversão para bypass gastrintestinal distal por perda insuficiente de peso - experiência em 41 pacientes [Gastric bypass Roux-en-Y gastrojejunostomy – conversion to distal gastrojejunoileostomy for weight loss failure – experience in 41 patients]. Arq Gastroenterol 2005; 42: 196-200.

[11]　Sarwer Db, Cohn NI, Gibbons LM, *et al.* Psychiatric diagnoses and psychiatric treatment among bariatric surgery candidates. Obes Surg 2004; 14: 1148-56.

[12]　Garrido Junior AB. Cirurgia em obesos mórbidos – experiência pessoal. Arq Bras Endocrinol Metab 2000; 44: 106-10.

[13]　Segal A, Fandiño J. Indicações e contra-indicações para realização das operações bariátricas [Bariatric surgery indications and contraindications]. Rev Bras Psiquiatria 2002; 24 (suppl 3): S68-S72.

[14]　de Zwaam M, Lancaster KL, Mitchell JE, *et al.* Health-related quality of life in morbidly obese patients: effect of gastric bypass surgery. Obes Surg 2002; 12: 773-80.

[15]　Dymek MP, Le GrangeD, Neven K, *et al.* Quality of life after gastric bypass surgery: a cross-sectional study. Obes Res 2002; 10: 1135-42.

[16]　Dymek MP, le Grange D, Neven K, *et al.* Quality of life and psychosocial adjustment in patients after Roux-en-Y gastric bypass: a brief report. Obes Surg 2001; 11: 32-39.

[17]　Maddi SR, Ross Fox S, Khoshaba DM, *et al.* Reduction in psychopathology following bariatric surgery for morbid obesity. Obes Surg 2001; 11: 680-185.

[18]　Sawrer DB, Wadden TA, Fabricatore AN. Psychosocial and behavioral aspects of bariatric surgery. Obes Res 2005; 13: 639-48.

[19]　van Hout GCM, Boekestein P, Fortuin FA, *et al.* Psychosocial functioning following bariatric surgery. Obes Surg 2006; 16: 787-94.

[20]   Delin CR, Watts J, Bassett DL. An exploration of outcomes of gastric bypass surgery for morbid obesity: patient characteristics and indices of success. Obes Surg 1995; 5: 159-70.

[21]   Buchwald H. Consensus Conference Statement Bariatric surgery for morbid obesity: health implications for patients, health professionals, and third-party payers. J Am Coll Surg 2005; 200: 593-604.

[22]   Hsu LK, Benotti PN, Dwyer J, *et al.* Non-surgical factors that influence the outcome of bariatric surgery: a review. Psychosom Med 1998; 60: 338-46.

[23]   Godfrey JR, Brownell K. Toward optimal health: The influence of the environment on obesity. J Women's Health 2008; 17: 325-30.

[24]   van Hout GCM, Vreeswijk CMJM, van Heck GL. Bariatric Surgery and bariatric psychology: evaluation of Dutch approach. Obes Surg 2008; 18: 321-25.

[25]   van Hout GCM, Van Oudeheusden I, Krasuska AT, *et al.* Psychological profile of candidates for vertical banded gastroplasty. Obes Surg 2006; 16: 67-74.

[26]   Herpertz S, Kielmann R, Wolf AM, *et al.* Do psychological variables predict weight loss or mental health after obesity surgery? A systematic review. Obes Res 2004; 12: 1554-69.

[27]   Larsen F. Psychosocial function before and after gastric banding surgery for morbid obesity. Acta Psychiatr Scand Suppl 1990; 359: 1-57.

[28]   Vallis MT, Ross MA. The role of psychological factors in bariatric surgery for morbid obesity: identification of psychological predictors of success. Obes Surg 1993; 3: 346-59.

[29]   Fandiño J, Benchimol AK, Coutinho WF, *et al.* Cirurgia bariátrica: aspectos clínico-cirúrgicos e psiquiátricos. Rev Psiquiatr Rio Gd Sul 2004; 26: 47-51.

[30]   DeMaria EJ. Bariatric surgery for morbid obesity. N Engl J Med 2007; 356: 2176-83.

[31]   Magdaleno R Jr, Chaim EA, Turato ER. Understanding the life experiences of Brazilian women after bariatric surgery. A qualitative study. Obes Surg 2008 Oct 2. [Epub ahead of print], PMID: 18830785.

[32]   Mitchell JE, Lancaster KL, Burgard MA, *et al.* Long-term follow-up of patients' status after gastric bypass. Obes Surg 2001; 11: 464-8.

[33]   Dixon JB, Dixon ME, O'Brien PE. Depression in association with severe obesity. Arch Intern Med 2003; 163: 2058-65.

[34]   Roberts RE, Kaplan GA, Shema SJ *et al.* Are the obese at greater risk for depression? Am J Epidemiol 2000; 152: 163-70.

[35]   Glinsky J, Wetzler S, Goodman E. The psychology of gastric bypass surgery. Obes Surg 2001; 11: 581-8.

[36]   Magdaleno Jr R, Chaim EA, Pareja JC, *et al.* The psychology of bariatric patient: what replaces obesity? A qualitative research with Brazilian women. Obes Surg. 2009 Mar 21. [Epub ahead of print], PMID: 19306052.

[37]   Kinzl JF, Schrattenecker M, Traweger C, *et al.* Quality of life in morbidly obese patients after surgical weight loss. Obes Surg: 2007; 17: 229-35.

[38]   Dixon JB, Dixon ME, O'Brien PE. Quality of life after Lap-Band placement: Influence of time, weigth loss, and comorbidities. Obes Res 2001; 9: 713-21.

[39]   Kolotkin RL, Head S, Hamilton M, *et al.* Assessing impact of weight on quality of life. Obes Res 1995; 3: 49-56.

[40]   Friedman KE, Reichmann SK, Costanzo PR, *et al.* Body image partially mediates the relationship between obesity and psychological distress. Obes Res 2002; 10: 33-41.

CHAPTER 5

Contour Deformities After Massive Weight Loss

Steven Rueda, Mari Rebane and Seth Thaller[*]

The DeWitt Daughtry Family Department of Surgery, University of Miami School of Medicine, Miami, Florida, USA

Abstract: Popularity of bariatric surgery has increased exponentially over the past several years. As a result, a new patient's population has emerged seeking plastic surgery: the Massive Weight Loss (MWL) patient. Patients who experience MWL present with bi-dimensional skin excess and are left looking "deflated," with disfiguring skin laxity circumferentially around the torso, including the breasts, redundant tissue on the upper arms, buttocks, and thighs leading to poor social acceptance and quality of life.

The strength with which the superficial fascia attaches to the underlying muscular fascia varies throughout the body. Areas of high strength are called zones of adherence and usually excessive tissue laxity are especially evident in these zones of decreased adherence. Post-bariatric weight loss contour deformities well exceed contour deformities plastic surgeons have ever encountered previously. These deformities are diverse and often severe in nature. It is not possible however to predict where body contouring deformities will materialize because they may be present anywhere on the body. Nevertheless their accurate classification can assist the surgeon in operative planning.

Keywords: Body contour deformity, superficial fascia, skin laxity.

1. INTRODUCTION

Obesity has become a universal healthcare condition. It has been estimated that close to 30% of Americans are obese as measured by the Body Mass Index. An additional 5% have been estimated to be morbidly obese and more than 65% of Americans are over their ideal body weight [1]. A large number of co-morbidities have been associated with obesity including hypertension, hyperlipidemia, type 2 diabetes, cardiovascular disease, osteoarthritis, and sleep apnea among others [2]. Lifestyle management of obesity remains very challenging and has been shown to lead to a weight reduction of no more than 5 to 10% if successful [1, 3].

Bariatric surgery is increasing in frequency and is offering obese patients at present the greatest degree of sustained weight loss. Once the final alternative for only the morbidly obese (body mass index >40), bariatric surgery is increasingly being performed in people who are only severely obese (body mass index >35) [4]. Subsequently, with an increase in number of successful weight loss patients, contouring of all regions of the body after bariatric surgery is growing in frequency and is rapidly undergoing modification and refinement growing to a new speciality in plastic surgery [4].

After shedding greater than 50 percent of their excess weight, patients are often left "deflated" with loose ptotic skin envelopes and oddly shaped protuberances well exceeding usual contour deformities most plastic surgeons have encountered previously [1, 4]. The excess skin that hangs from the torso, abdomen and extremities is not only extremely unsightly, but can be painful and susceptible to recurrent intertriginous infections [1]. Post bariatric deformities are vast in scope and variety and have yet to be adequately classified. They can only be loosely predicted based on preoperative appearance, degree of weight loss, and patient age [4].

---

*Address correspondence to Seth Thaller: The DeWitt Daughtry Family Department of Surgery, University of Miami School of Medicine, Miami, Florida, USA; Tel:+1-305-243-4500, Fax:+1-305-243-4535; E-mail: SThaller@med.miami.edu

Bishara S. Atiyeh and M. Costagliola (Eds)

2. UNDERSTANDING THE ANATOMY OF BODY CONTOURING [5]

As post-bariatric body contouring surgery enters a new era of sophistication, improving patient care and optimizing results entails a comprehensive understanding of the superficial fascia system. Superficial fascia system consists of a connective tissue network interwoven between the dermis and the muscular fascia. Superficial fascia has two main functions: anchoring the skin to the underlying musculature and tissues, and encasing the fat in the trunk and extremities. It achieves this through various horizontal membranous sheets that interconnect with one another and with the dermis and muscular fascia; encased between these membranous sheets lies the body fat (Fig. 1).

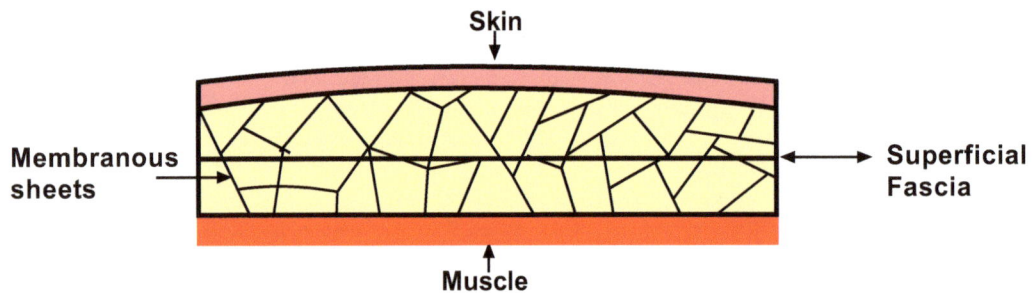

Fig. 1. Superficial Fascia System.

The strength with which the superficial fascia attaches to the underlying muscular fascia varies throughout the body. Areas of high strength are called zones of adherence. On the body, these areas of tight adherence are visible as the skin creases and plateaus. Skin creases are the groin, the inframammary, the gluteal, and the joint creases. Plateau areas are the anterior and posterior midline of the chest and back. Areas high in fat deposit, on the other hand, have low strength of fascial attachment.

The superficial fascia provides anatomic support to the fat and the skin. Utilizing the fascial system in body contouring surgery has several advantages. Advantages include better support to tissues undergoing contouring, equalizing the skin tension lines, lifting areas of ptosis, assisting in predicting final location of scars, and preventing excessive tissue displacement.

3. FACTORS THAT INFLUENCE THE SUPERFICIAL FASCIA AND PATHOLOGIC CHANGES

Variations in the superficial fascia are mainly due to adiposity, body regions, and gender. Age, amount of sun damage, and obesity has also been shown to significantly weaken fascial attachments. Adverse effects of adiposity on the fascia are even seen in people with BMI inconsistent with obesity.

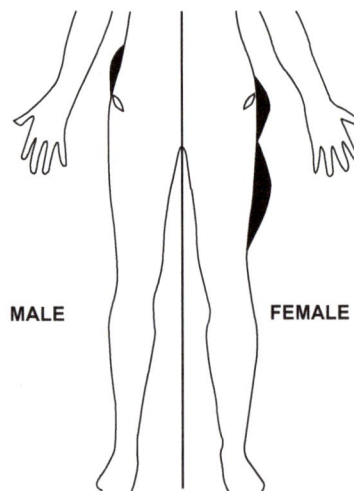

Fig. 2. Zones of Decreased Adherence in Males and Females. Areas in Black Denote Locations of Reduced Adherence.

Increasing quantities of fat separate the layers of fascia, making them significantly weaker. Anatomical nature of the fascial system varies in different regions of the body. In the majority of the trunk, fascia lies in a single sheet as in Scarpa's and Colle's fascia. Fascial layer is also much more prominent on the posterior half of the trunk and thighs than on the anterior half.

Male or female gender affects the zone of adherence of the fascia. As seen in Fig. **2**, the fascia is strongly adhered to the periosteum of the iliac crest in males and to the gluteal depression in females. Superficial fascia also varies between males and females in the breast. In women, the fascia splits into anterior and posterior lamellae forming the fibrous stroma of the breasts.

Female fat deposition is circumferentially concentrated around the flanks, thighs, lower abdomen, and intra abdominally. Men tend to accumulate fat more centrally. After these patients undergo radical weight loss, excess skin tends to develop in areas of maximum fat accumulation [6].

Skin Laxity

The etiology of skin laxity after rapid weight loss has been speculated upon. Microscopic studies have shown that the subdermal to aponeurosis fibroelastic spans, overflowing with adipocytes in the obese, with rapid weight loss has fractured elastin fibers, which subsequently translates into sagging of the abdominal, breast, and buttock skin and of the inner portions of the arms and thighs [7]. With increasing age, the entire complex of skin, fat, and fascia undergoes further stretching and relaxation. Excessive tissue laxity is especially evident in the zones of decreased adherence.

Despite its usually poor tone due to the prolonged period during which it was under tension and despite the frequent history of "yo-yo" dieting, the skin has the ability to tighten to some degree following MWL depending on the patient's age and the amount of weight loss. Although genetics can play a role in the quality of the skin, it is often noted that the greater the drop in the body mass index, the greater the degree of tissue and skin laxity. The age of the patient also plays a role, as younger patients may have preserved skin elasticity while older patients tend to exhibit a greater degree of skin laxity. However, whatever tightening occurs, it is usually achieved during the first postoperative year, and after this time any additional improvement is virtually impossible [4, 8-10] (Fig. **3**).

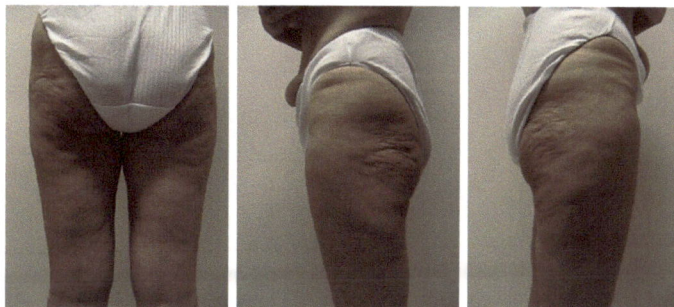

Fig. 3. Poor Skin Retraction in a Patient following Bariatric Weight Loss.

Ptosis

Ptosis occurs in areas directly above zones of adherence. Ptosis is seen as lax skin consisting of fat bulging over a tight zone of adherence. Gravity will continue to pull on the fat in the lax tissue. Increased adherence below will help accentuate the resultant anatomical problem. Increased fat or demands on the tissue eventually overcome its modulus of elasticity. This is why ptosis in a massive weight loss patient resembles that of pregnancy induced ptosis. The regions of the body most commonly affected by excess skin tissue are the medial part of the arms, the breasts, the thoracic and abdominal wall, especially in the lateral areas, and the inner and outer thigh [8]. Tissue deflation is often, but not always, present in the massive weight loss patient. Patients with a similar body mass index and total weight loss can present as deflated, mildly deflated, or non-deflated (minimum loss of fat) [10].

Cellulite

There are two types of cellulite: primary and secondary. Primary cellulite arises from hypertrophy of fat cells. It is visible whether the patient is supine or standing. Primary cellulite is not correctable by surgery.

Secondary cellulite, on the other hand, is due to increased skin laxity and the effects of gravity. Membranous portions of the fascia pull on the dermis, giving it the appearance of an orange skin (peau de orange). Secondary cellulite is only seen with the patient standing.

4. CONTOUR DEFORMITIES FOLLOWING MASSIVE WEIGHT LOSS AND THEIR CLASSIFICATION

During weight gain, the anatomical changes that occur in soft tissues and the pattern of adipose deposition are determined by the patient's gender, age, caloric intake, level of activity, and genetic predisposition. After massive weight loss, a broad spectrum of deformities can result in unusual body contour deformities. The extent of the deformities in the arms, upper thorax, abdomen, waist, back, buttocks, thighs, and legs is similarly influenced by the patient's age, gender, original weight, the lost weight, and genetic predisposition [11] (Figs. **4-6**).

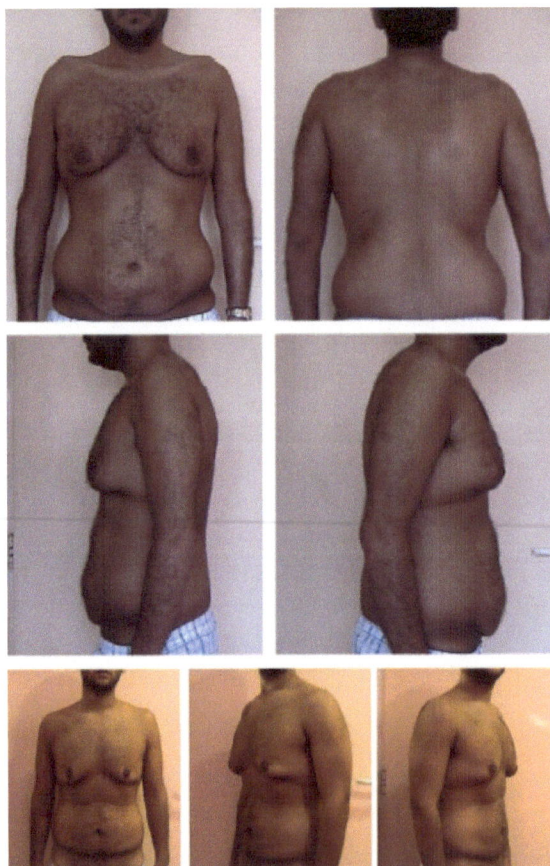

Fig. 4. Chest and Abdominal Deformity in Young Male Patients following Weight Loss.

More than just excising excess skin and subcutaneous tissue, post-bariatric body contouring requires unique insights into the nature of the post-bariatric deformities [1]. A classification system is a valuable tool in systematically assessing a patient and describing the deformities in a manner that is translatable from surgeon to surgeon. The best classification systems categorize preoperative appearance and formulate appropriate treatment guidelines in a succinct and structured manner. Unfortunately, post–bariatric weight loss deformities are diverse, disordered, and often unpredictable [4].

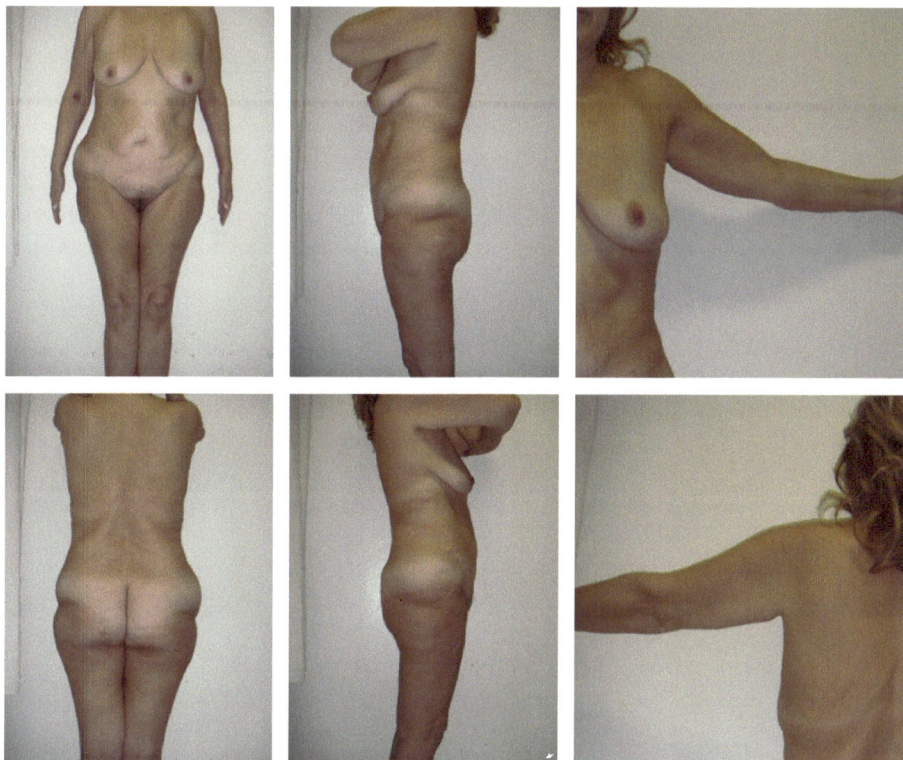

Fig. 5. Post Bariatric MWL of 30 Kg in a Female Patient. Deflation Deformities of Breasts, Abdomen, Arms, and Thighs with Asymmetrical Residual Fat Deposits around the Flanks and Hips.

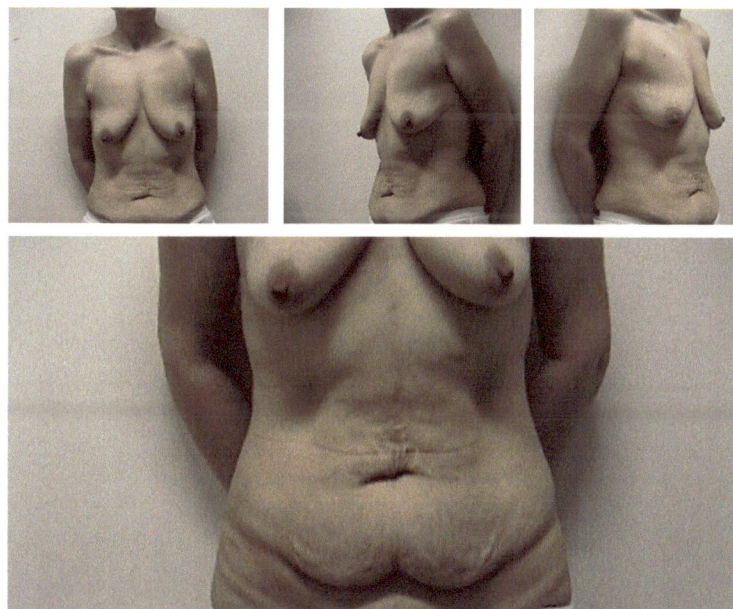

Fig. 6. Deflation Deformity of Breasts and Abdomen in a Middle Aged Female Patient after Loss of 50 kg following Gastric Banding.

Excluding the face and neck region, ten anatomical areas have been delineated by the Pittsburgh Rating Scale. The ten regions are: arms, breasts, abdomen, flank, mons, back, buttocks, medial thighs, hips/lateral thighs, and lower thighs/knees. A four-point grading scale ranging from 0-3 describes the common deformities found in each region. A grade of 0 indicates an appearance within a normal range. A grade of 1 indicates a mild deformity; a grade of 2, moderate deformity; and a grade of 3, the most severe deformity.

The rating scale is customized for each region of the body to account for the variety of deformities. Generally, non-excisional or a minimally invasive procedure is required for correction a deformity considered mild and a moderate deformity would require an excisional procedure. For most anatomical regions, significant ptosis of adipose-filled rolls or obvious skin folds representing the most severe level of deformity can only be rectified by combinations of excisional, lifting, and noninvasive procedures and would frequently involve large areas of undermining. The Pittsburgh Rating Scale may be useful in both classifying the individual deformities in a specific region and performing a comprehensive assessment for proper surgical planning. It could display the potential to classify and rank massive weight loss deformities in a clinically useful manner [4, 12] (Table 1).

Following massive weight loss, the abdomen is usually the first body area that draws attention. Like other regions of the body, hypertrophy of adipose tissue results in a three-dimensional expansion of subcutaneous tissues of the abdomen causing stretching of Scarpa's fascia and the superficial fascial system as well as dermal breakage of skin to a varying degree. Consequently, zones of adherence and demarcations become loose and the skin develops striae. With MWL, the damaged tissue is deflated and the abdominal tissue becomes lax in both vertical and transverse directions [11]. Most MWL patients present with epigastric fullness and a larger suprapubic pannus consisting of loose hanging skin and subcutaneous tissue that may be demarcated by a variable transverse fascial attachment between the abdominal dermis and superficial muscular fascia at the vicinity of the umbilicus (Fig. 7).

In some MWL patients, the suprapubic pannus often extends inferiorly over the groin region and posteriorly as a roll that overhangs the hips. It can be the source of significant discomfort, recurrent infection and difficulty with personal hygiene. In most patients, however, the excess skin and subcutaneous tissue is not confined solely to the anterior abdomen but rather extends in a form of "cone-like" circumferential excess with a ptotic mons pubis. MWL patients often have also associated abdominal wall laxity or abdominal hernias not always evident on pre-operative physical exam that need to be addressed as well [1, 11] (Fig. 8).

Table 1: Classification of Deformities in 10 Body Regions Assessed on a Scale Ranging from 0 (normal) to 3 (most severe). Adapted from Song A.Y. *et al.* (*Plast Reconstr Surg* 116: 1535, 2005).

AREA	DEFORMITY
ARMS	Normal Adiposity with good skin tone Loose, hanging skin without severe adiposity Loose, hanging skin with severe adiposity
BREASTS	Normal Ptosis grade I/II or severe macromastia Ptosis grade III or moderate volume loss or constricted breast Severe lateral roll and/or severe volume loss with loose skin
BACK	Normal Single fat roll or adiposity Multiple skin and fat rolls Ptosis of rolls
ABDOMEN	Normal Redundant skin with rhytids or moderate adiposity without overhang Overhanging pannus Multiple rolls or epigastric fullness
FLANK	Normal

	Adiposity
	Rolls
	Ptosis or rolls
BUTTOCKS	Normal
	Mild to moderate adiposity and/or mild to moderate cellulite
	Severe adiposity and/or severe cellulite
	Skin folds
MONS	Normal
	Excessive adiposity
	Ptosis
	Significant overhang below symphysis
HIPS/LATERAL THIGHS	Normal
	Mild to moderate adiposity and/or mild to moderate cellulite
	Severe adiposity and/or severe cellulite
	Skin folds
MEDIAL THIGHS	Normal
	Excessive adiposity
	Severe adiposity and/or severe cellulite
	Skin folds
LOWER THIGHS/KNEES	Normal
	Adiposity
	Severe adiposity
	Skin folds

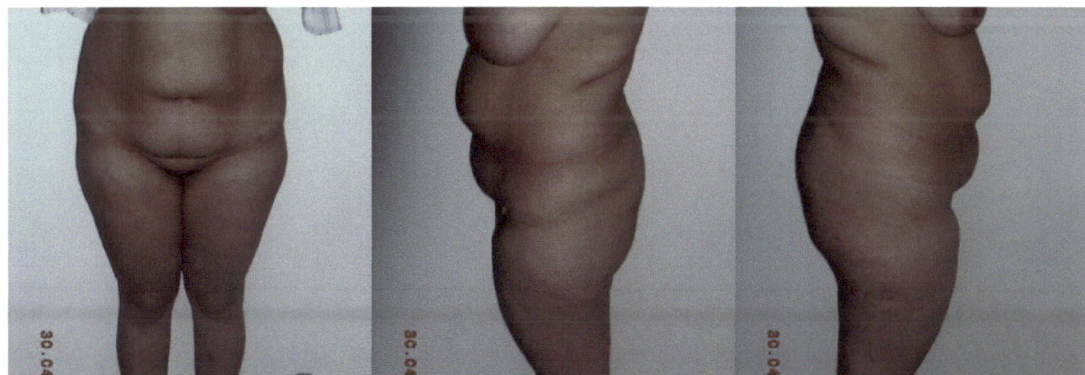

Fig. 7. Abdominal Deformity with Supraumbilical Fullness following Weight Loss.

The major deformity of the breast after massive weight loss in both men and women is ptosis or sagging of the breast which leaves the Nipple Areola Complex (NAC) below its appropriate anatomic position [5, 6] (Figs. **9, 10**). Moreover, with breast deflation and ptosis, the Inframammary Fold (IMF) often becomes loose and descends inferiolaterally [1, 11]. In both males and females, upper-torso laxity presents as cascading rolls of tissue, extending from the sternum to the vertebral column which in addition to the contour deformities of the upper extremities can be quite significant and extremely unsightly. On the back, up to five back rolls may develop. These include the axillary roll, the breast roll, the scapular roll, the lower thoracic roll, and the hip roll. The breast roll can result in further IMF descent, which appears as lower thoracic excess. The scapular and lower thoracic rolls can extend over the lateral chest and upper abdomen between the IMF and the umbilicus and present as upper abdominal excess [1, 11].

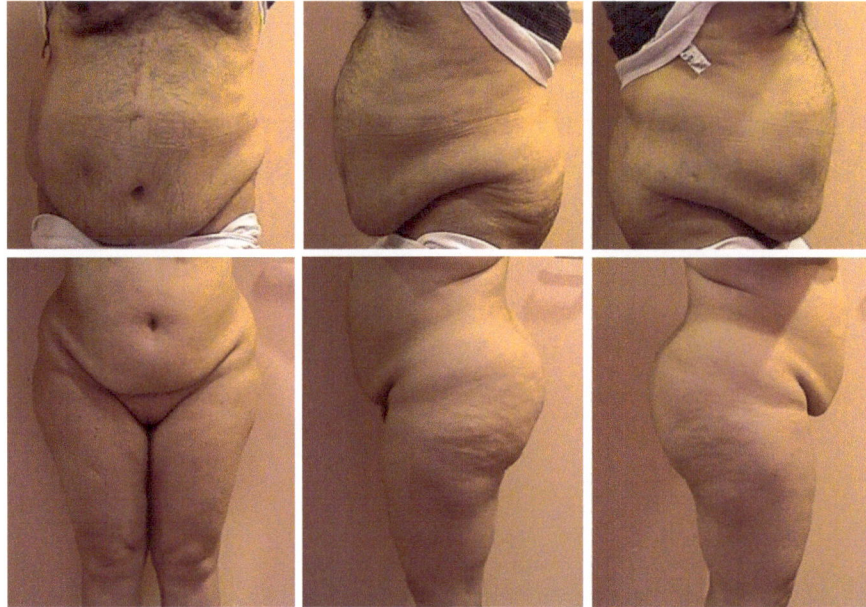

Fig. 8. Abdominal Pannus with Marked Abdominal Wall Laxity. Abdominal Deformity with Supra-Umbilical Fullness following Weight Loss.

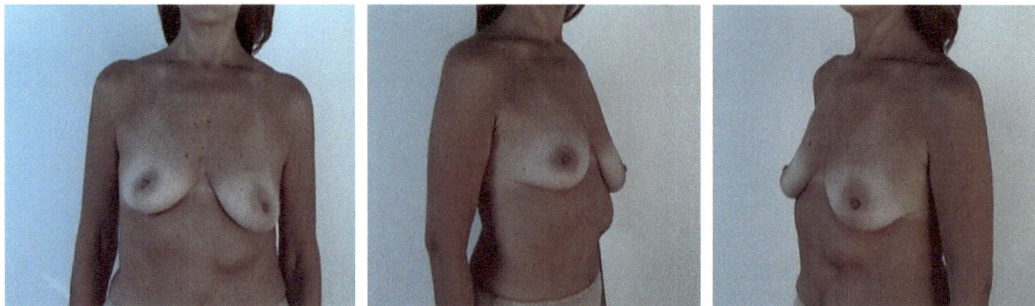

Fig. 9. Breast Ptosis with Accentuation of Asymmetry following Bariatric Weight Loss. Location of the Gastric Band Inflation Valve Well Apparent in the Left Submammary Area over the Lower Chest Wall Ribs.

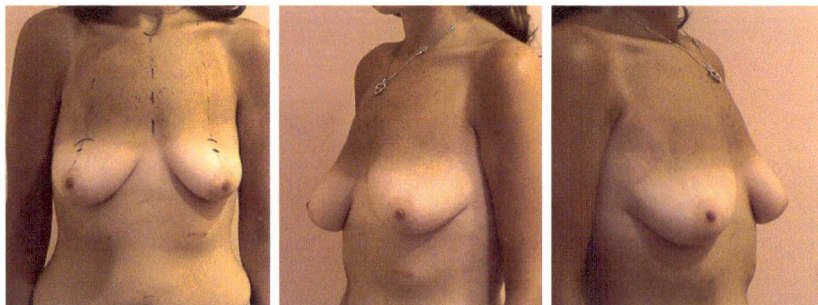

Fig. 10. Deflated Ptotic Breasts with Stretch Marks following Bariatric Weight Loss in a Young Patient.

Fat localization to limbs is prevalent in women. Following MWL they mostly complain of functional rather than cosmetic problems. In both upper and lower extremities, the most affected area by MWL is the medial aspect of the limbs. In the thigh, this means difficulties in walking as well as irritation, dermatitis and sensation of heaviness. Likewise for the arm, skin redundancy causes a reduced range of movements in addition to the same dysfunctions mentioned for the thigh interfering with the various actions performed during the day [13]. Buttock ptosis, decreased projection and flattening of lower back curse are also common contour deformities following MWL (Fig. **11**). Large amounts of excess skin draped from the proximal half of the arm gives the appearance of a "bat wing" (Fig. **12**). Excess skin hanging from the thigh is also unsightly and prevents patients from wearing bathing suits and shorts [1,11].

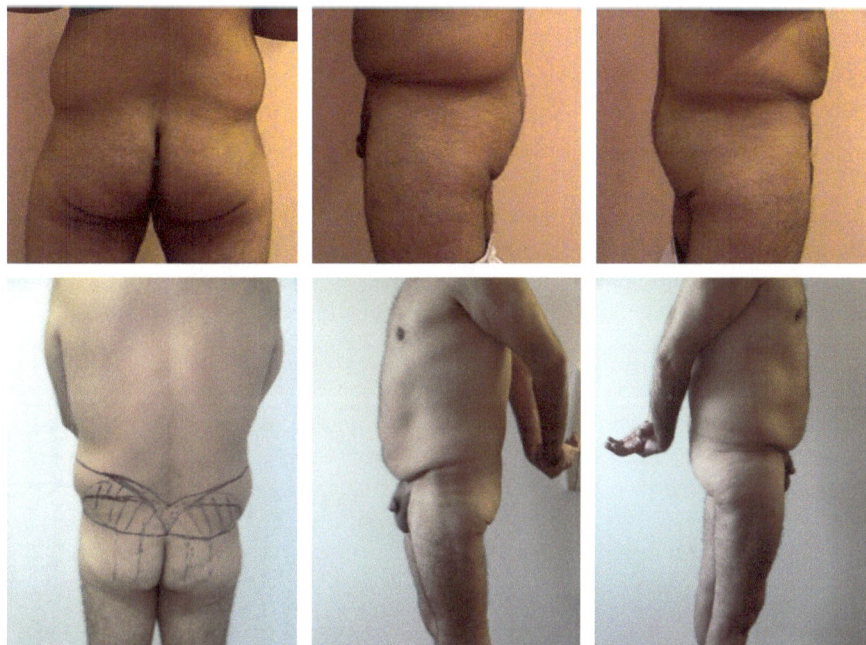

Fig. 11. Buttock Ptosis, Decreased Projection and Flattening of Lower Back Curve.

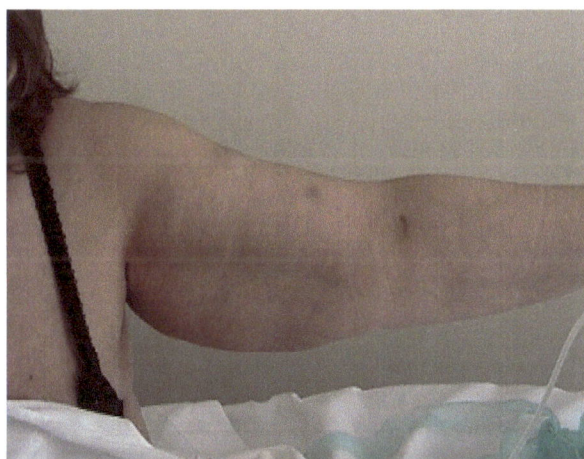

Fig. 12. "Bat Wing Deformity".

5. BODY CONTOURING TECHNIQUES

The appropriate surgical intervention for body contouring depends upon the amount and distribution of excess skin and subcutaneous tissue and satisfying results of the procedure are mainly dependent on the ability of the skin to retract as well as on resultant scars that respect the boundaries of aesthetic units. In general, as the amount of weight loss increases, so does the aggressiveness of the surgical intervention [1].

Panniculectomy

Panniculectomy involves excising excess skin inferior to the umbilical line and closing the wound over suction drains. Panniculectomy has proven to be clinically successful at treating skin irritation. However, it does not address the issues of excess horizontal skin, skin laxity, or aesthetic positioning of the umbilicus [14].

Abdominoplasty

The procedure begins with an incision around the umbilical stalk. To allow for greater skin resection the flap is elevated towards the xiphoid process and costal margins. Plication in a transverse or oblique direction is done to close the rectus diasthesis defect. This procedure works well for anterior abdominal excess. However, abdominoplasty is not sufficient to address posterior skin excess. Instead, belt lipectomy may be required to correct this deformity.

Belt Lipectomy

This procedure resembles abdominoplasty. However, the skin flap is elevated farther laterally to allow for larger resection of lateral tissue. The patient is then positioned in the lateral decubitus and a resection from the anterior axillary line to the midline of the back is completed. Closure is done over suction drains to the back and flank. This procedure is intended for management of circumferential skin excess.

Lower Body Lift

This procedure requires changing the patient's position from prone to supine during the surgery. While in the prone position, incisions are made inferiorly and superiorly on the back and hips. A closed suction drain is used for midback closure. Patient is then placed in the supine position and abdominoplasty is completed. Many post-bariatric patients have a flat buttock. To address this issue it is possible to use autologous augmentation through de-epithelialized rotation flaps based on the superior gluteal artery.

Brachioplasty

This procedure starts with a circumferential liposuction to decompress the arm. Skin excess is later excised. Resection is based on a vertical incision in the bicipital groove of the upper arm. Wound closure is accomplished with absorbable 3-0 sutures (2 layers), and compressive dressing.

Breast Contouring

Breast deformities are also common following massive weight loss. Simple removal of excess skin can lead to an undesirable flat breast profile in females. To reconstruct normal breast appearance, implants or mastopexy are often necessary. In some patients, autologous volume augmentation can be achieved by recruiting local folds of tissue that might be otherwise excised in the course of body contouring [1]. Frequently a free nipple graft is also indicated.

Suction Assisted Lipectomy (SAL)

Although some patients may experience an evenly distributed MWL resulting in a relatively thin layer of subcutaneous tissue throughout their body, for most patients this is not the case. Suction-Assisted Lipectomy (SAL) may be required in order to address recalcitrant adipose deposits which are common in the upper abdomen, back, flanks, arms and legs. Liposuction is used to "sculpt" areas of the body that need further refinement in their contour [1].

Generally more than one procedure may be performed during one setting. Mastopexy, for example, can be performed simultaneously with abdominoplasty. Yet, combining the procedures potentially increases the risk of complications. Risk is increased due to the larger amount of blood loss and increased operating room time. Most worrisome complication is deep vein thrombosis. Combined procedures are contraindicated in smokers, in patients with history of thrombosis, and in other medical co-morbidities.

6. CONCLUSION

The wide breadth and variety of deformities seen following bariatric surgery and MWL are an indication that surgical options are also endless. Unexpected folds and creases materialize in areas where there had been previously large amounts of adipose tissue, and previously smooth contours may cave into pleats and puckers. These transformations can take place anywhere on the body, and it is unpredictable where these deformities will materialize [4]. Brachioplasty, mastopexy, reduction mammaplasty, augmentation mammaplasty, abdominoplasty, suction-assisted lipectomy, circumferential and noncircumferential lower body lift, upper body lift, medial and lateral thighplasties, buttock lipectomy, and lower leg lipectomy all have a role in this patient population [4]. To achieve an aesthetic body contour, patients must accept, however, numerous long scars. They must understand that recurrent laxity and scar migration is not necessarily a failure of the operation but rather an expected outcome [15].

Body contouring in MWL patients is intrinsically different from traditional body contouring in normal weight patients. Even though many of the basic plastic surgery principles apply to this particular group of patients, a paradigm shift in approach and technique is necessary in order to achieve the desired and expected optimal results [16]. Before undergoing any of these procedures, it is important for the plastic surgeon to be aware that post bariatric patients require a thorough evaluation. Preoperative planning and appropriate treatment selection are tantamount to successful outcome. A systematic approach is ideal in addressing each area of the patient's body and quantifying the level of deformity in each particular region [4]. Patients' priorities should serve as a guide to staging the procedures and assessing the risk of multiple operations is crucial as the complications of massive body contouring can be very serious. Significant accommodations must be made if these operations are to be performed safely, effectively, and aesthetically. Patient education is also crucial to ensure that patients understand the major nature of these procedures and potential sequelae [16].

REFERENCES

[1]   Spector JA, Levine SM, Karp NS. Surgical solutions to the problem of massive weight loss. World J Gastroenterol 2006; 12: 6602-07.

[2]   Borud LJ, Warren AG. Body contouring in the postbariatric surgery patient. J Am Coll Surg 2006; 203: 82-92.

[3]   Salem L, Jensen CC, Flum DR. Are bariatric surgical outcomes worth their cost? a systematic review. J Am Coll Surg 2005; 200: 270-8.

[4]   Song AY, Jean RD, Hurwitz DJ, et al. Classification of contour deformities after bariatric weight loss: the Pittsburgh Rating Scale. Plast Reconstr Surg 2005; 116: 1535-44.

[5]   Lockwood TE. Superficial fascial system (SFS) of the trunk and extremities: A new concept. Plast Reconstr Surg 1991; 87: 1009-18.

[6]   Nemerofsky RB, Oliak DA, Capella JF. Body lift: An account of 200 consecutive cases in the massive weight loss patient. Plast Reconstr Surg 2006; 117: 414-30.

[7]   Hurwitz DJ, Rubin JP, Risin M, et al. Correcting the saddlebag deformity in the massive weight loss patient. Plast Reconstr Surg 2004; 114: 1313-25.

[8]   Fotopoulos L, Kehagias I, Kalfarentzos F. Dermolipectomies following weight loss after surgery for morbid obesity. Obes Surg 2000; 10: 452-59.

[9]   Atiyeh BS, Hayek S, Dibo S. Contouring of male breast after bariatric surgery and massive weight loss – A case report. Aesthet Surg J 2008; 28: 688-96.

[10]   Examination of the massive weight loss patient and staging considerations. Plast Reconstr Surg 2006; 117S: 22S-30S.

[11]   Agha-Mohammadi S, Hurwitz DJ. Management of upper abdominal laxity after massive weight loss: reverse abdominoplasty and inframammary fold reconstruction. Aesthetic Plast Surg 2009 Nov 21. [Epub ahead of print].

[12]   Song, AY, O'Toole JP, Jean RD, et al. A classification of contour deformities after massive weight loss: Application of the Pittsburgh Rating Scale. Semin Plast Surg 2006; 20: 24-29.

[13]   Bruschi S, Datta G, Bocchiotti MA, et al. Limb contouring after massive weight loss: functional rather than aesthetic improvement. Obes Surg 2009; 19: 407-11.

[14]   Manahan MA, Shermak MA. Massive panniculectomy after massive weight loss. Plast Reconstr Surg 2006; 117: 2191-7.

[15]   Heitmann C, Germann G. Body contouring surgery after massive weight loss. Part I: abdomen and extremities. Chirurg 2007; 78: 273-84.

[16]   Ali AS. Preface. Accessed Jan. 4, 2010, *www.qmp.com/uploads/pdfs_preface/Aly_Preface.pdf.*

CHAPTER 6

Post-Bariatric Body Contouring Surgery and Patient Safety

Bishara S. Atiyeh* and Amir Ibrahim

American University of Beirut Medical Center, Beirut, Lebanon

Abstract: Massive Weight Loss (MWL) patients differ from other patients in several parameters. They are at an increased risk for complications secondary to potential nutritional deficiencies, persistent obesity, venous varicosities, poor quality and inelastic tissue. As the numbers of patients who seek body contouring after extreme weight loss increase, surgeons must be able to inform patients of the risks and complications of body-contouring surgery as they relate to the specific co-morbidities of the particular patient. Likewise, surgeons will need to alter the aggressiveness of the procedure according to the risk versus benefit in patients who fall into higher-risk groups. The foundation for a safe body contouring practice involves a combination of good patient selection and managing patient expectations. Appropriate pre-operative assessment intra- and post-operative care as well as selection of the appropriate timing, type, and magnitude of surgery are all essential.

Keywords: Body contouring, patient selection, patient safety.

1. INTRODUCTION

The advent of bariatric surgery has provided a means of large, rapid, and sustained weight reduction to bring patients closer to their ideal body weight. A persistently poor body image despite the patient's enormous accomplishment with successful weight loss remains however a significant issue. Since it is virtually impossible to correct the excess skin left after extreme weight loss by diet or exercise, the increase in bariatric surgery has been paralleled by a similar increase in body contouring procedures to bring patients closer to their ideal body shape. Body contouring procedures, in a properly selected group, can be immeasurably rewarding to the patient as well as to the plastic surgeon [1-3]. Although post-bariatric body contouring procedures are associated with high rates of patient satisfaction, post-operative complications continue to negatively impact initially satisfying results [1, 4].

As with all surgical procedures, patient safety is of paramount importance. Massive Weight Loss (MWL) patients differ from other patients on several parameters and may have special needs [5]. Patients who have undergone bariatric surgery and have lost a massive amount of weight are healthier than when they were obese, but still present risks that should be addressed. They are an inherently difficult patient population whose co-morbid conditions must be addressed to perform safe and cautious procedures. They may be at increased risk for complications secondary to potential nutritional deficiencies, persistent obesity, venous varicosities, and poor quality, inelastic tissue [1, 2, 6, 7]. Appropriate pre-operative assessment and selection of the appropriate timing for surgery is essential [2].

2. PATIENT SELECTION AND RISK FACTORS

The foundation for a safe body contouring practice involves a combination of good patient selection and managing patient expectations [6]. In light of increased demand for body contouring procedures after MWL, plastic surgeons naturally look for predictors of poor outcomes/complications to help guide their patient selection [2]. Major body contouring surgery can involve prolonged operative times, substantial blood loss and fluid shifts, and greater physiologic stress than even the bariatric procedure [1]. A thorough pre-operative evaluation of the MWL patient is critical to ensure safety and a good aesthetic outcome [8] (Table **1**).

Address correspondence to Bishara S. Atiyeh: American University of Beirut Medical Center, Beirut, Lebanon; Tel: +961-3-340032; E-mail: batiyeh@terra.net.ll

Table 1: Important Factors to be Considered in Patient Selection.

Factors to be considered in patient selection
BMI
Associated co-morbidities
Nutritional status
Smoking
Psychosocial condition

Body Mass Index - BMI

When considering body contouring procedures, obesity serves as an independent risk factor. The poor outcomes attributable to patient obesity and secondary effects, such as hypertension, diabetes, sleep apnea, cardiovascular disease, and poor healing, have long been recognized [2]. Despite some claims to the contrary [9], the literature is replete with data documenting a correlation between obesity and increased occurrence of complications after body contouring and Body Mass Index (BMI) at the time of body contouring surgery is a predictor not only of aesthetic outcome but also of complication profile [1, 10].

Without further identifying a specific BMI over which there is a significant increase in complications, and although there is no consensus on BMI criteria for this surgery, a number of studies have shown that there is an increased surgical risk in obese patients. Though at this time, there is no clear BMI threshold above which surgery should be refused, it is clear that higher BMIs have been associated with increased complications and poorer outcomes [2, 5, 6]. Most authors advise against body contouring in patients with a BMI greater than 32 and recommend functional panniculectomy only for those with a BMI greater than 35 [1, 2, 9, 11, 12]. The risk of surgical complications decreases as the patient moves out of the morbidly obese category and approaches ideal body weight and complication risk, in particular risk of Deep Vein Thrombosis (DVT), increases significantly with increasing BMI. Thus risks of surgery need to be weighed against functional benefit in obese (BMI - 30) and severely obese (BMI - 40) individuals [1, 11, 13, 14]. The best candidates for extensive body contouring surgery after weight loss reside in the BMI range of 25 to 30 [14].

It is unknown exactly how much weight loss will reduce surgical risk, but a reasonable approach is to avoid operating on patients who are morbidly obese unless there is strong indication, such as chronic panniculitis or a pannus that severely limits ambulation. In such a patient, panniculectomy is a purely functional procedure and should be performed with no direct undermining of tissues and little regard to aesthetic contour [14]. Proceeding with surgery for these patients with greater caution, less aggressive surgery and more aggressive peri-operative care make typically bad prognostic factors less predictive of problems [9].

Associated Co-Morbidities

Patient safety starts with an emphasis on preoperative planning and strategy with special attention to the current medical status of the patient who has lost a massive amount of weight [14]. Along with MWL comes a tremendous improvement, if not complete resolution, of many medical disorders associated with obesity. Even moderate weight loss has a significant impact on medical co-morbidity in obese patients. A 10-kg weight loss is associated with a 30% reduction in diabetes-related mortality and has major benefits with regard to blood pressure, plasma glucose, angina, serum lipids, hemostatic factors, and even reduces cancer risk. Weight loss associated with gastric restrictive surgery reduces blood pressure in a linear fashion, with 50% to 60% of patients becoming normotensive [14]. There are concerns however that not all means of weight reduction are equal in long-term outcomes. Bariatric patients differ from patients who used exercise as a means of weight reduction in that their cardiopulmonary status has not had the strengthening over time compared with those who used exercise and lifestyle modification as their means of weight reduction [3].

Although any amount of weight loss can be beneficial in improving obesity-related disorders, complete resolution of these co-morbidities may not occur. Therefore, the plastic surgeon must be wary of unresolved medical problems [14]. The bariatric group commonly has underlying hypertension, undiagnosed diabetes, and heart disease with such manifestations as atrial fibrillation. Given the complexity of various diseases and co-morbidities affecting the patient who has lost massive amounts of weight, a thorough preoperative clearance and cardiac evaluation is recommended. Residual medical problems should be assessed before surgery with appropriate consultation to specialists as needed to optimize the patient for surgery [7].

Nutritional Status

Even with a good BMI, the patient's nutritional status may not be adequate [6, 15]. Patients who undergo surgical intervention for weight loss must be monitored according to the type of procedure performed. The type of procedure, mal-absorptive or restrictive bariatric surgery, has implications for what type of nutritional deficiencies may be encountered [3, 7]. Restrictive procedures including gastric stapling and banding procedures (GB) are less likely to result in nutritional complications. Mal-absorptive procedures involve bypassing a segment of small intestine to decrease the available absorptive surface, or redirecting the biliary and pancreatic juices so that digestion and absorption occur farther down the small intestine. The prototypical mal-absorptive procedure, jejuno-ileal bypass, is rarely performed today because nearly 50% of patients require reversal due to life-threatening metabolic complications [7, 14].

Anemia is a common side-effect of both bariatric surgery and of the altered anatomy. Nutritional disorders are also common. They are inherent to the nature of bariatric surgery. MWL patients may become as well noncompliant with the post-bariatric nutritional program [1]. Often MWL patients consume poorly balanced diets. This leads to thiamine, folate, B_{12}, calcium, albumin, and potassium deficiencies. These deficiencies adversely affect wound healing and can lead to other complications during body contouring procedures. A rational approach to nutritional screening before body contouring surgery starts with the collection of historical data about eating habits and current nutritional supplements. Any patient with persistent nausea and vomiting may require evaluation for anastamotic stricture or other anatomic problems. If the patient is intolerant for meat or dairy products, inquiries should be made about alternative sources of protein intake. Protein malnourishment can negatively impact wound healing [1, 7, 13]. Protein levels need to be monitored in patients who have undergone a distal gastric bypass or Bilio-Pancreatic Diversion (BPD) procedure due to the amount of protein that can be lost in the stool. Particular attention should be focused as well on vitamins with known mal-absorption following bariatric surgery including thiamine, vitamin D, calcium, iron, vitamin B_{12}, and folate. Proximal gastric bypass patients usually have deficiencies of iron, folate, calcium, vitamin B_{12}, and vitamin D. Nutrient deficiencies specific to BPD include as well mal-absorption of fat-soluble vitamins in particular vitamins A and D [1, 3, 7, 14].

Although routine laboratory screening for vitamin deficiencies is not mandatory and probably not cost effective, most authors recommend a preoperative laboratory work-up, including a complete blood count, electrolytes, albumin, and pre-albumin [1, 6] as well as Prothombin/Partial Prothrombin Time (PT/PTT) and uric acid level. All patients considering elective body contouring should be on a regimen of multivitamins, iron, and calcium supplements before surgery. When there is concern about protein or nutrient deficiencies in a patient who seeks body contouring surgery, he or she should be referred to the bariatric nutritional team for further evaluation and treatment [3, 7, 14].

Smoking

Smoking significantly decreases local cutaneous blood flow, and it should be documented, addressed, and strongly discouraged with the patient. Though the effect of massive weight gain and loss on the subcutaneous adipose layer is analogous to a delay, with the vascular network expanding to meet the metabolic demands of the increased tissue mass and then remaining in place after body mass decreases, great caution still needs to be exercised with tobacco smokers. Major body contouring surgery involves the creation of large tissue flaps. Nicotine causes vasoconstriction, which can result in flap necrosis, infection, and major wound healing complications. Smokers are required to stop smoking 1 month before surgery and urine Cotinine tests can be used to ensure cessation of smoking at the time of surgery. If smoking cessation is not possible or cannot be guaranteed, then the extent of the procedure performed, especially the amount of tissue undermining, is limited [1, 6, 7, 14].

Psychological Condition and Patients' Expectations

Body contouring patients want surgery immediately; they want perfect results, no downtime, no risk, and no out-of-pocket charge. Their expectations are an important guide for selection and must be well assessed at the initial consultation. Patients are told that body contouring procedures are bigger than bariatric procedures, with bigger scars, longer recovery times, and that wound healing problems are common. When patients present for post-bariatric reconstruction, they should be aware that the procedures are functional and that the results may not be equivalent to a cosmetic operation [6]. The "trade off" of skin excision for acceptance of scars must be emphasized. Patients must understand that scars are inevitable, not invisible, and that the quality of the scar is controlled by many factors, including genetic, ethnic, and natural biologic processes [6].

The incidence of depression is very high among bariatric patients. A majority report the use of antidepressants. Provided that patients have appropriate expectations and motivation for body contour procedures, this condition is not a contraindication for surgery and does not require clearance from a psychiatrist. If patients have just started the medication, they should be allowed to stabilize for several weeks before scheduling surgery. Any patient with a psychiatric diagnosis beyond simple depression should follow up with their mental health provider and obtain a letter of clearance documenting the patient's realistic expectations. Bipolar disorder or schizophrenia should be approached with great caution. These conditions may lead to unreasonable expectations and noncompliance [6, 14]. Men who fit the "SIMON" criteria of the single, immature, overly expectant, and narcissistic male may be difficult to satisfy even after major transformations [6].

Special Risk Factors

Male gender, hypothyroidism, and Ehlers-Danlos syndrome are definite risk factors for wound-healing problems and must be considered in determining the prognosis for potential failure in wound healing. Body contouring techniques for these patients must aim toward conservatism [9]. Asthmatic patients are also at increased risk for requiring blood transfusions as well as having three or more procedures [9].

3. PROPER TIMING AND STAGING OF BODY CONTOURING SURGERY AND STAGING

Following bariatric surgery, weight loss typically decreases exponentially and then subsequently stabilizes after a period of 15 to 18 months. Body contouring surgery should only be considered after the patient has achieved a stable weight plateau for 3 months or longer. A period of rapid weight loss may not be the best time for good wound healing [1, 14, 16]. This delay is advocated for three reasons. First, it allows the patient to achieve a metabolic and nutritional homeostasis. Second, the risk of surgical complications decreases as the patient moves out of the morbidly obese category and approaches ideal body weight. Third, aesthetic outcomes tend to be better for patients who are close to their ideal body weight [14].

As a general rule, it seems logical to assume that single-stage procedures of multiple body parts should be avoided. They may expose the patient to prolonged operative times, increasing the likelihood of morbidities such as hypothermia, anemia, thromboembolism, and wound-healing complications. Staging of body contouring procedures for MWL patients helps to ensure optimal outcomes as well as enhance patient safety. The staging process can be as simple as one procedure or as complex as a series of procedures that spans 12 months or more. While there is no one algorithm, a number of commonalties have emerged for staging body contouring for MWL patients [8]. Several variables influence the decision of how to stage procedures, including the size of the operating team, operative setting, surgeon experience, patient financial resources, insurance coverage, and most importantly the patient's personal goals [6]. Stages should also be dictated by both the physician's and the patient's comfort level. Considerations should include a comfortable operating time that is patient and practice specific as well as the patient's ability to tolerate longer surgery. Final decision is dictated by the patient's preoperative physical condition, whether proposed surgery is to be performed by a solo plastic surgeon or by a team, and by the BMI [8].

Because one procedure will most likely have an effect on adjacent areas of the body, the surgeon must carefully plan the sequence of body contouring procedures. A lower body procedure can affect the thighs and anterior chest, while a brachioplasty can affect the chest and back. This concept must be understood by the patient, especially those who are looking for a "quick fix" [8]. An additional benefit of staging is the planned opportunity to make any minor revisions that may be necessary following a previous stage [6, 17].

As a general principle, the plastic surgeon should make every effort to elicit from the patient which anatomic areas are of greatest concern and then divide the procedures into as many stages as necessary to help ensure patient safety. The patient's highest priorities should be addressed first. It is often recommend combining procedures such as abdominoplasty and mastopexy, a very powerful combination that can reshape the entire trunk in the first stage. Another common strategy is lower body lift, abdominoplasty, and an upper body procedure (mastopexy or brachioplasty), followed by a second stage consisting of medial thigh lift and another upper body procedure for a comprehensive 2-stage approach [6].

It has been shown, however, that operative duration is not an independent predictor of complications [1, 11]. Moreover, there are no data to support a definite "time limit in surgery" rule; therefore, it is inappropriate to set arbitrary boundaries on time spent in the operating room. There are many variables to consider, such as the type of surgery; 6 hours spent performing a rhytidectomy cannot be considered the same as 6 hours of liposuction, because the metabolic and physiological effects of the procedures are significantly different. Nor are there significant data to suggest that 7 hours in surgery is any more dangerous than 5 hours [5]. Although combining 3 or more procedures in one operative setting is associated with an increased hospital length of stay greater than 2 days, it does not increase risk of wound dehiscence, seroma, or blood transfusion [1, 9, 16].

Single-stage total body lift procedures offer many advantages for patients by decreasing cost, facilitating time off work, and decreasing the number of general anesthesia inductions [1, 18]. Multiple stages, however, mean more operations with more recoveries, more time off work, and added expense to the patient [6, 17]. There is also an unproven but evident feeling that surgical rehabilitation that is costly and time-consuming over many stages is less likely to be completed. Little to nothing is thus gained when a MWL patient with low self-esteem has a flat abdomen and no thigh saddlebags, but still hides sagging inner thighs [16]. These single-stage procedures can safely be performed in healthy patients by a plastic surgeon and team experienced in body contouring surgery at a tertiary academic medical center. While the guidelines for single-stage surgery concern optimal age, BMI, health, and motivation, the operative plan and its implementation reflect the surgeon's training, experience, patient evaluation, intra-operative progress, and performance of the assembled team. The patients' factors are balanced, so that younger patients may have a higher BMI and older patients a lower BMI. There are, however, no specific literature-based recommendations for surgeons with less experience, fewer assistants, or those who choose to perform outpatient procedures [1, 16, 18].

It is advisable, nevertheless, to apply general surgical guidelines when performing body contouring surgery for MWL patients. As with other surgical procedures, the healthier the patient and the fewer the procedures performed, the lower the complication rate. Naturally, three or four procedures performed in one operative stage can be more demanding on the patient and the surgeon, and tend to lengthen the recovery time. As a general recommendation, it is critical to gain more experience before attempting multiple procedures in one stage and though no objective data have yet been published, it is recommended to limit surgery in one setting from 6 to 7 hours and wait 3 months between staged procedures [1, 5, 19] (Table **2**).

Table 2: Single Stage Total Body Contouring Surgery: Advantages and Disadvantages.

Advantages	Disadvantages
Decreased cost	More demanding to surgeon
Less time off work	Lengthy recovery
Decreased number of general anesthesia inductions	Difficult to perform in solo at a peripheral hospital

4. ANESTHESIA PREOPERATIVE EVALUATION

An appropriately trained anesthesiologist or nurse anesthetist is necessary to manage the complex physiology associated with the major procedures of post-bariatric body contouring. A thorough preoperative evaluation needs to address the American Society of Anesthesiologists (ASA) classification, nicotine use, maximum and current BMI, history of Deep Vein Thrombosis (DVT) or Pulmonary Embolism (PE), nutritional status, and psychosocial health. The ASA classification, however, is as an imprecise guide to clinical health of the patient and predictor of anesthetic and surgical patient risks [1].

A complete history and physical examination are the cornerstones of the preoperative evaluation. Laboratory tests should be ordered on these patients on a routine basis, well in advance, so that the tests may be interpreted and the patient treated accordingly, if necessary. Blood work should include routine electrolyte analysis, a complete blood count, and pre-albumin and albumin levels. In some cases, the patient's gastric bypass surgery may contribute to either protein deficiency or anemia. In these cases, correction should be achieved before surgical intervention [5, 7, 14]. It should be realized that 50% of MWL patients are anemic. This should be recognized preoperatively and treated appropriately. Moreover, the large amount of well-vascularized excess skin that is excised during body contouring can lead to excessive blood loss and worsen an existing anemia, especially when multiple procedures are combined in one setting [1, 5].

On occasion, patients may continue to take diet medications such as phentermine or herbal supplements that have been associated with intraoperative events or may delay wound healing. By alternating serotonin levels, phentermine and fenfluramine, have been linked to aortic valve injury and pulmonary hypertension. Patients who have chronic illnesses or medial conditions should obtain proper clearance from their internist before surgery [5, 7, 14]. A consultation with the anesthesiology department before body contouring surgery is recommended, especially if the anesthesiology team is inexperienced with these patients and procedures. As anesthesiology teams become more familiar, they will develop protocols that take into account the nuances of the MWL patient [5].

Anesthetic issues that may also require special consideration include intubation, hypothermia, and intravenous lines. It is not uncommon for MWL patients to have anatomy that makes intubation difficult and to develop hypothermia with a large exposed percentage of the patient's body surface area during a lengthy body contouring surgery [5].

5. INTRA-OPERATIVE CONSIDERATIONS

When working with the MWL patient undergoing body contouring surgery, there are a number of considerations that the operating room team must understand with regard to patient positioning and support [5].

Patient Positioning

Whether the prone-supine or supine-lateral decubitus position is used, it is important for the operating room team to be very aware of the potential undesirable effects that can occur in these positions, including, but not limited to, hemodynamic changes caused by impaired venous return or alterations in ventilation and perfusion and compression injuries at bony prominences or nerves. Regardless of the position the patient is placed in, it is imperative to keep the cervical spine neutral to prevent stress and torsion of the spinal cord [5].

Padding and Patient Support

As many of these patients are still large, it is important to pad pressure points to reduce the potential for skin necrosis. Important anatomical structures that need support include shoulders, hips, arms, face, eyes, and breasts [5].

Hypothermia

Large amounts of uncovered skin surface during body-contouring procedures can lead to precipitous decreases in core body temperature, contributing to patient instability and post operative wound infection. Warming equipment such as warming devices (Bear Hugger®, Augustine Medical, Inc., Eden Prairie, Minnesota) and forced air warming blankets as well as prewarmed intravenous fluids are extremely important to avoid hypothermia [7].

6. POST-OPERATIVE CARE

Postoperative care for the MWL patient following body contouring surgery includes, in addition to ambulation as early as possible, the use of steroids and/or diuretics, antibiotic prophylaxis, the use of compression garments, and the timing of the next intervention. Pain management is also an important consideration in the immediate post-operative period [5].

Steroids and/or Diuretics

Short-term steroid therapy helps reduce swelling, postoperative pain, and postoperative nausea and vomiting. However, if a patient has a history of gastric or stomal ulcers, or if there are complications related to the patient's bypass surgery, steroids should be avoided. Some surgeons also use diuretics to treat patients, especially in cases of severe swelling. Good surgical judgment should be used, and diuretics should not be considered "routine" in the management of these patients [5].

Compression Garments

Compression garments produce beneficial effects in the immediate postoperative period however do not appear to affect long-term outcome. Gentle compression with light spandex exercise-type clothing is probably recommended. Most surgeons do not use garments on the arms and lower extremities, because of the increased incidence of blistering, pain, and tissue necrosis. If liposuction of the outer thighs is part of the surgical procedure, some believe the use of a light compressive garment helps to reduce swelling in this area. Others believe that binders are useful in abdominal procedures to help hold the dressings in place and support wound closure. However, the use of binders immediately postoperatively could compromise skin flap blood supply if already made tenuous by extensive undermining and tension closures and increase intra abdominal pressure [5, 7].

Pain Management

Narcotics and the duration of their use vary depending on the procedure and the patient's individual pain tolerance. Epidural analgesia is an extremely effective method of controlling pain, it requires however the cooperation of the anesthesia team, which may or may not be well equipped for it. Pain pumps with catheters threaded either into the pre-fascial space or in a sub-fascial position away from drains are useful for MWL patients who undergo body contouring procedures. Injecting Marcaine® (bupivacaine HCl; Sanofi-Aventis, Bridgewater, N.J.) into the deep fascia helps also to control postoperative pain [5].

Antibiotic Prophylaxis and Infection Control

For longer, multistage procedures, the risk for infection increases. In part, this can be a consequence of nutritional deficiencies leading to impaired immune states, as well as underlying skin infections. To decrease skin bacterial counts, washing twice daily with a topical broad-spectrum soap, such as chlorhexidine, starting 3 days before a procedure is recommended. According to standard operative guidelines, prophylactic intravenous antibiotics are infused 30 minutes to 1 hour before the start of the surgical procedure and continued for an additional 24 hours postoperatively. Although there are no conclusive data to support prolonged administration of antibiotics for this surgical procedure, it is common practice to keep patients on a first-generation cephalosporin until their drains have been removed. If wound dehiscence or infection occurs, basic principles of wound management dictate therapy [1, 7, 14].

7. COMPLICATIONS

Even though the majority of complications after MWL body contouring surgery are minor and treatable in the office setting, they are very common including, besides poor or less than optimal aesthetic outcome, wound-healing problems, seromas, and thromboembolic complications. Other complications include, deep suture abscess requiring surgical removal, postoperative bleed requiring return to the operating room, drain removal under anesthesia, hematoma, lymphocele, pain requiring consultation, and bleeding ulcer within weeks of surgery requiring transfusion. Complications are not mutually exclusive. For all body-contouring procedures, complications have been estimated at around 14.4% for wound problems, 12.9% for seromas, and 2.9% for skin infections [7, 9]. To reduce the potential for these problems, understanding potential risks helps development of better preoperative, intraoperative, and postoperative care plans [5].

Thromboembolism

Obesity is a well-documented risk factor for DVT and Pulmonary Embolism (PE) necessitating a sound protocol for prophylaxis. Other risk factors include increased age, history of spontaneous miscarriages, pregnancy, use of oral contraceptives, previous history of DVT and PE, immobility, recent surgery, and coagulation abnormalities [3]. Venous thromboembolism is an important source of morbidity and the most common cause of mortality following body contouring procedures [1, 12]. Most MWL patients are classified as high risk according to criteria defined by the American College of Chest Physicians (ACCP). The risk of DVT is highest in the first 1 to 2 weeks following surgery and it is necessary to initiate thromboembolism prevention preoperatively and to continue well into the postoperative setting [1]. There is, however, no consensus statement or specific recommendations for patients undergoing body contouring procedures. Different opinions on the dosage and when to use Heparin (Wyeth-Ayerst Pharmaceuticals, a division of American Home Products Corporation, Philadelphia, Pa.) or low-molecular-weight Heparin products and when chemoprophylaxis is to be combined with mechanical prophylaxis using intermittent compression devices or graded compression stockings have been expressed. The risk of venous thromboembolism has to be weighed against the relatively high risk of postoperative bleeding [1, 5].

Seroma

Seroma is the most common complication of surgery in many clinical series. The incidence of seroma in the MWL patient is however unknown, but it may occur in up to 50 % of some patient groups. Seromas may result in large cavities that can lead to dehiscence and chronic wounds. Various interventions to reduce the incidence of seroma have been described such as placement and maintenance of drains until output is less than 30 mL per day, progressive tension sutures, fibrin sealant, and preservation of a thin layer of fat. However, little data currently support any one technique [1, 5] (Fig. **1**).

Wound Dehiscence

Wound healing complications are common after body contouring surgery, with rates from 8 to 66 percent in the MWL population. The cause of aberrant healing is likely multifactorial. Given their function in other populations prone to similar problems, abnormal cytokine, matrix metalloproteinase, and tissue inhibitor of metalloproteinase levels likely play a major role. As seen in cancer, burn, transplant, and obese wounds, increased inflammatory cytokines in the MWL wound may shift the Matrix Metalloproteinase (MMP)–to–tissue inhibitor of metalloproteinase balance toward greater proteinase activity. However, reasons for this abnormal healing remain unclear; malnutrition, increased skin tension, impaired blood flow, prolonged operative time, intraoperative hypothermia, and multiple operators may all contribute to the pathogenesis [20].

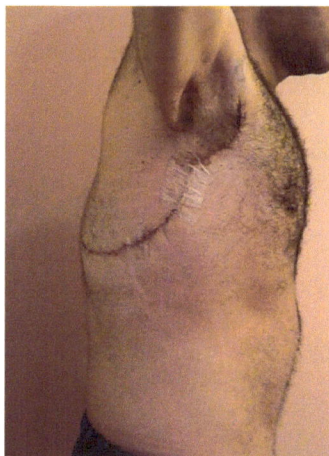

Fig. 1. Large Seroma following Torsoplasty.

Nutrition certainly plays a key role. However, wound healing problems may be secondary to causes beyond nutrition. Following MWL body contouring, significant wound tension and reduced blood flow as a result of wide undermining definitely impair healing. Hypoxia significantly impacts extracellular matrix turnover, leading to tissue ischemia and elevated MMP-2 and tissue inhibitors of metalloproteinase (TIMP)-2 levels, augmenting matrix breakdown. Operative and postoperative positioning may contribute also to poor healing, as skin dehiscence commonly occurs in regions of greatest tension, including the buttocks and hips. Wound complications increase as well with operative times, but it is unclear whether this reflects prolonged operative hypothermia, incision length, surgeon fatigue, the effect of multiple operators, or other surgical variables. Hypothermia itself is known to increase susceptibility to wound infections and delay healing. Another source of stress associated with a prolonged operative time may be increased incision length. Perhaps longer incisions generate a traumatic stress response and exaggerate a MMP–to–tissue inhibitor of metalloproteinase imbalance, leading to further wound instability [20].

Wound dehiscence can occur either early (*i.e.,* immediately after surgery) or later in the postoperative period. Early dehiscence is typically seen with patient movement by either the staff or the patient him- or herself. Wound dehiscence from inappropriate movement can be minimized in MWL patients through sound education of the patient and the nursing care team. Dehiscence, later in the healing process, can be secondary to motion but is most often due to an underlying seroma, which will manifest as an open wound with drainage [5] (Fig. **2**).

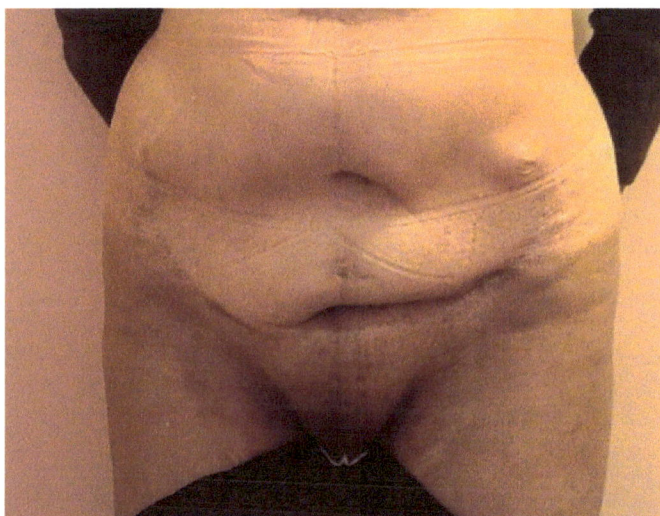

Fig. 2. Poor Aesthetic Result following Wound Dehiscence and Secondary Wound Healing.

8. CONCLUSION

The increasing number of patients with MWL has led to a growing responsibility placed on body-contouring surgeons to perform safe and cautious procedures in this inherently difficult patient population [7]. Body contouring after MWL has evolved from "amputative" procedures, such as panniculectomy, to more thoughtful "body lift" procedures; the name itself implies a restoration of form and cosmesis that transcends simple removal of excess skin and fat [21]. As experience is accumulating in this subspecialty, technical post-bariatric body contouring procedures are being refined and dramatic aesthetic benefits have been seen in postoperative results. However, few objective guidelines exist on the optimization of patient care and safety in this complex patient population [2].

The ongoing metabolic changes after significant weight loss, nutritional challenges, psychological effects of bariatric surgery, and likely persistent secondary effects of obesity (diabetes, cardiovascular disease, and pulmonary disease) can make these patients poor surgical candidates [2]. However, as the numbers of patients who seek body contouring after extreme weight loss increase, surgeons must be able to inform patients of the risks and complications of body-contouring surgery as they relate to the specific co-morbidities of the particular patient. Likewise, surgeons will need to alter the aggressiveness of the procedure according to the risk versus benefit in patients who fall into the higher-risk groups [3].

It must be realized that massive weight loss affects the body from head to toe, not just focal body parts. The surgeon and patient therefore need to determine what is problematic either functionally or aesthetically, prioritize, and map a course for treatment. Procedures may need to be staged if multiple body regions will be addressed to improve safety. Stages may also be necessary to refine surgical results from an earlier procedure to tighten residual laxity. "Measure twice, cut once" is an important principle, as one would not want to convert tissue excess to tissue deficiency [9]. Moreover, the concept of vascular supply driven selective undermining is being increasingly recognized as one of the most important steps to improve both aesthetics and safety of post MWL body contouring procedures. By a selective approach to the extent of dissection with greater respect for and maintenance of an increasingly rich blood supply, a more liberal contouring by liposuction and ultimately enhanced aesthetics can be achieved [21]. Limiting both undermining and adjunctive liposuction reduces the risk of wound-healing problems and seroma formation. Awareness of hernia potential with abdominal scarring is also necessary when addressing the abdomen. Autogenous tissue augmentation, using tissue that would otherwise be disposed, is helpful for addressing involutional changes associated with MWL, particularly in the breast and buttock area. Finally, aggressive patient care is necessary from the beginning, at the time of consultation, to the end, into the remote postoperative period, at which time the patient has reached his or her goals [9].

Although there is no exact maximal amount of time for body contouring procedures, a prudent and conservative approach should be taken when deciding how much surgery is safe to perform. There is no single accepted algorithm, but in general, a smaller body mass allows for more procedures to be performed safely at one time. More procedures can be undertaken at one time if a team is used [7]. With a team approach, procedures commonly combined include circumferential body lift with liposuction of thighs, upper body lift with arms, or thigh lift with minor revisions [7]. Surgeries should be kept short enough to prevent significant blood loss and significant hypothermia. A commonly used threshold is to limit surgery to 6 hours, although this certainly is not universally accepted as a limit. It should be thought of as a guideline, not a standard, which is modifiable by physician and patient comfort levels [7].

To date it remains unclear whether the particular method of weight reduction (dietetic vs. bariatric) may additionally predispose for increased complication rates in the secondary reconstructive body contouring procedures. Bariatric surgery is known to achieve successfully and rapidly massive weight reduction but may induce nutritional imbalance through malabsorption and intake restriction. In comparison, dietetic and exercise weight reduction occurs slower with a postulated lower risk for nutritional imbalance. As might be expected, metabolic assessments and protein intake are better in the patients who have lost weight through diet and exercise and bariatric surgery patients present with a significantly lower preoperative hemoglobin level than patients following dietetic weight reduction. Despite major physiologic differences between the two groups of MWL patients, well-selected post bariatric MWL patients are likely to have complication rates comparable to diet and exercise patients. Careful patient selection and nutritional optimization may equilibrate the risks between the diet and exercise and bariatric procedure patients [22, 23].

Independent of the particular method of weight reduction, selecting patients with a high excess body mass index loss (EBMIL) $\geq 30\%$, indicative of a lower necessary soft tissue resection weight, followed by a stable weight plateau maintained for at least 12 months prior to the body contouring procedure has been shown to successfully reduce complication rates to only 26.9%. High complication rates exceeding 40% occur in patients with low EBMIL [23].

Table 3: Reporting of Weight Loss: % BMI Loss and % Excess BMI Loss (EBMIL). Adapted from Deitel and Greenstein [24].

% BMI Loss = [(Starting BMI - Follow-up BMI) / Starting BMI] X 100
Excess weight defined as BMI>25 BMI in excess = Patient's BMI – 25 **% Excess BMI Loss = 100 - [(Follow-up BMI - 25 / Starting BMI - 25) X 100]** *** If an individual has an initial BMI of 45, then the 20 BMI units above the upper limit of the normal of 25 BMI units, represents a %EBMIL of 100; a loss of 10 BMI units (to a BMI of 35) would be a %EBMIL of 50.**

Similar to microsurgery, bariatric plastic surgery requires a skilled, coordinated team approach pre-, intra-, and postoperatively [10]. Centers of Excellence in body contouring that provide a team approach combining comprehensive patient evaluation, outcomes research, and surgical training may be the optimal approach for treating the MWL patient [6]. In properly selected patients body contouring following MWL procedures can be immeasurably rewarding to both the patient as well as the surgeon. For optimal results, it is thus critical to recognize the risk factors for poor outcomes and higher complication rates [2]. Literature general consensus guidelines to ensure patient safety are summarized in Table **4**.

Table 4: Consensus Guidelines for Post Bariatric Body Contour Surgery.

Operate only patients after stable weight plateau for 3 months or longer
Avoid body contouring surgery in patients with a BMI greater than 32
Ensure that patients have adequate nutritional status
Assess residual medical and psychosocial problems
Ensure that patients have stopped smoking 1 month before surgery
Make thorough evaluation of anatomic deformities and plan specific procedures
Elicit which anatomic areas are of greatest concern and then divide the procedures into as many stages as necessary to help ensure patient safety
Address patients' highest priorities first
Limit surgery in one setting to 6 to 7 hours and wait 3 months between staged procedures
Initiate mechanical prophylaxis before surgery to be continued postoperatively and consider strongly

chemoprophylaxis with unfractionated or low molecular weight heparin
Prevent intraoperative hypothermia
Pay attention to anemia and blood loss and transfuse whenever indicated
Male patients must be treated more conservatively than female patients
Perform body contouring only in accredited inpatient or outpatient facilities

REFERENCES

[1]   Colwel AS, Borud LJ. Optimization of patient safety in postbariatric body contouring: A Current review. Aesthet Surg J 2008; 28: 437-42.

[2]   Au K, Hazard SW III, Dyer AM. Correlation of complications of body contouring surgery with increasing body mass index. Aesthet Surg J 2008; 28: 425-29.

[3]   Sanger CA, David LR. Impact of significant weight loss on outcome of body contouring surgery. Ann Plast Surg 2006; 56: 9-13.

[4]   Aly AS, Cram AE, Chao M, *et al.* Belt lipectomy for circumferential truncal excess: The University of Iowa experience. Plast Reconstr Surg 2003; 111: 398-413.

[5]   Safety considerations and avoiding complications in the massive weight loss patient. Plast Reconstr Surg 2006; 117S: 74S-81S.

[6]   Gusenoff JA, Rubin JP. Plastic surgery after weight loss: Current concepts in massive weight loss surgery. Aesthet Surg J 2008; 28: 452-55.

[7]   Davison SP, Clemens MW. Safety first: Precautions for the massive weight loss patient. Clin Plastic Surg 2008; 35: 173-83.

[8]   Examination of the Massive Weight Loss Patient and Staging Considerations. Plast Reconstr Surg 2006; 117S: 22S-30S.

[9]   Shermak MA, Chang D, Magnuson TH, *et al.* An outcomes analysis of patients undergoing body contouring surgery after massive weight loss. Plast Reconstr Surg 2006; 118: 1026-31.

[10]   Matarasso A. Commentary: A clinical review of total body lift surgery. Aesthet Surg J 2008; 28: 304-5.

[11]   Arthurs ZM, Cuadrado D, Sohn V, *et al.* Post-bariatric panniculectomy: Pre-panniculectomy body mass index impacts the complication profile. Am J Surg 2007; 193: 567-70.

[12]   Nemerofsky RB, Oliak DA, Capella JF. Body lift: An account of 200 consecutive cases in the massive weight loss patient. Plast Reconstr Surg 2006; 117: 414-30.

[13]   Shermak MA, Chang DC, Heller J. Factors impacting thromboembolism after bariatric body contouring surgery. Plast Reconstr Surg 2007; 119: 1590-96.

[14]   Rubin JP, Nguyen V, Schwentker A. Perioperative management of the post–gastric-bypass patient presenting for body contour surgery. Clin Plastic Surg 2004; 31: 601-10.

[15]   Gravante G, Araco A, Sorge R, *et al.* Wound infections in post-bariatric patients undergoing body contouring abdominoplasty: the role of smoking. Obes Surg 2007; 17: 1325-31.

[16]   Hurwitz DJ, Agha-Mohammadi S, Ota K, *et al.* A clinical review of total body lift surgery. Aesthet Surg J 2008; 28: 294-303.

[17]   O'Toole JP, Rubin JP. Evaluation of the massive weight loss patient who presents for body-contouring surgery. In: Rubin JP, Matarasso A, Eds. Aesthetic Surgery After Massive Weight Loss. New York: Elsevier Health Sciences, 2008; pp. 13-20.

[18]   Hurwitz DJ. Single-staged total body lift after massive weight loss. Ann Plast Surg 2004; 52: 435-41.

[19]   Aly A, Downey SE, Eaves FF III, *et al.* Panel discussion—Evolution of body contouring after massive weight loss. Plast Reconstr Surg 2006; 118 (Abstract Suppl): 55.

[20]   Albino FP, Koltz PF, Gusenoff JA. A comparative analysis and systematic review of the wound-healing milieu: implications for body contouring after massive weight loss. Plast Reconstr Surg 2009; 124: 1675-82.

[21]   Kolker AR, Lampert JA. Maximizing aesthetics and safety in circumferential-incision lower body lift with selective undermining and liposuction. Ann Plast Surg 2009; 2: 544-48.

[22]   Gusenoff JA, Coon D, Rubin JP. Implications of weight loss method in body contouring outcomes. Plast Reconstr Surg 2009; 123: 373-76.

[23]   De Kerviler S, Hüsler R, Banic A, *et al.* Body contouring surgery following bariatric surgery and dietetically induced massive weight reduction: a risk analysis. Obes Surg 2009; 19: 553-59.

[24]   Deitel M, Greenstein RJ. Recommendations for Reporting Weight Loss. Obes Surg 2003; 13: 159-60.

CHAPTER 7

Abdominal Contour Surgery for the Massive Weight Loss Patient

Ulrich M. Rieger[*]

Department of Plastic, Aesthetic & Reconstructive Surgery, Innsbruck Medical University, Anichstrasse 35, 6020 Innsbruck, Austria

Abstract: Patients requiring surgical skin excision after massive weight loss are challenging and require an individualized and structured approach. The characteristic abdominal deformity includes a draping apron of panniculus often extending to the back, gluteal and thigh areas. Occasionally these deformities are associated with previous surgical scars in the upper abdomen resulting from open gastric bypass surgery or from other procedures such as open cholecystectomy. These scars can compromise the blood supply of the abdominal skin. For adequate and safe abdominal contouring both excess skin and fat as well as the remaining perfusion of the remaining tissues must be addressed to achieve satisfactory results. The key to satisfactory results is a thorough analysis of horizontal and vertical skin and fat excess of abdominal area keeping in mind the torso, buttock, flank areas and choosing an adequate and safe procedure addressing the respective areas of skin and fat excess while preserving the blood supply of the abdominal area in a scarred abdomen.

Keywords: Abdominal contouring, panniculectomy, abdominoplasty, mid-body-contouring.

1. BACKGROUND

Massive Weight Loss

The obesity pandemic in western countries is becoming a major health problem. In recent years a growing number of obese patients are seeking a surgical solution for their weight problem. Bariatric surgery is an effective treatment modality resulting in substantial and long-term weight reduction [1]. Consecutively, the number of patients being referred to plastic surgeons for body contouring procedures after Massive Weight Loss (MWL) following gastrointestinal bypass surgery is steadily increasing [2].

Abdominal Contour Surgery

Abdominoplasty often is among the first procedures to be performed in these patients. Depending on whether patients have lost weight spontaneously or whether some kind of bariatric surgery has been carried out the selection of the abdominal contouring procedure varies. The plastic surgeon has to carefully consider the incisions used for bariatric surgery by the visceral surgeon because abdominoplasty procedures do not only have to be adapted according to patients' excess tissue but also to the remaining blood supply of the abdominal skin flap.

2. PATIENT SELECTION AND EVALUATION

Patient Selection

Abdominal body contouring after MWL is often the initial step of an extensive series of body contouring procedures that are needed to restore a functional and pleasing body contour. It is mandatory, however, for the patient to have achieved a stable body weight after bariatric surgery for a minimum of one year prior to any surgical body contouring. Although MWL leads to a significant decrease of medical co-morbidities such as heart disease, hypertension, diabetes or hypercholesterolemia [3], these medical problems should be under control before post-bariatric surgery is carried out. Moreover, in the long term, tobacco abuse has deleterious effects on microcirculation and blood supply leading to post-operative tissue necrosis. Since

*Address correspondence to Ulrich M. Rieger: Department of Plastic, Reconstructive & Aesthetic Surgery, Medical University Innsbruck, Anichstrasse 35, 6020 Innsbruck, Austria; Tel: +43-512-5040; e-mail: ulrich.rieger@uki.at or ulrich.rieger@i-med.ac.at

abdominoplasty procedures are associated with relatively high complication rates of up to 50% with obese patients being associated with the highest rates of complications [4], cessation of smoking is mandatory before carrying out extensive body-contouring in order minimize complications.

Patient Evaluation

Patients presenting for abdominal body contouring after gastro-intestinal bypass surgery or gastric banding may present with little stab wound incisions resulting from laparoscopic access ports or with vertical or horizontal scars from open bypass surgery. A vertical scar often does not limit blood supply of the abdominal flap in body contouring procedures at all, nor do the laparoscopic access ports. However, these vertical scars may be disfiguring since they tend to be broad or even hypertrophic and therefore these scars should be treated adequately in body contouring procedures to achieve satisfactory results. Patients with transverse abdominal scars, however, do need individualized approaches [5] for abdominal body contouring due to the limitations in blood supply of the abdominal flap.

In addition a patient who has had an open abdominal procedure has an increased risk of hernia formation [6, 7]. A massive weight loss patient may have an excess subcutaneous fat component, which can make palpation of a hernia difficult. Therefore, a hernia can remain occult until the time of surgery. Preoperative abdominal wall ultrasound can be of aid for its diagnosis.

Most of these MWL patients present with excess abdominal wall laxity as well as excess fat and skin in the torso, buttock and flank areas. Undoubtedly, in these patients adequate body contouring cannot be achieved by an abdominoplasty alone and therefore adjacent body areas need to be addressed additionally. As the weight loss deformities increase with additional excess tissue extending from the abdomen to the lateral and posterior trunk, more extensive procedures such as circumferential body-lifting or belt lipectomy procedures may be required.

Procedure Selection

The key to satisfactory results is a thorough analysis of the horizontal and vertical skin and fat excess of the abdomino-torso region and then choose an adequate and safe procedure addressing the respective areas of skin and fat excess while preserving the blood supply of the abdominal area in a scarred abdomen.

Anatomy

The abdominal fat and skin overlying the rectus muscles is mainly supplied by vessels derived from the deep superior and inferior epigastric vessels coursing within the rectus muscles. Peforators originating from these vessels run through the rectus fascia, the overlying fat and then supply the skin. Further blood supply is derived from the superficial superior and inferior epigastric vessels as well as from the superficial iliac circumflex vessels. Minor blood supply is derived from the lateral intercostals and subcostal vessels including the lumbar vessels running from back to front within the fatty tissue (Fig. 1). When performing a traditional full abdominoplasty with wide undermining up to xiphoid process including umbilical transposition as described by Pitanguy, the remaining blood supply of the abdominal flap consists only of these minor lateral vessels as well as the superficial superior epigastric vessels.

In the presence of transverse supraumbilical or subcostal scars, as we may encounter after open gastrointestinal bypass surgery or open cholecystectomy, scaring and subsequent subdermal fibrosis after these incisions represents a great limitation to the remaining blood supply to the abdominal flap as the blood supply by the superficial superior epigastric vessels is additionally jeopardized. Performing full abdominoplasty in presence of these scars is associated with a higher risk of postoperative complications [8] including partial or complete tissue necrosis.

In the following paragraphs, some examples of excess fat and skin distribution types after MWL are described together with the respective corresponding contouring procedure. MWL patients ideally fitting the described fat and skin distribution profiles are often theoretical however the classification enables the surgeon to address the deformities in a structured approach.

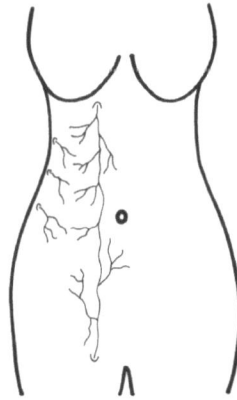

Fig. 1. Blood Supply of the Abdominal Fat and Skin overlying the Rectus Muscles: Perforator Vessels from the Deep Superior and Inferior Epigastric Vessels Coursing within the Rectus Muscles; the Superficial Superior and Inferior Epigastric Vessels as well as the Superficial Iliac Circumflex Vessels; Minor Blood Supply from the Lateral Intercostals and Subcostal Vessels including the Lumbar Vessels.

3. ABDOMEN WITHOUT SCARS

Full Abdominoplasty (Fig. 2, 3, and 4)

A traditional full abdominoplasty begins with an incision low in the inguinal folds, but can be varied as to patient needs and further consists of wide undermining up to the xiphoid process centrally and to the costal margin laterally. It includes umbilical transposition. The umbilicus is circumcised and reinserted in a triangular incision on the abdominal flap. In presence of true diastasis of the rectus muscle, midline suture plication of the fascia is performed, beginning below the xiphoid processus and continuing down to the pubis. The procedure has been described in detail by Pitanguy [9].

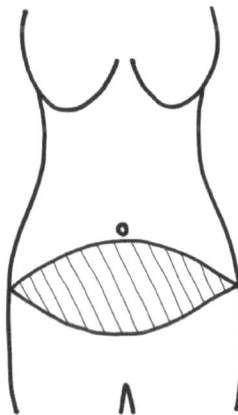

Fig. 2. Full Abdominoplasty: Traditional Resection Lines (Shaded Area).

Panniculectomy

Panniculectomy is indicated in patients who are still morbidly obese in terms of their BMI (>35 kg/ m2) even though they may have lost more than 50% of their initial body weight. Often these patients are severely disabled by physical limitations, pain, back strains or rashes caused by the massive abdominal pannus. Panniculectomy is performed with conservative undermining and can be assisted by suspending the skin and fat pannus from ceiling chains. Limited undermining leads to a preservation of blood supply below the horizontal abdominal scar and limits seroma formation due to minimal dead space. An in depth description of the procedure has been published by Manahan *et al.* [10].

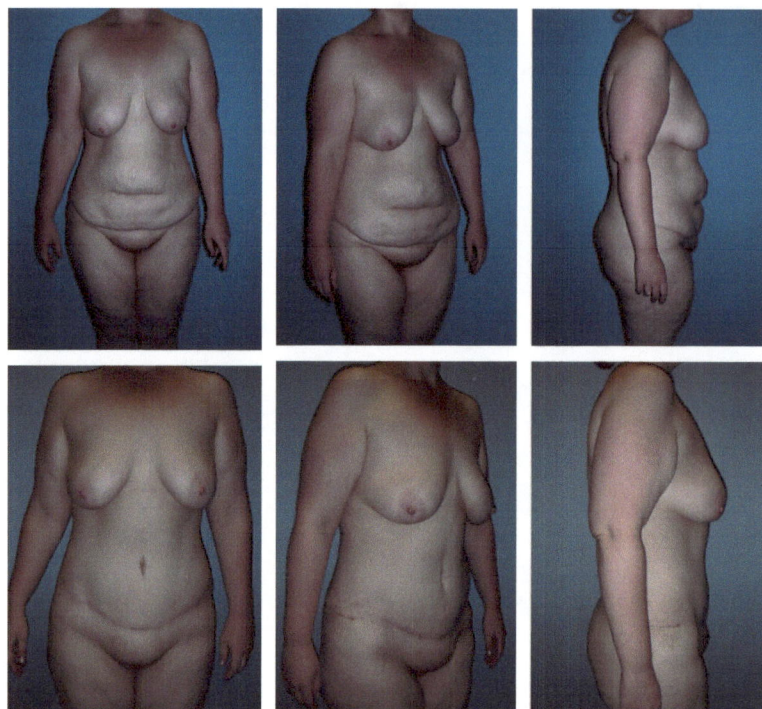

Fig. 3. Traditional Full Abdominoplasty after Laparoscopic Gastric Banding and MWL.

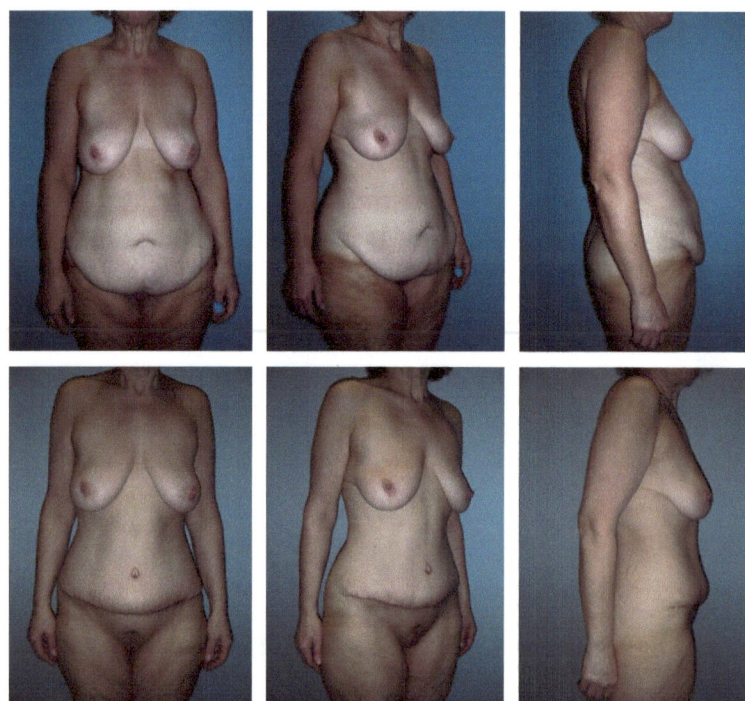

Fig. 4. Spontaneous Weight Loss, Traditional Full Abdominoplasty.

Fleur-de-Lys Abdominoplasty (Mixed Abdominopalsty: Horizontal and Vertical Scars) (Figs. 5, 6, and 7)

Fleur-de-Lys type procedures are ideal for patients presenting with vertical and horizontal abdominal skin excess in conjunction with variable fat excess. In case of additional buttock and flank laxity the Fleur-de-Lys abdominoplasty can be extended to a Fleur-de-Lys circumferential body lift, thereby addressing the gluteal areas and thighs as well.

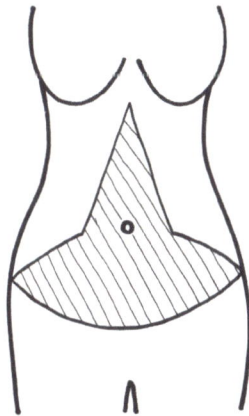

Fig. 5. Fleur-de-Lys Abdominoplasty, Resection Area (Shaded Area) Resulting in a T-Scar.

The technique has been described in different modifications by Cardoso, Dellon and Borud, among others [11-13]. It combines vertical and horizontal resections to restore the abdominal contour to the upper and lower abdomen and the mons pubis. The procedure is especially indicated for patients with a huge panniculus and supra-umbilical dermatochalasis. Essentially, after resection of a cloverleaf like panniculus lateral tissue excess is mobilized and transposed medially. Full plication of the rectus fascia is possible if needed.

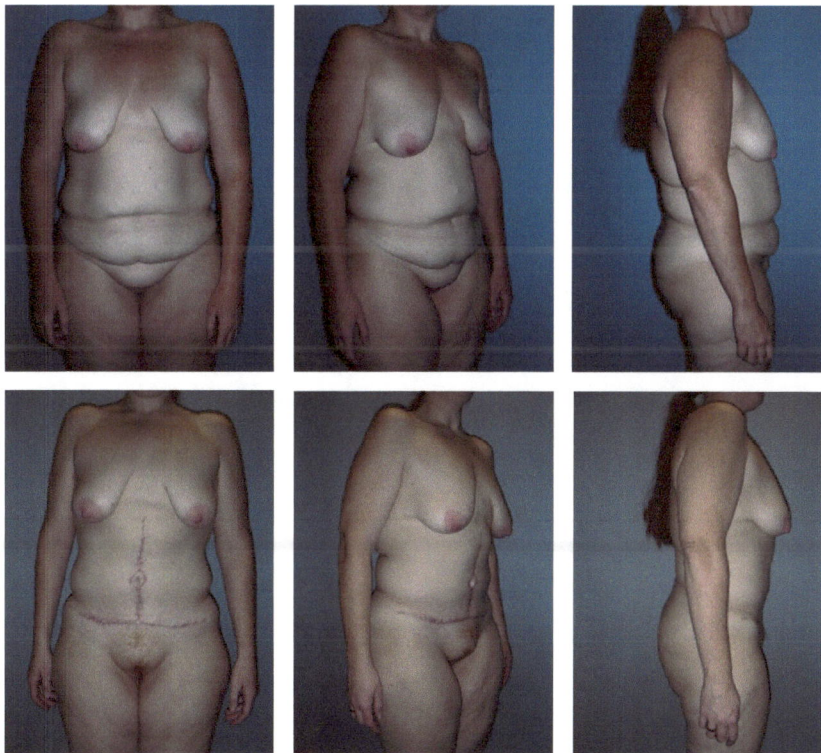

Fig. 6. Fleur-de-Lys Abdominoplasty After Laparoscopic Gastric Banding and MWL.

Reverse Abdominoplasty/IMF Reconstruction (Fig. 8)

Correction of the abdominal laxity involves both the abdominal region and its surrounding tissues. In those post-bariatric patients with significant lower thoracic/upper abdominal laxity, correction of excess abdominal tissue and laxity through a circumferential abdominoplasty underserves the upper abdominal region. In these patients, aesthetic treatment of the abdominal region will demand management of both the lower thorax/upper abdomen as well as the lower abdomen. This can be achieved with the use of reverse abdominoplasty/IMF reconstruction. It can be performed safely independently or combined with circumferential abdominoplasty when limited abdominal undermining is performed. For single-stage procedures, all skin between the umbilicus and pubis is excised with preservation of Scarpa's fascia about the groin and loose areolar tissue over the external oblique fascia. The umbilicus is then released and a supraumbilical tunnel of about 2–3 in. (5-7 cm) is dissected all the way to the xiphoid. Care is taken to preserve the lateral row of rectus abdominus muscle perforators. Following anterior rectus imbrication, the abdominal tissue is discontinuously undermined using vascular dissector/dilators. The abdominoplasty closures are then concluded. The reverse abdominoplasty/neo-IMF reconstruction is then undertaken by excising the upper abdominal tissue over the muscular fascia in between the IMF and the reverse abdominoplasty markings. Tissues to be excised may be de-epithelialized as the epigastric component of the spiral flap to augment the breast. The upper abdominal tissue is again discontinuously undermined and then raised to the new appropriately elevated IMF position [14].

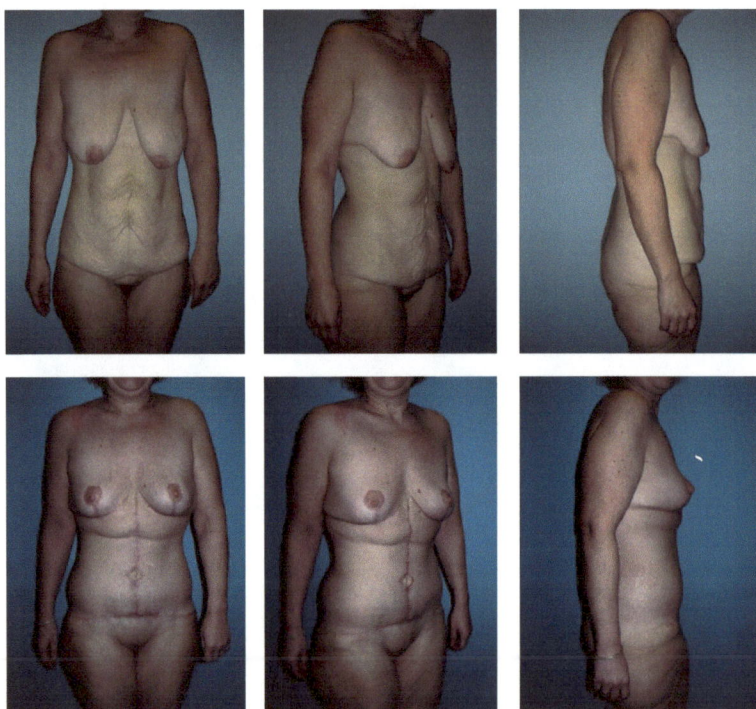

Fig. 7. Fleur-de-Lys Abdominoplasty Combined with Mastopexy after Laparoscopic Gastric Banding and MWL.

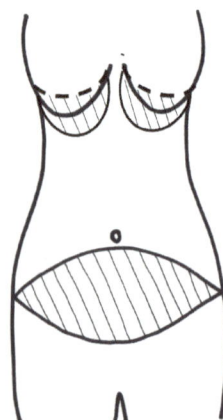

Fig. 8. Reverse Abdominoplasty/IMF Reconstruction.

4. ABDOMEN WITH SCARS

Some surgeons consider the left oblique or left and right oblique incision to be standard incisions for bariatric surgery due to a very low rate of incisional hernia compared to vertical midline incisons [15-17]. As MWL patients often carry co-morbidities such as cholecystolithiasis they may also present with a transverse scar in the upper right abdomen resulting from open cholecystocytectomy or with chevron incisions resulting from other more extensive procedures.

Due to the limited blood supply of the abdominal flap [8], often a limited abdominoplasty of the low transverse type with limited undermining only up to the level of the umbilicus is performed in order not to compromise blood supply in the zone between the old transverse scar (resulting from the gastric bypass) and the new transverse scar (resulting from the abdominoplasty). While limitation of mobilization of the abdominal flap itself leads to preservation of vascular zones [18] and therefore is a safe procedure, it does not yield an aesthetically acceptable result. These patients following gastrointestinal bypass surgery need full abdominoplasty by extensive lifting and undermining well above the umbilicus.

Perforator Sparing Abdominoplasty [19] (Figs. 9, 10, and 11)

This procedure is suitable for patients with excess fat and skin very much limited to the abdomen and flank areas or for patients who do not wish more extensive procedures. It suits best for horizontal excess skin with variable amounts of excess fat.

The operative procedure consists of an abdominoplasty with wide undermining up to the xiphoid process as described by Pitanguy [9]. To prevent hypoperfusion of the flap due to the transverse scar, selective dissection of periumbilical abdominal wall perforator vessels is performed to secure flap blood supply. These perforators can be visualized pre-operatively by color duplex imaging [20, 21]. Vessel tunnelling through the rectus sheaths and rectus muscles and ligation of the cranial perforator branches provides sufficient flap mobility without perforator tension or traction. Flap undermining is done around those perforator vessels. Through extensive perforator preparation sufficient flap mobility without perforator tension or traction is achieved. The abdominal flap is then further elevated to the xiphoid process centrally and the costal margins laterally and excess of skin and subcutaneous tissue is excised. Hereby the lifting effect can be extended all over the abdominal flap up to the xiphoid. The umbilicus is circumcised and reinserted in a triangular incision on the abdominal flap. Full plication of the rectus fascia can be performed as needed. The perforator sparing abdominoplasty procedure has been described in detail before [19]. Adjunct procedures such liposuction of the flanks can be performed in the same procedure for additional touch-up [22, 23].

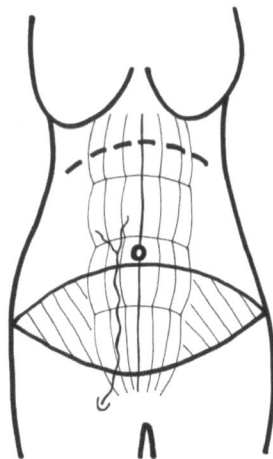

Fig. 9. Perforator Sparing Abdominoplasty. Area of Resection (Shaded Area), the Rectus Abdominis Muscles including One Exemplary Dissected Perforator Vessel.

From our experience, actual perforator dissection through the fascia combined with muscle tunneling and ligation of the ascending branch of the perforator as done in perforator sparing abdominoplasty [19] may yield favorable aesthetic results, but requires basic microsurgical skills. In cases with extensive supra-umbilical excess panniculus, only insufficient flap mobility may be achieved despite extensive perforator dissection. The limiting factor hereby is represented by the actual lengths of the dissected perforators which may be disrupted by too much traction when pulling the abdominal flap caudally for good tightening. The resulting new scar is positioned low in the inguinal folds, however the old scar from the gastric bypass moves caudally and often to the central part of the abdomen. The tightening effect of this method is excellent however the bypass scar can become even more accentuated. The method yields excellent results in terms of abdominal skin sensibility.

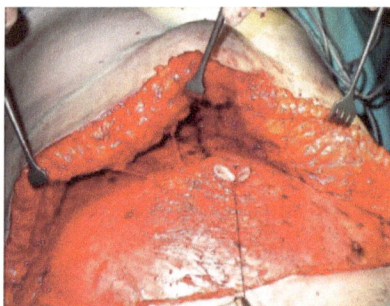

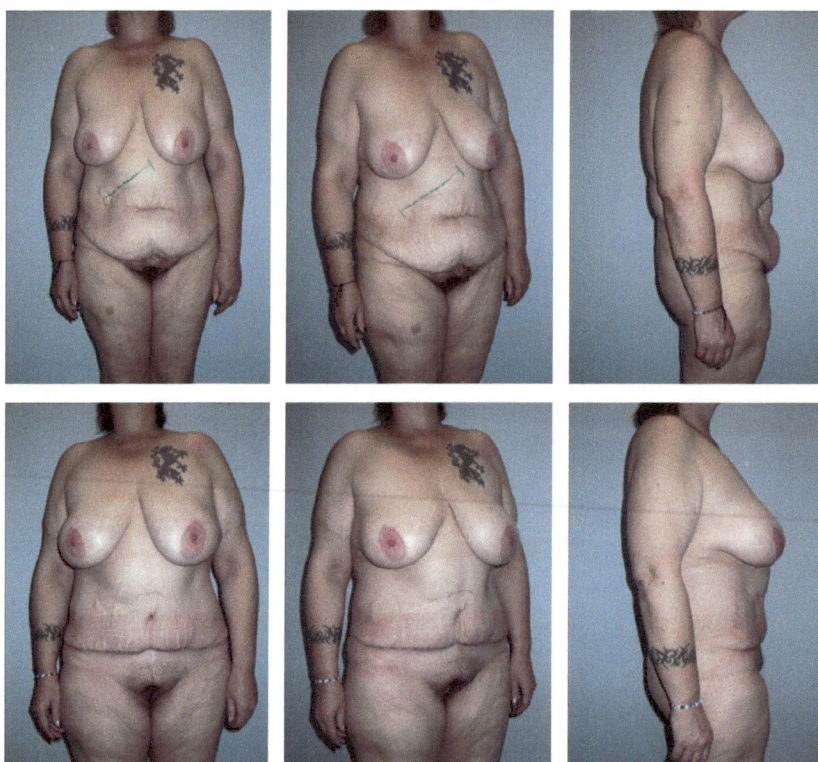

Fig. 10. Perforator Sparing Abdominoplasty for MWL, Patient with Scar Resulting from Open Cholecystectomy (Subcostal Incision).

Abdominoplasty - Cranial and Caudal Incision - Two-Staged Procedure (Figs. 12 and 13)

Staged abdominoplasty procedures and/or limited undermining procedures are suitable for patients presenting with horizontal and/or vertical abdominal excess skin combined with severe excess fat.

In a first procedure a limited low transverse abdominoplasty with undermining up to the umbilicus and excess skin resection is performed. Six months later, at the soonest, a second procedure consisting of an inverted abdominoplasty with scar positioning in the submammary folds and excision of the bilateral subcostal scars from the gastric bypass surgery is performed. Midline suture plication of the rectus fascia in case of diastasis can be performed in each of the steps if needed.

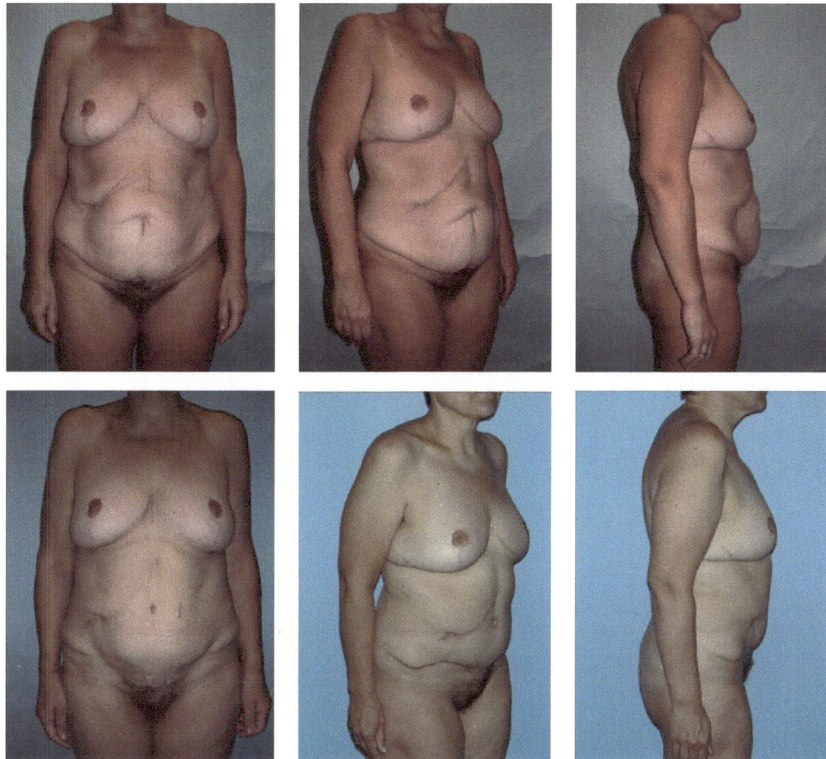

Fig. 11. Perforator Sparing Abdominopasty Technique for MWL, Patient with Scar Resulting from Open Cholecystectomy (Subcostal Incision).

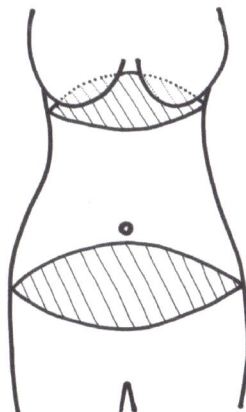

Fig. 12. Abdominoplasty - Cranial and Caudal Incision - Two-Staged Procedure; Displayed Are Areas of Resection (Shaded Sreas); Note: Allow at Least 6 Months Between the Two Resections; Significant Undermining Has to be Performed Both in Regular as well as Inverted Abdominoplasty.

To yield a good tightening effect disruption of most of the abdominal wall perforators has to be performed. Suture plication of the rectus muscles is possible in two separate steps, first in the lower abdomen then later in the upper part. The scars can be positioned discretely in the sub-mammary folds as well as low in the inguinal folds. Naturally, hiding of the sub-mammary scar is only possible in females. The surgeon should make sure that excess skin resection can be performed in such way that the original sub-costal scar is included in the resected excess skin. However, the main disadvantage of the two-staged procedure is a severely compromised sensitivity of the abdominal skin between the two transverse scars. Nevertheless, a very good tightening effect can be achieved by this technique.

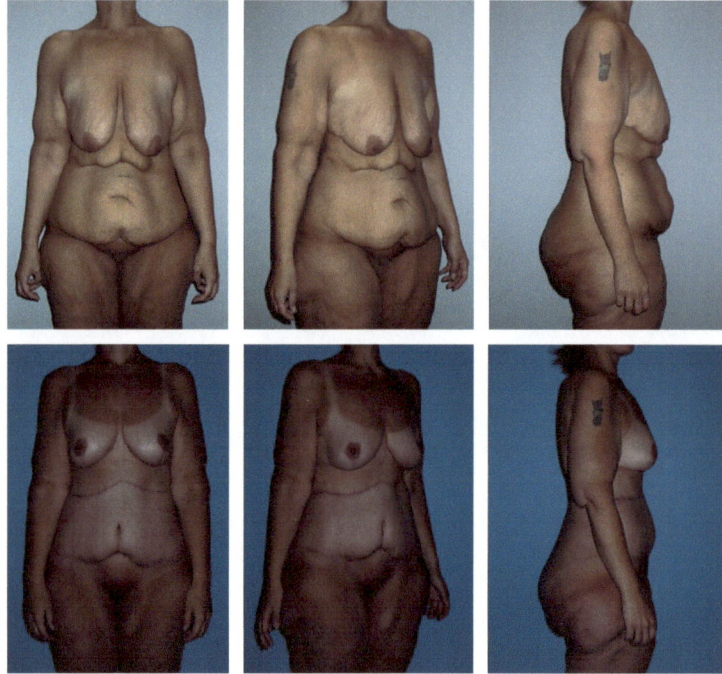

Fig. 13. Skin Island with Cranial and Caudal Incision and Mastopexy.

Abdominoplasty - Cranial and Caudal Incision - One-Staged Procedure (Fig. 14)

Pre-operatively several paraumbilical perforators are identified by color duplex imaging. The paraumbilical area is circumcised including the preoperatively marked perforators. Then an incision pursuant to a low transverse abdominoplasty is made low in the inguinal folds and suprapubic region. Through another incision placed along the submammary folds excision of the old transverse scar is performed. Excess skin resection follows from cranial as well as from caudal by preservation of a central island flap whose blood supply is ensured by the paraumbilical perforators. The options for suture plication of the rectus fascia are limited by the central skin island.

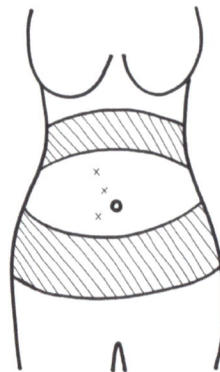

Fig. 14. Abdominoplasty - Cranial and Caudal Incision - One-Staged Procedure: Asterisks Mark Paraumbilical Perforator Vessels Supplying the Periumbilcal Skin Island, which Should not be Undermined, Shaded Areas Mark Resected Tissues.

This single stage technique is very safe and quick to perform. In order to be sure to include at least one perforator in proximity to the umbilicus, pre-operative marking by color duplex imaging can be performed. This way a well-perfused skin island around the umbilicus is left without having to dissect perforators microsurgically. However, the tightening effect of the abdominal wall is limited and plication of rectus fascia cannot be performed to a full extent due to the overlying skin island in the central part of the abdomen. In most cases the old bypass scar can be included in the cranial excess tissue resection. Positioning of the new scars is performed in a similar way as in the two-staged procedure. The sensibility of the skin island is severely compromised.

Fleur-de-Lys Abdominoplasty with Horizontal to Vertical scar Transposition [12, 24, 25] (Figs. 15, 16, 17, 18, and 19)

This technique is performed as described by Dellon *et al.* [12]. It combines vertical and horizontal resections to restore the abdominal contour to the upper and lower abdomen and the mons pubis. The procedure is especially indicated for patients with a huge panniculus and supra-umbilical dermatochalasis. After resection of a cloverleaf like panniculus, lateral tissue excess is mobilized and transposed medially thus converting the transverse subcostal scar into a vertical scar. Blood supply of the flap is ensured by intercostal vessels. Full plication of the rectus fascia is possible if needed.

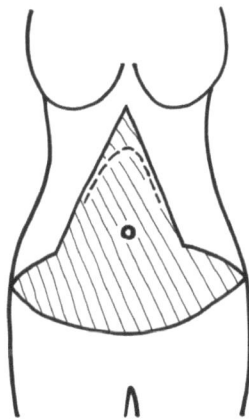

Fig. 15. Fleur-de-Lys Abdominoplasty with Horizontal to Vertical Scar Transposition after Open Gastric Bypass *via* Bilateral Subcostal Incision (Shaded: Area of Resection).

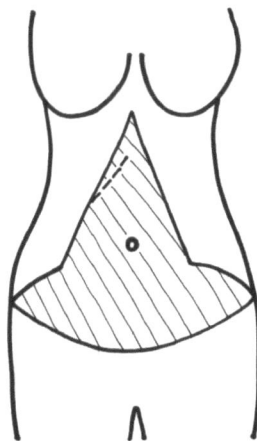

Fig. 16. Fleur-de-Lys Abdominoplasty with Horizontal to Vertical Scar Transposition after Kocher Incision from Open Cholecystectomy (Shaded: Area of Resection).

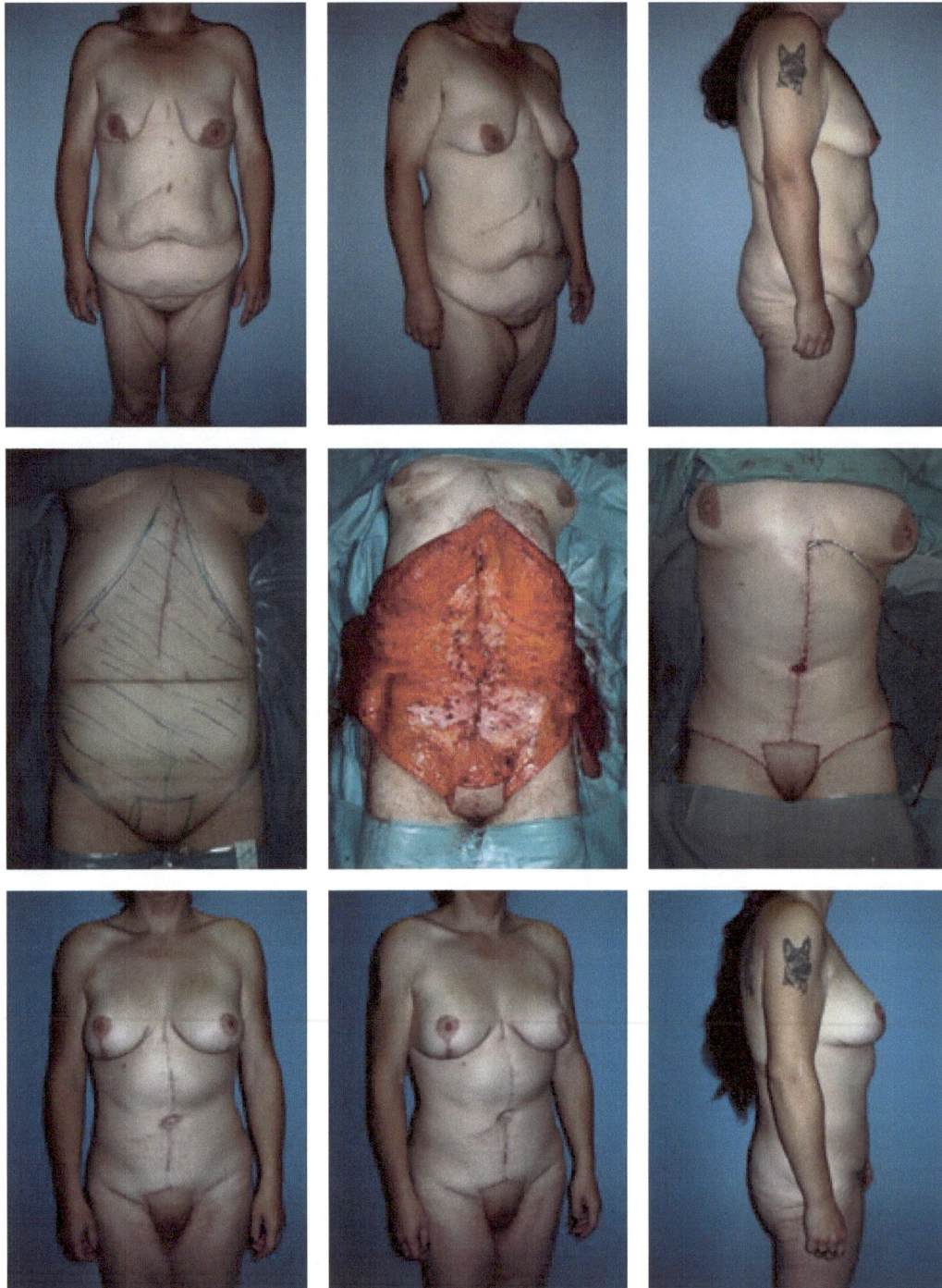

Fig. 17. Fleur-de-Lys Abdominoplasty with Scar Transposition (Open Cholecystectomy) after Laparoscopic Gastric Banding.

Modified Fleur-de-Lys abdominoplasty is very suitable to resect huge excess panniculus tissue in a single stage procedure. A skilled surgeon can transform the old transverse subcostal scar resulting from the open gastric bypass into a vertical scar, which would have resulted anyway due to the nature of the technique. To the author, the vertical scar seems less conspicuous than the transverse scar across the abdominal wall. The tightening effect is excellent and plication of the rectus fascia is easy to perform in case of true diastasis of the rectus muscles. In addition excess tissue of the flanks can be removed as well. The sensibility of the abdominal skin is impaired but still better than after techniques with cranial and caudal incisions.

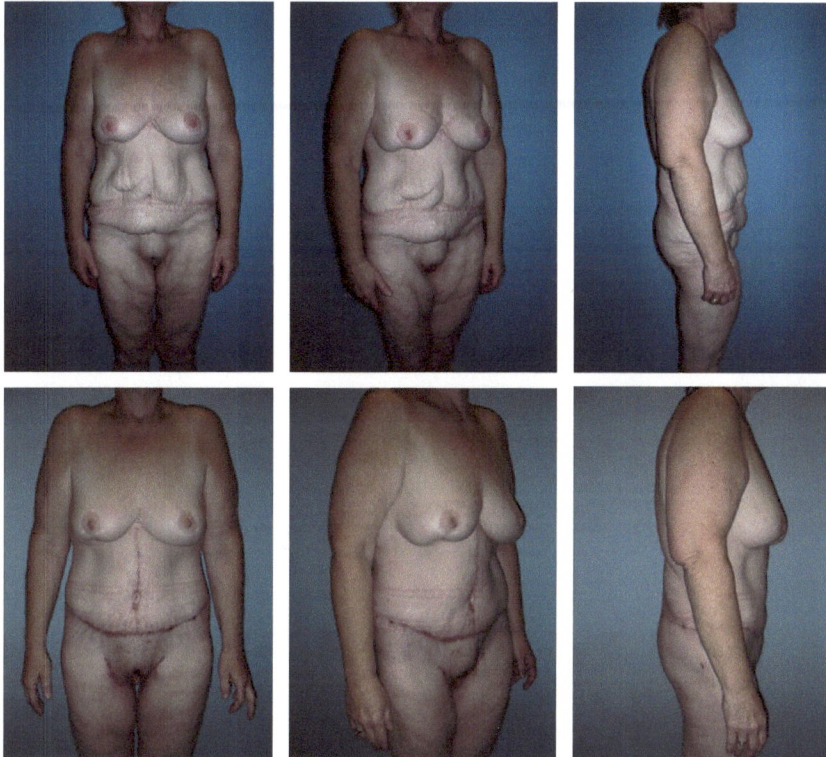

Fig. 18. Abdominal Body Lift after MWL (Note Subcostal Scar from Open cholecystectomy, Mastopexy Had Been Performed Elsewhere).

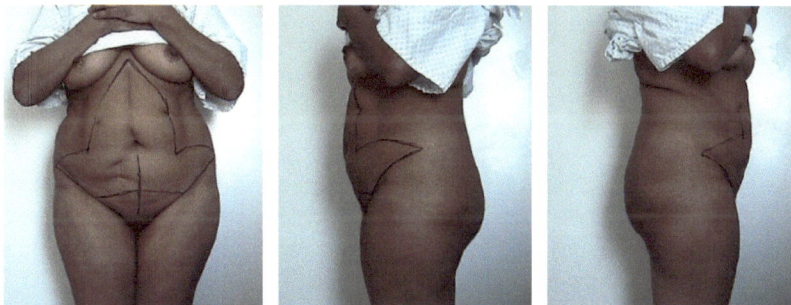

Fig. 19. Planning of Fleur-de-Lys Abdominoplasty in a patient with Multiple Abdominal Scars.

5. POST OPERATIVE CARE

Post-operative care is similar for all described procedures. Ambulation is usually begun on the first post-operative day. Suction drains are removed within 48 to 72 hours. Custom-made compression garments should be worn for 8 weeks post-operatively. Out-patient visits are scheduled around day 14 post-operatively (suture removal), then regular follow-up visits at 3 and 6 months as well as one year postoperatively.

6. DISCUSSION OF ALTERNATIVE ABDOMINAL CONTOURING CONCEPTS

Abdominal body contouring implies a significant impairment of the vascular supply of a considerable part of the abdominal flap as well as the adjacent areas [26]. Long transverse supra-umbilical or sub-costal scars particularly compromise blood supply from the superior epigastric arteries. Therefore these patients are especially demanding and as reported by De Castro *et al.* they have an increased risk for partial flap necrosis when body contouring is performed [8].

Different operative techniques have been proposed in order to preserve abdominal flap blood supply and reduce postoperative complications [18, 27, 28]. Among these, techniques preserving the peri-umbilical rectus abdominis perforator vessels proved to reduce postoperative wound complications [29, 30] and are suitable as well for patients following open gastrointestinal bypass surgery [19].

The technique of "Mid-Body-Contouring" has been described by Strauch *et al.* [31]. This technique enables the surgeon to achieve satisfactory contour results in patients with excess horizontal fat and skin excess. However, in patients with transverse upper abdominal scars this excellent technique is inapplicable due to severe limitations to abdominal flap perfusion. With the modification of perforator sparing abdominoplasty, patients with transverse abdominal scars can be treated equally with similar results.

Persichetti *et al.* [32] have described an "anchor line abdominoplasty" which is in principle a modification of a fleur-de-Lys approach with special emphasis on enhancing perfusion of the abdominal zone 3 [18] by preserving segmental lateral abdominal wall perforator vessels. We feel that this concept is important, especially when a Fleur-de-Lys circumferential abdominoplasty technique is applied because excessive lateral mobilization may disrupt these lateral perforators and the surgeon needs to carefully evaluate the amount of circumferential excess skin and fat preoperatively in order to leave perfusion intact. Other variations such as the "four blade propeller pattern" described by Carwell *et al.* [33] describe similar excision patterns in circumferential torsoplasty. Wallach has reported that the Fleur-de-Lys approach is especially suitable in patients presenting with significant upper midline fullness [34]. The authors feel that, indeed, upper midline fullness can be corrected excellently with this technique even in presence of subcostal scars. Other modifications include the "Modified Circumferential Torsoplasty" described by Davison *et al.* [35] with extensions of excess fat and lateral wedge skin excisions to the lateral thighs. This concept can be applied equally in most other body lift techniques, *e.g.* as well in the Fleur-de-Lys circumferential body lift.

Adjunct contouring procedures such as liposuction have been described in different variations for indications such as "Marriage abdominoplasty" [36] with extensive flap thinning or for touch up in selected areas [37]. The authors feel that liposuction can only be an adjunct and not be a substitute for direct excision. Shrinkage of skin as described by Walgenbach *et al.* [36] is often not observed in massive weight loss patients with extensively stretched skin.

When performing post-bariatric surgery for MWL the surgeon is most often aiming for functional improvement in the first line and aesthetic improvement in the second line as has been mentioned by Bruschi *et al.* [38]. All of the above described techniques except panniculectomy can yield both acceptable aesthetic and functional results. With panniculectomy the improvement is most often only functional. In addition to functional improvement Cintra W Jr. *et al.* have described a significant quality of life improvement after abdominoplasty following MWL, especially in domains such as "social/ cultural performance", "self esteem" and "good self-image" [39], a finding that has been confirmed by Song *et al.* for post-body contouring MWL patients in general [40]. Needless to mention that post-bariatric body contouring complications, especially major complications such as flap necrosis and consecutively worse contouring results may interfere with this improved quality of life. Therefore a thorough analysis of both the remaining perfusion as well the contouring needs before abdominal body contouring especially in patients with scared abdomen may help to improve patients' condition in many aspects including physical and psychosocial well-being.

7. CONCLUSION

For optimal contouring of the truncal region following post bariatric MWL, most patients benefit from excision of as much excess skin and subcutaneous tissue as possible. For such patients, abdominoplasty is a safe and effective method of body contouring which can be optimally refined through both horizontal and vertical excisions of tissue resulting in an inverted T scar. The presence of a prior vertical upper abdominal scar makes contouring procedures that include a vertical component less onerous. However, if a patient does not already have an upper abdominal scar, he or she may choose to accept less reduction in tissue volume with less than optimal body contouring in exchange for a limited scar [33]. Regardless, a BMI as close to the ideal as possible is necessary for the complication rate to approach that of the general population undergoing abdominoplasty [41, 42]. However, in those patients who remain in the 'obese' category (BMI>30) maximal resection of both the excess horizontal and vertical abdominal tissues may still be performed safely provided no or extremely limited undermining is performed for closure [43].

ACKNOWLEDGEMENT

Thanks to Ilonka Heider, MD, Cantonal Hospital St. Gallen, Switzerland for providing excellent anatomic illustrations.

REFERENCES

[1]   Spivak H, Hewitt MF, Onn A, *et al.* Weight loss and improvement of obesity-related illness in 500 U.S. patients following laparoscopic adjustable gastric banding procedure. Am J Surg 2005; 189: 27-32.

[2]   Savage RC. Abdominoplasty following gastrointestinal bypass surgery. Plast Reconstr Surg 1983; 71: 500-509.

[3]   Taylor J, Shermak M. Body contouring following massive weight loss. Obes Surg 2004; 14: 1080-85.

[4]   Neaman KC, Hansen JE. Analysis of complications from abdominoplasty: a review of 206 cases at a university hospital. Ann Plast Surg 2007; 58: 292-98.

[5]   Rieger UM, Erba P, Kalbermatten DF, *et al.* An individualized approach to abdominoplasty in the presence of bilateral subcostal scars after open gastric bypass. Obes Surg 2008; 18: 863-69.

[6]   Podnos YD, Jimenez JC, Wilson SE, *et al.* Complications after laparoscopic gastric bypass: a review of 3464 cases. Arch Surg 2003; 138: 957-61.

[7]   Hesselink VJ, Luijendijk RW, de Wilt JH, *et al.* An evaluation of risk factors in incisional hernia recurrence. Surg Gynecol Obstet 1993; 176: 228-34.

[8]   de Castro CC, Aboudib JJ, Salema R, *et al.* How to deal with abdominoplasty in an abdomen with a scar. Aesthet Plast Surg 1993; 17: 67-71.

[9]   Pitanguy I, Mayer B, Labrakis G. Abdominoplasty--personal surgical guidelines. Zentralbl Chir 1988; 113: 765-71.

[10]   Manahan MA, Shermak MA. Massive panniculectomy after massive weight loss. Plast Reconstr Surg 2006; 117: 2191-97.

[11]   Cardoso dC, Salema R, Atias P, *et al.* T abdominoplasty to remove multiple scars from the abdomen. Ann Plast Surg 1984; 12: 369-73.

[12]   Dellon AL. Fleur-de-lis abdominoplasty. Aesthet Plast Surg 1985; 9: 27-32.

[13]   Borud LJ, Warren AG. Modified vertical abdominoplasty in the massive weight loss patient. Plast Reconstr Surg 2007; 119: 1911-21.

[14]   Agha-Mohammadi S, Hurwitz DJ. Management of upper abdominal laxity after massive weight loss: reverse abdominoplasty and inframammary fold reconstruction. Aesthet Plast Surg 2009 Nov 21. [Epub ahead of print].

[15]   Alvarez-Cordero R, Aragon-Viruette E. Incisions for Obesity Surgery: a brief report. Obes Surg 1991; 1: 409-11.

[16]   Jones KBJ. The superiority of the left subcostal incision compared to mid-line incisions in surgery for morbid obesity. Obes Surg 1993; 3: 201-05.

[17]   Jones KBJ. The left subcostal incision revisited. Obes Surg 1998; 8: 225-28.

[18]   Huger WEJ. The anatomic rationale for abdominal lipectomy. Am Surg 1979; 45: 612-17.

[19]   Rieger UM, Aschwanden M, Schmid D, *et al.* Perforator-sparing abdominoplasty technique in the presence of bilateral subcostal scars after gastric bypass. Obes Surg 2007; 17: 63-67.

[20]   Chang BW, Luethke R, Berg WA, *et al.*Two-dimensional color Doppler imaging for precision preoperative mapping and size determination of TRAM flap perforators. Plast Reconstr Surg 1994; 93: 197-200.

[21]   Rand RP, Cramer MM, Strandness DEJ. Color-flow duplex scanning in the preoperative assessment of TRAM flap perforators: a report of 32 consecutive patients. Plast Reconstr Surg 1994; 93: 453-59.

[22]   Matarasso A. Liposuction as an adjunct to a full abdominoplasty. Plast Reconstr Surg 1995; 95: 829-36.

[23]   Rieger UM, Erba P, Wettstein R, *et al.* Does abdominoplasty with liposuction of the love handles yield a shorter scar? An analysis with abdominal 3D laser scanning. Ann Plast Surg 2008; 61: 359-63.

[24]   Ramsey-Stewart G. Radical "Fleur-de-Lis" Abdominal after Bariatric Surgery. Obes Surg 1993; 3: 410-14.

[25] Duff CG, Aslam S, Griffiths RW. Fleur-de-Lys abdominoplasty--a consecutive case series. Br J Plast Surg 2003; 56: 557-66.

[26] Mayr M, Holm C, Hofter E, *et al.* Effects of aesthetic abdominoplasty on abdominal wall perfusion: a quantitative evaluation. Plast Reconstr Surg 2004; 114: 1586-94.

[27] Matarasso A. Liposuction as an adjunct to a full abdominoplasty revisited. Plast Reconstr Surg 2000; 106: 1197-1202; discussion 1203-05.

[28] Saldanha OR, Souza Pinto EB, Mattos WN, Jr., *et al.* Lipoabdominoplasty with selective and safe undermining. Aesthet Plast Surg 2003; 27: 322-27.

[29] Graf R, de Araujo LR, Rippel R, *et al.* Lipoabdominoplasty: liposuction with reduced undermining and traditional abdominal skin flap resection. Aesthet Plast Surg 2006; 30: 1-8.

[30] Schoeller T, Huemer GM, Kolehmainen M, *et al.* Management of subcostal scars during DIEP-flap raising. Br J Plast Surg 2004; 57: 511-14.

[31] Strauch B, Herman C, Rohde C, *et al.* Mid-body contouring in the post-bariatric surgery patient. Plast Reconstr Surg 2006; 117: 2200-11.

[32] Persichetti P, Simone P, Scuderi N. Anchor-line abdominoplasty: a comprehensive approach to abdominal wall reconstruction and body contouring. Plast Reconstr Surg 2005; 116: 289-94.

[33] Carwell GR, Horton CE, Sr. Circumferential torsoplasty. Ann Plast Surg 1997; 38: 213-16.

[34] Wallach SG. Abdominal contour surgery for the massive weight loss patient: the fleur-de-lis approach. Aesthet Surg J 2005; 25: 454-65.

[35] Davison SP, Clemens MW, Chang S. Modified circumferential torsoplasty for the massive-weight-loss patient. Ann Plast Surg 2007; 59: 453-58.

[36] Walgenbach KJ, Shestak KC. "Marriage" abdominoplasty: body contouring with limited scars combining mini-abdominoplasty and liposuction. Clin Plast Surg 2004; 31: 571-81.

[37] Kim J, Stevenson TR. Abdominoplasty, liposuction of the flanks, and obesity: analyzing risk factors for seroma formation. Plast Reconstr Surg 2006; 117: 773-79.

[38] Bruschi S, Datta G, Bocchiotti MA, *et al.* Limb contouring after massive weight loss: functional rather than aesthetic improvement. Obes Surg 2009; 19: 407-11.

[39] Cintra W Jr., Modolin ML, Gemperli R, *et al.* Quality of life after abdominoplasty in women after bariatric surgery. Obes Surg 2008; 18: 728-32.

[40] Song AY, Rubin JP, Thomas V, *et al.* Body image and quality of life in post massive weight loss body contouring patients. Obesity (Silver Spring) 2006; 14: 1626-36.

[41] Larsen M, Polat F, Stook FP, *et al.* Satisfaction and complications in post-bariatric surgery abdominoplasty patients. Acta Chir Plast 2007; 49: 95-98.

[42] Leahy PJ, Shorten SM, Lawrence WT. Maximizing the aesthetic result in panniculectomy after massive weight loss. Plast Reconstr Surg 2008; 122: 1214-24.

[43] Moya AP, Sharma D. A modified technique combining vertical and high lateral incisions for abdominal-to-hip contouring following massive weight loss in persistently obese patients. J Plast Reconstr Aesthet Surg 2009; 62: 56-64.

CHAPTER 8

Thigh Lift

Bishara S. Atiyeh[*], Shady Hayek and Amir Ibrahim

American University of Beirut Medical Center, Beirut, Lebanon

Abstract: Thigh lifting for excess soft tissue and skin laxity became an essential part of the lower body contouring after massive weight loss. Since early 1950's, thigh deformity and different techniques of thigh lift have been addressed. In this chapter, we address and review the anatomical considerations of the different procedures of medial thigh lift and lateral thigh lift as well as the combined techniques.

Keywords: Thigh lift, thigh deformity after massive weight loss, medial thigh lift, lateral thigh lift, vertical thigh lift, horizontal thigh lift, spiral lift.

1. INTRODUCTION

Excess thigh soft tissue and skin laxity secondary to Massive Weight Loss (MWL) requires contouring as an essential step in body re-sculpturing [1]. This area is commonly treated in combination with circumferential abdominoplasty and lower body lift but unfortunately it is difficult to manage [2]. Despite preservation and tight approximation of the subcutaneous fascial system, proper and complete correction of saddlebag deformity and mid-thigh laxity in the massive weight loss patient is not easy to achieve [2]. Factors contributing to postoperative laxity and recurrent ptosis are the diseased skin collagen and elastin, the inability to transmit mechanical forces effectively beyond a short distance from the line of closure, and adherence of the skin to underlying fascia preventing tightening beyond the adherence without extensive undermining which may severely compromise blood supply. Leaving behind excessively heavy and redundant distal thigh skin adds to the distraction forces as well [2].

2. HISTORICAL BACKGROUND

Aesthetic improvement of the thigh by tissue excision through an oblique incision starting in the anterior superior iliac spine, traversing the groin, and ending on the posterior medial thigh was first reported by Lewis in 1957. He proposed also a vertical excision starting at the posterior medial thigh ending proximal to the medial border of the knee [3]. In 1964, Pitanguy described the thigh-buttock lift with an oblique incision across the buttocks extending around the thigh circumference (Fig. **1**). The technique results, however, in noticeable scars and is associated with significant wound healing complications, unnatural contours, early recurrence of deformities, and painful prolonged post operative disability. Later modifications of this technique included dermal flap suspension to muscle fascia or periosteum, however, these modifications, did not address the basic problems of the Pitanguy incision design which produces along with undermining a devascularized inferior skin flap. Tight wound closure adds to the risk of poor wound healing and tends to produce a flattened, aged buttock contour and accentuates supratrochanteric depression [4, 5].

In an attempt to address concerns raised by previous techniques, Lockwood [5] proposed a different surgical design for circumferential lifting and recommended anchoring the dermal tissue of the distal medial thigh tissue to the Colles' fascia to allow for a more stable and long-term result. These changes have improved results and decreased complications but still could not resolve the fundamental problem of poor tissue fixation of ptotic thigh skin to proximal rigid tissue [6].

*****Address correspondence to Bishara S. Atiyeh:** American University of Beirut Medical Center, Beirut, Lebanon; Tel: +961-3-340032; E-mail: batiyeh@terra.net.lb,

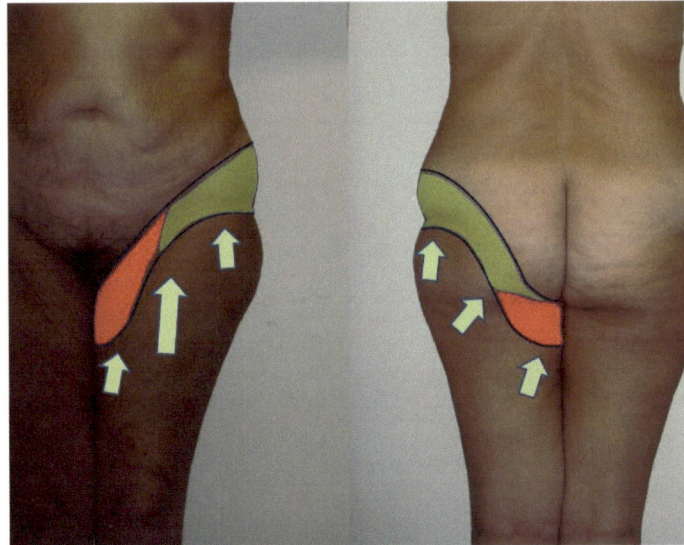

Fig. 1. Pitanguy Thigh-Buttok Lift to Achieve Circumferential Tightening of Tissues.

On the whole, with current techniques, the lower body lift typically addresses the lateral thigh. Fat is removed by liposuction and direct excision and the skin laxity is improved with a combination of excision and discontinuous undermining. The medial thigh is addressed with the thighplasty [6]. Though numerous modifications and refinements have been described regarding skin incisions, tissue excision, Suction Assisted Lipectomy (SAL), level and extent of undermining, and suspension, two main surgical techniques are primarily employed: horizontal excision with superior superficial fascial suspension as described by Lockwood to achieve a lifting effect or vertical excision and skin approximation without tension to reduce thigh circumference. Both may be performed alone or in combination with each other or with a more extensive lower body lift as required (Fig. **2**).

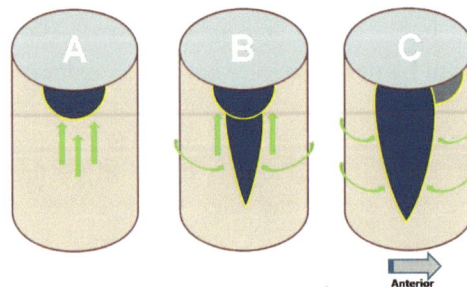

Fig. 2. Diagramatic Representation of Horizontal and Vertical Designs for Medial Thighplasty. (A) Standard Medial Thighplasty with Vertical Vector Superior Pull. (B) Combined Horizontal and Vertical Design. (C) Pure Vertical Design with Anterior and Posterior Horizontal Vectors. Excision of Superior Dog Ears Results in a Horizontal Scar.

3. THIGH DEFORMITY FOLLOWING MWL

Sagging skin deformity after rapid weight loss is diffuse, involves primarily the subcutaneous tissue, and is complicated by underlying adherences to deep fascia [2]. Generally, the most affected area is the medial aspect of the thighs, however, a significant excess of skin showing an unattractive fat area is usually observed both anteriorly and medially [4]. Redundant soft tissues create overlapping folds of skin that are subject to increased rates of infection, an unsightly body image and poor-fitting clothing [3]. Patients present usually with very specific problems consisting in difficulties in deambulation as well as irritation due to chafe between thighs, dermatitis and sensation of heaviness [1].

When evaluating a patient for thigh rejuvenation, a critical assessment must be made of the quality, location, and degree and extent of skin excess. Mathes and Kenkel [7] proposed a classification system of thigh deformities following MWL patients. The typeI MWL patients demonstrate skin laxity over the entire thigh but do not have significant residual lipodystrophy (often denoted as deflated). These patients are treated with a horizontal vector thigh lift. TypeII patients demonstrate both skin laxity and significant lipodystrophy (denoted as non-deflated) and require SAL in addition to a thigh lift either in one or two stages. SAL is usually performed at the first stage combined with a lower body lift to be followed in 3 to 6 months by a second-stage horizontally based medial thigh lift [6, 7].

4. ANATOMICAL CONSIDERATIONS

In contradistinction to other areas undergoing surgical rejuvenation, such as the abdomen or the lateral thigh, the medial thigh has a relatively thin dermal component [6]. As described by Lockwood, underneath the dermis there is a fatty layer within which the superficial fascial system is sandwiched. The superficial fascia consists of one to several thin, horizontal membranous sheets of connective tissue separated by varying amounts of fat with interconnecting vertical and oblique fibrous septae [5, 7]. It provides the major structural support for the skin and subcutaneous tissue of the trunk and extremities [5].

Colles' fascia, originally described by Colles in 1811, is a distinct connective tissue layer. It provides the anatomic shelf that defines the perineal thigh crease and can be visualized by dissecting the skin and superficial fat off the vulva [6]. Lockwood [8] described how Scarpa's layer and the posterior part of the urogenital diaphragm are continuous with the deep layer of the superficial perineal fascia in the medial thigh. Laterally, Colles' fascia attaches to the ischiopubic rami. Anteriorly, it is continuous with Scarpa's fascia of the abdominal wall, and posteriorly it is fused with the posterior border of the urogenital diaphragm. The medial thigh skin is indirectly supported by this subcutaneous fascial network, and accurate reconstruction of this anatomic relationship during medial thighplasty is necessary to limit untoward results [8].

5. MEDIAL THIGH LIFT

One of the areas of greatest concern for both the surgeon and the MWL patient is the medial thigh [3] which remains a troublesome region for body contouring in patients with generalized lipodystrophy and skin flaccidity [6]. SAL of the medial thigh is effective in patients with lipodystrophy without skin laxity. It fails, however, to remodel and tighten the inner thigh where the skin is often thin and inelastic; on the contrary it may result in conspicuous contour abnormalities [6]. In many patients, and in particular in MWL patients where skin laxity can be quite severe and extend down to, and even below, the knee, rejuvenation of the medial thigh requires excision and re-draping of the medial thigh skin as well as removal of fat deposits [6].

The medial thigh lift proposed by Lewis did not gain widespread acceptance because of the many complications associated with the technique including delayed healing, wound dehiscence, wound infection, necrosis, effusions, pain, inferiorly displaced and wide scars, lateral traction vulvar distortion, and unreliable results with early recurrence of ptosis [4, 9, 10]. Lockwood, by describing a fascial anchoring technique attempted to address these issues [6, 7, 11]. Presently, medial thighplasty may be performed either with the Lockwood horizontal medial thigh lift [8] or with the vertically oriented thigh lift [7]. Deciding which technique is indicated depends on the extent of the skin excess. In general, if the skin laxity is limited to the proximal third of the thigh, it can be addressed with a transverse excision of the upper medial thigh skin with lifting and suspension. Further skin excess beyond the upper one-third of the thigh requires a vertical component resulting in a T-shaped or L-shaped scar. Regardless of which technique is used, associated excess fat can be excised or better removed with SAL [6, 7, 11]. Both, horizontal or vertical techniques are subject, however, to untoward results such as scar widening, wound migration, deformity of the vulva, pubic hair migration and recurrent ptosis. Careful pre-operative planning is required [6, 9, 12].

Lockwood's Medial Thighplasty

If only a transverse excision is required, a horizontally oriented ellipse is drawn preoperatively in the standing position demarcating the area of medial thigh skin that can be securely resected. The superior incision consists of a line starting at the ischio-crural line along the labia majora in the perineal crease and proceedes along the groin towards the anterior superior iliac spine if necessary [1, 9]. To avoid excessive resection, the point of maximum resection width is better determined with the patient in the standing position by moderately pulling the redundant medial skin excess towards the planned incision in the perineal crease [9]. If further excision is needed for circumferential excess, a vertical component may be added for both horizontal and vertical excisions along the medial thigh [1] (Figs. **3, 4**).

Positioning of the patient depends on the surgeon's preference. The procedure can be successfully performed either in the supine or in the frog-leg position. Incision down to fascia lata is made laterally and the anterior medial thigh flap is undermined only lateral to the femoral triangle [3]. Medially, in the area of the triangle, the incision is carried only to the superficial fascia and careful dissection is performed at the level of the fascia between the fatty layers to preserve the lymphatics and prevent the formation of seromas or lymphoceles [1, 3]. Colles' fascia is identified after traction is applied laterally to the dissected tissues [13, 14]. After broad undermining of the thigh flap and excision of redundant skin, the flap is stretched cephalad and anchored securely with strong sutures to Colles' fascia for support [1, 8, 10]. Closing the tissue is usually performed in three layers (the fascia, dermis, and epidermis). 2-0 anchoring suture is used on the fascia and 3-0 Sutures are used on the dermal and epidermal layers. Drains should be placed at the end of the procedure.

The standard Lockwood medial thighplasty technique has its vector of pull vertically toward the groin. Unfortunately this pull is responsible for much of the associated morbidity such as recurrent ptosis and traction deformity of the labia. In the MWL patient, this risk is increased as greater demands are placed in an effort to raise more of the distal thigh [6]. The technique necessitates also a very long rehabilitation and can result in significant complications. It causes scars in the inguinal fold and on the posterior buttock fold which are rather painful over a long period. Skin resection and lipectomy in these regions sometimes cause lymphorrea and delayed healing. The procedure may entail also lipoaspiration in addition to significant undermining, sometimes necessitating blood transfusions [4, 12]. Moreover, the superior pull and surgical resection pattern rotates the medial thigh tissues anteriorly. This way the scar becomes usually too long extending to the anterior superior iliac spine and frequently is of bad quality. Furthermore, the vertical traction on the anterior thigh skin can give an unnatural aspect with vertical folds [9]. A posteriorly rotated excision pattern on the other hand leads to a significant scar in the buttock fold, directly visible and rarely symmetrical [9]. To avoid these pitfalls, a concentric surgical excision pattern has been proposed and is characterized by a shorter scar located in the perineal fold extending superiorly towards the mons pubis (Fig. **5**). The resultant scar is hidden from both the anterior and the posterior views [9].

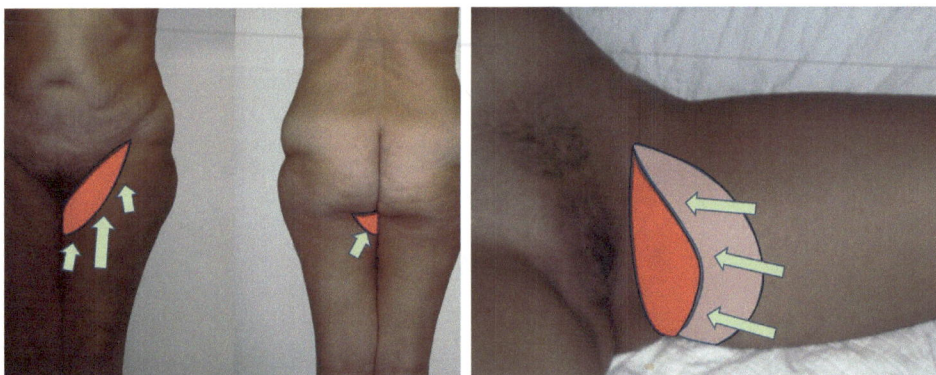

Fig. 3. Lockwood's Medial Thighplasty. Superior Pull with Anterior Rotation of Medial Thigh Skin. Excision Area in red, Undermining Area in Pink. Incision in Inguinal Crease and Posteriorly Extends into the Infragluteal Fold.

It has been recently stressed that the non-use of liposuction or insufficient liposuction associated with extensive undermining in classical horizontal medial thighplasty techniques is a frequent cause of failure and scar migration. It seems that lightening of the inferior flap is mandatory to make the anchoring sutures more reliable. SAL thus must be performed most importantly at the determined elliptical resection area where almost all the fat needs to be eliminated. This is the key to avoid subsequent extensive undermining [9]. With liposuction, the skin to be resected is simply dissected at the subdermal level and the thin liposucted tissues located between the lower medial thigh horizontal skin incision and the anchoring point are no longer thick and thus and are left untouched. Consequently, all the lymphatic and other vessels are preserved. Key sutures are then placed anchoring the medial thigh superficial fascia and dermis of none undermined skin to Colles' fascia and skin excision is completed without much tension and over resection [9].

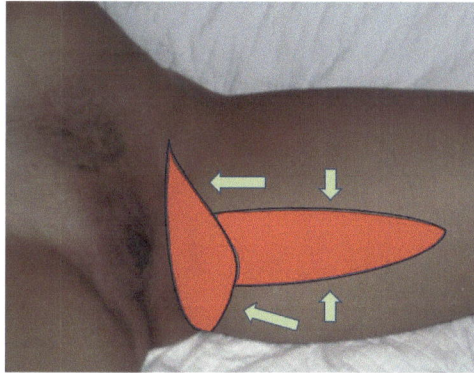

Fig. 4. Additional Vertical Excision to Standard Lockwood's Medial Thighplasty Can Be Performed When Required.

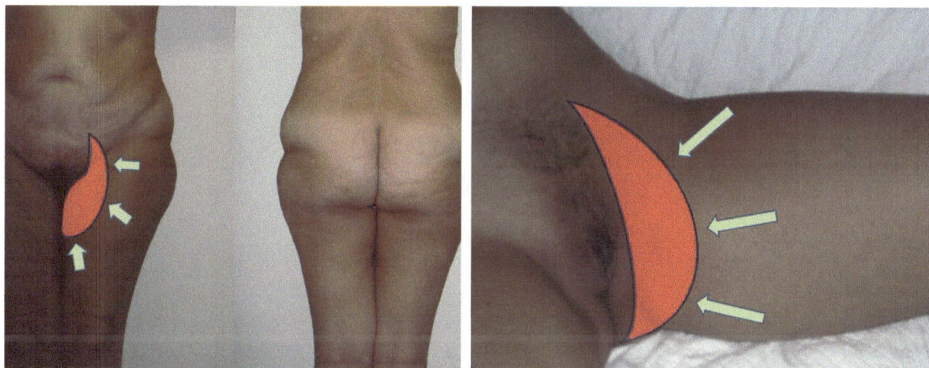

Fig. 5. Concentric Medial Thighplasty. Anterior Part of the Incision Line Begins Along the Labia Majora, then Extends Anteriorly and Vertically Along the Lower Part of the Mons Pubis Lateral Border. Posteriorly, the Incision Is Not in the Infragluteal Fold But Goes Into the Buttock Toward the Ischiatic Bone and Anus.

Vertically Oriented Medial Thigh Lift

The horizontal laxity seen in the MWL patient is often quite severe and difficult to completely address with the classic vertical vector pull. Furthermore, this pull cannot address the middle distal third of the thigh [6]. Circumferential contouring may be combined with a classical Lockwood's medial thighplasty. In this case, an additional incision is performed vertically along the median line of the thigh inner aspect, and excessive adipose–dermoid tissue is then removed. The resulting scar will therefore have a T shape, with an inguinal and an axial component [1]. A typical vertical medial thigh lift, however, uses only an anterior and posterior horizontal vector to accomplish thigh contouring. It does not rely on any type of vertical elevation or pull and there is no need for tension closure to the Colles' fascia. The fundamental principle of this technique focuses on a reorientation of the vertical vectors horizontally producing a strong horizontal pull medially, both anteriorly and posteriorly, and allowing for both skin and thigh circumference reduction by closing the thigh as a cylinder from the knee to the groin. All of the tension is focused along the medial aspect of the thigh and is thus transferred from Colles' fascia and the surrounding important structures to the medial aspect of the thigh, decreasing the likelihood of labial distortion. The horizontal incision in the groin is limited to removal of the "dog ear" and does not contribute to the actual lift [6, 7] (Fig. **6**).

As with any procedure, marking of the elliptical medial thigh excision is critical. The scar is planned to run along the medial aspect of the thigh in an effort to minimize visibility taking into consideration the fact that the posterior thigh is more mobile proximally, whereas the anterior thigh is more mobile in the mid and distal thigh regions. The proximal dog-ear is planned to be removed anteriorly along the groin. To prevent anterior visibility of the scar superiorly, more anterior skin is left proximally [6, 7].

Anterior incision is made first down to the deep fascia which is subsequently elevated. Maintenance of the plane of dissection is critical. It helps avoid the saphenous vein and preserves the deep lymphatics. Of particular importance is the dissection proximally near the groin. All fat in this area must be preserved and the skin is merely undermined to avoid the lymphatic network [7]. Venous branches are usually ligated [15]. Quilting sutures can help prevent the formation of seromas [16] and the skin can be closed in 2 layers similar to the Lockwood lift. Drains can be used especially if the patient has a high BMI or required liposuction. Vertically oriented medial thighplasty allows for predictable improvement of the redundant medial thigh commonly seen in patients following MWL [7].

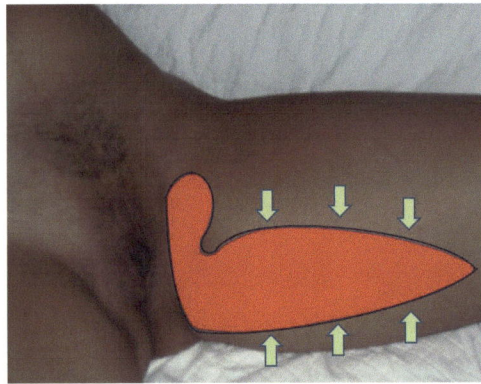

Fig. 6. Vertically Oriented Medial Thigh Lift. Superior Dog Ear Excised Anteriorly Along the Groin Crease.

6. LATERAL THIGH LIFT AND COMBINED MEDIAL AND LATERAL THIGHPLASTY – THE SPIRAL LIFT

Bulging, sagging, and loose skin of the lateral thigh can be vastly improved by lower body lift, however, deformities usually partially reappear within 1 year due to inadequate removal of enough lateral skin. To improve long term results, full abduction of each operated thigh before closure in the prone position has been proposed to allow an increase in the width of skin excision and closure under moderate tension causing the lateral thigh skin to be taut upon leg adduction [2]. Undermining the inferior skin flap beneath the superficial fascial system to the level of the trochanteric fat deposit releases the superficial fascial system zone of adherence and allows optimal transmission of the lifting forces to the distal thigh [5]. Though liposuction reduces excess fat and mobilizes the distal flap, further discontinuous undermining with the Lockwood Underminer (Byron Medical, Inc., Tucson, Ariz.) beneath the superficial fascia and over the fascia lata must be performed to just above the knees, except in the patient with very lax skin, until the entire lateral thigh skin can be pulled to the desired location without the restriction of vertical septa [2]. Superior incision is then performed followed by limited undermining of the superior flap. Flank and lateral thighs are approximated along the midlateral lines over the external oblique muscle as tightly as possible with the leg fully abducted. The subcutaneous fascial system of the lateral thigh and lower torso is stout and the ultimate closure tension must be as strong as the tissue tolerates. Leaving behind the appropriate amount of flank fat is artistry that considers sex, nearby fat deposits and bone structure, and the anticipated thinning of the overlying tightly approximated subcutaneous tissues [2].

A combination of SAL and direct excision of excess skin and fat at the epicenter of ptotic tissues followed by pursestring closure of the skin defect has been described for selected patients with a severe, focal, lateral thigh lipodystrophy resulting in dramatic improvement of saddlebag deformity [17]. Potential drawbacks of pursestring closure are the open areas that must heal secondarily, the initial unpleasant appearance of gathered and pleated skin, and the potential for secondary scar revisions [17]. An M-shaped thigh lift has also been described to correct similar lateral thigh deformities. The traditional supragluteal circular incision is complemented by bilateral downward extensions in the lateral thighs allowing easy and adequate skin and fat removal. The adequacy of the resection, leading to superior aesthetic results, seems to compensate for the resultant long and visible thigh scar and should be considered in selected patients [18] (Fig. **7**).

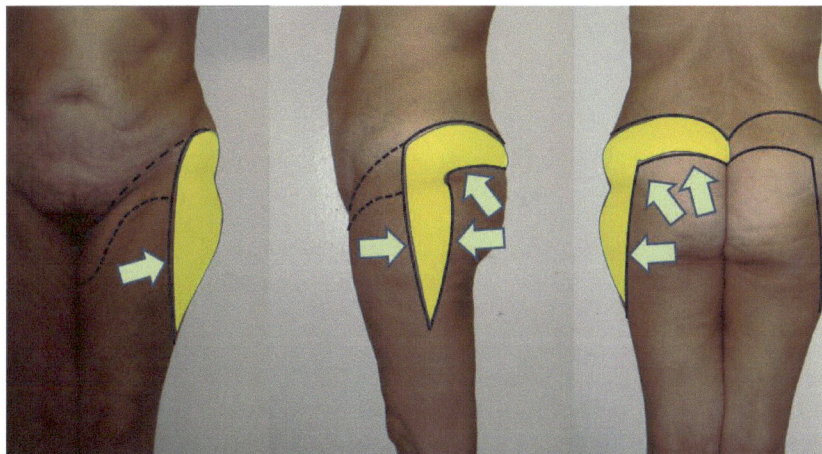

Fig. 7. M-Shaped Thigh Lift.

Patients with a pear- or guitar-shaped body contour deformity involving lipodystrophy and excess skin in the thighs, buttock, and waist traditionally have been treated with a combination of liposuction, flankplasty, belt lipectomy, lower body lift, and/or medial thigh lift [19]. The spiral lift composed of an extended flankplasty together with some modifications, accompanied by a buttock lift and autologous augmentation can be accomplished with a single spiral incision easily concealed by underwear, resulting in removal of excess skin and fat [19]. An incision extending from the medial infra-gluteal fold and upper medial thigh to the inguinal fold anteriorly then to the iliac crest and down to the intergluteal fold posteriorly [4] allows tissue excision and reconfiguration of a natural silhouette [19]. Superficial fascial anchoring to the Colles' fascia, the inguinal ligament, and the anterior superior iliac spine, aggressive liposuction, lateral and anterior thigh lift with undermining and high lateral tension, and a buttock lift with autologous augmentation are all essential steps in the technique to achieve optimal results and avoid scar migration and hypertrophy [19]. This technique results in improved contour and tension of the medial and lateral thigh, trochanteric area, and gluteal region [19] (Fig. **8**).

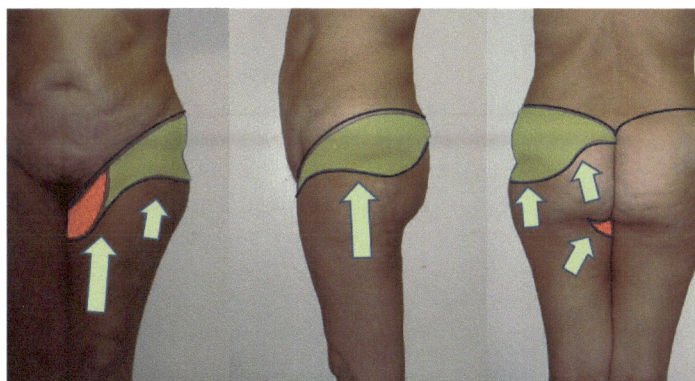

Fig. 8. Combined Lateral and Medial Thighplasty.

A somewhat similar two steps procedure has been also suggested. In step 1, abdominal lipectomy is performed in association with lipoaspiration of the thighs medial and lateral aspect to regularize the subcutaneous fat layer in this area. In step 2, after 6 months, the buttock thigh lift is performed using a heart-shaped incision extending from the perianal region on the inguinal fold anteriorly to the iliac crest and down to the intergluteal fold posteriorly [4]. Prior superficial lipoaspiration of the inner aspect of the thigh makes the second intervention of a groin lift easier because dissection and subdermic lipectomy in this region becomes not necessary [4]. The intervention removes localization of thick fibrotic fat and badly vascularized tissue. The superficial liposuction as well helps with the retraction of skin. With the heart-shaped technique, skin in the dorsal region is resected and the buttock, anchored on the dorsal aponeurosis, is pulled upward [4]. Gluteal autoaugmentation, usually desirable, is possible with this incision. For a circumferential excess of thigh skin, the technique may be combined with a medial thigh vertical skin resection [4].

7. CONCLUSION

The problems faced by patients following MWL are not only relative to esthetics but mainly to vital functions. This concept justifies these procedures as functional surgery in pathologic individuals. Such concept has to be clear when discussing about legitimacy of limb contouring as a mutualistic, state-financed procedure [1].

Most of the recent techniques of medial thigh lift are based on Lockwood's concept of vertical vector pull supporting the thigh lift with sutures to create a superficial fascia-like suspension [6]. All recent modifications attempt to increase the fascial support and decrease tension on the healing scars. No studies, however, have examined the superiority of one technique over another. Regardless, there is clearly merit in limiting the size of the tissue flaps and the degree of undermining [6]. Anyway, the Lockwood medial thigh lift does not address many of the anatomic issues encountered in the MWL patient. The vertical lift seems to be more appropriate to this group of patients. There is no need for tension closure to Colles' fascia and the groin incision in made only anterioly avoiding a T- incision and is limited to remove the ''dog ear'' [6].

REFERENCES

[1]   Bruschi S, Datta G, Bocchiotti A, *et al.* Limb contouring after massive weight loss: functional rather than aesthetic improvement. Obes Surg 2009; 19: 407-11.

[2]   Hurwitz DJ, Rubin JP, Risin M, *et al.* Correcting the saddlebag deformity in the massive weight loss patient. Plast Reconstr Surg 2004; 114: 1313-25.

[3]   Borud LJ, Cooper JS, Slavin SA. New management algorithm for lymphocele following medial thigh lift. Plast Reconstr Surg 2008; 121: 1450-55.

[4]   Cannistra C, Valero R, Benelli C, *et al.* Thigh and buttock lift after massive weight loss. Aesthetic Plast Surg 2007; 31: 233-7.

[5]   Lockwood TE. Transverse flank-thigh-buttok lift with superficial fascial suspension. Plast Reconstr Surg 1991; 87: 1019-27.

[6]   Mathes DW, Kenkel JM. Current concepts in medial thighplasty. Clin Plast Surg 2008; 35: 151-63.

[7]   Kenkel JM, Eaves FF 3rd. Medial thigh lift. Plast Reconstr Surg 2008; 122: 621-22.

[8]   Lockwood TE. Fascial Anchoring Technique Advances in medial thighplasty. Plast Reconstr Surg 1988: 299-304.

[9]   Le Louarn C, Pascal JF. The concentric medial thigh lift. Aesthetic Plast Surg 2004; 28: 20-23.

[10]   Candiani P, Campiglio GL, Signorini M. Fascio-fascial suspension technique in medial thigh lifts. Aesthetic Plast Surg 1995; 19: 137-40.

[11]   Hurwitz DJ. Medial thighplasty. Aesthetic Surg J 2005; 25: 180-91.

[12]   Hodgkinson DJ. Medial thighplasty, prevention of scar migration, and labial flattening. Aesthetic Plast Surg 1989; 13: 111-14.

[13]   Leitner DW, Sherwood RC. Inguinal lymphocele as a complication of thighplasty. Plast Reconstr Surg 1983; 72: 878-81.

[14]   Moreno CH, Neto HJ, Junior AH, *et al.* Thighplasty after bariatric surgery: evaluation of lymphatic drainage in lower extremities. Obes Surg 2008; 18: 1160-64.

[15]   Schultz RC, Feinberg LA. Medial thigh lift. Ann Plast Surg 1979; 2: 404-10.

[16]   Shermak MA, Rotellini-Coltvet LA, Chang D. Seroma development following body contouring surgery for massive weight loss: patient risk factors and treatment strategies. Plast Reconstr Surg 2008; 122: 280-88.

[17]   Garcia JA, Driscoll DN, Donelan MB. Pursestring thigh lift direct approach for a problematic deformity. Ann Plast Surg 2006; 57: 330-2.

[18]   Gomes HC, Ribeiro RB, Bizerra MM, *et al.* The M-shaped thigh lift. Plast Reconstr Surg 2008; 122: 100e-101e.

[19]   Sozer SO, Agullo FJ, Palladino H. Spiral lift: medial and lateral thigh lift with buttock lift and augmentation. Aesthetic Plast Surg 2008; 32: 120-5.

CHAPTER 9

Gluteal Contouring and Enhancement for the Massive Weight Loss Patient

Bishara Atiyeh*, Shady Hayek and Amir Ibrahim

American University of Beirut Medical Center, Beirut, Lebanon

Abstract: Gluteal deformities encountered in patients who have lost a massive amount of weight are unprecedented in body-contouring surgery and plastic surgeons need uncommon techniques when contouring and augmenting the gluteal region in these patients. Following massive weight loss, a lot of patients show atrophy in their buttocks. In addition to the skin laxity and the volume loss relaxation of the superficial fascial apron contributes to gluteal ptosis as well. To restore a youthful gluteal contour, part of the tissue normally resected in circumferential lower body lift can be used enhancing gluteal contour with autologous tissue augmentation. Gluteal aesthetics in massive weight loss patients can be achieved also with large volume autologous fat transfer or alloplastic implants. Adjunctive techniques such as resection and tightening of the gluteal superficial fascia system (gluteal SMAS), posterior thigh lift and infragluteal diamond lift can refine the results.

Keywords: Gluteal contouring, buttock ptosis, lower body lift, belt lipectomy.

1. INTRODUCTION

Loss of fat improves overall health and self-esteem of patients however; it results in excessive skin laxity and deformities that may have a counterproductive effect. Exaggerated fat loss in the face, breasts, and buttocks, specifically, produces flatness in areas where fullness is desirable. There is at present a growing demand for surgical correction of these deformities. The population that has sustained massive weight loss (MWL), with its severe deformities in the buttocks region, is no exception to this trend [1, 2].

The first signs of buttocks ptosis with redundant tissue crossing caudally the lower gluteal crease at the midline of the posterior thigh appear as the buttock enlarges from weight gain or from normal body changes in adolescence. The gluteal tissue becomes more prominent and begins to form an angle at the posterior thigh. At the same time, the crease becomes more apparent and extends laterally [3]. With aging, anatomic changes that occur in the gluteal region contribute to platypygia (flattening of the buttocks). However, several changes may coexist, including accumulation of subcutaneous adipose tissue around the gluteal region, loss of adipose volume in the buttock, postmenopausal accumulation of intra-abdominal fat coupled with rectus diastasis, skin laxity and buttock ptosis, increased hip width, and lengthening of the infragluteal fold. All of these natural changes contribute to decreased buttock projection and ptosis. Dramatic weight loss may cause significant adipose tissue loss in the buttock and exacerbates skin laxity and platypygia seen with normal aging [2].

Gluteal deformities encountered in MWL patients are unprecedented in body-contouring surgery, and plastic surgeons need uncommon techniques when contouring and augmenting the gluteal region in these patients [2]. Varying degrees of buttocks platypygia and decreased projection may be encountered. Many develop deflated buttocks that look unnaturally flat and lack shape and definition with lack of an attractive curve in the lower back which may appear totally straight [4-6] (Fig. **1**). In addition to skin laxity and volume loss, relaxation of the superficial gluteal suspension system (fascial apron or superficial facial system – SFS) contributes to gluteal ptosis as well [2, 3, 5, 7]. MWL patients develop also skeletal changes that may contribute to platypygia. These changes seem to be permanent and tend to exacerbate any pre-existing primary or secondary platypygia caused by loss of adipose tissue in the gluteal region after MWL [2].

*Address correspondence to Bishara Atiyeh:** American University of Beirut Medical Center, Beirut, Lebanon; Tel: +961-3-340032; E-mail: batiyeh@terra.net.lb

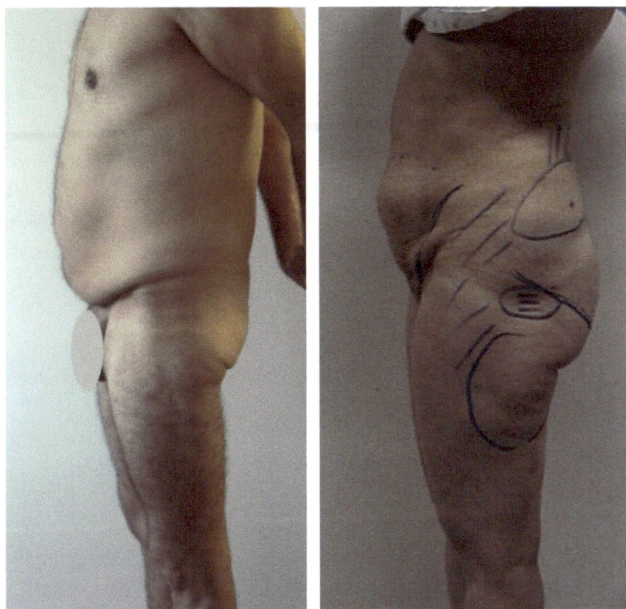

Fig. 1. Buttock Ptosis, Decreased Projection and Flattening of Lower Back Curve.

Aesthetic surgeons regularly perform procedures on the torso, gluteal region, and lower extremities that have immense impact on the aesthetic perception of the buttock [8]. Most of the deformities following MWL involving the abdomen, thighs, and buttocks can be corrected with single-stage belt lipectomy (circumferential body lift - CBL) with lateral thigh and buttock lift as well as liposuction. This treats truncal contour as a unit and may effectively correct contour deformities of the abdomen, the lateral highs, the buttocks, the supragluteal area, the lateral supragluteal area, and the lower back, leaving only one circular scar [1, 2, 9]. A classic CBL, however, has the potential to either impair or enhance the gluteal aesthetics. Alone it fails to address deficient projection and definition of the buttocks [1, 2, 8]. Because of its powerful lifting effect, as the buttock skin is raised, CBL can significantly flatten the buttock [6, 8]. Loss of gluteal projection and resulting flattened contour is directly proportional to the extent of lift achieved [9] and aggressive lifting can exacerbate platypygia and may erode patient satisfaction among those who have experienced MWL [2, 5]. In the buttocks as in the breasts, a "lift" alone does not reliably improve contour in an area that requires volume [10].

Due to genetic programming, adipose tissues in certain body regions are more resistant to weight loss than are others. Following weight loss, the ''apple'' somatotype seems to have less adipose tissue in the gluteal region, whereas the ''pear'' tends to retain more tissue in this area [2]. Regardless of body type, race, or ethnicity, many MWL patients tend to lose gluteal volume and want to have this deformity addressed. Most desire additional volume and expressly wish to avoid and reverse buttock flattening [1, 2]. These patients can benefit from augmentation, which produces a more natural and aesthetically pleasing shape [6]. Gluteal autologous augmentation can be achieved by preserving and repositioning part of the tissue normally resected in CBL [4, 5, 6, 8]. Large volume autologous fat transfer or alloplastic implants are also options to be considered for gluteal aesthetics in MWL patients [5, 6, 8]. Adjunctive techniques such as resection, tightening and suspension of the gluteal superficial fascia system (gluteal SMAS), posterior thigh lift, and infragluteal diamond lift can refine the results [5].

For children and teens, on the other hand, the interpretation of BMI is both age- and sex-specific and is often referred to as BMI-for-age [11]. After BMI is calculated for children and teens, the BMI number is plotted on the CDC BMI-for-age growth charts (for either girls or boys) to obtain a percentile ranking. Percentiles are the most commonly used indicator to assess the size and growth patterns of individual children. The percentile indicates the relative position of the child's BMI number among children of the same sex and age [12] (Fig. **1**).

2. ANATOMICAL AND AESTHETIC CONSIDERATIONS

Buttock projection is created from a combination of the gluteus maximus muscle and fat deposits and the buttock is divided into three sections: upper, middle, and lower. Each area must be evaluated and addressed independently for optimal aesthetic result. In determining the most appropriate procedure, three components of the buttock must be evaluated: volume, shape, and skin quality/laxity [3] and to achieve the best aesthetic shape, buttock contouring and gluteal augmentation need to be addressed concomitantly because they are interrelated but one should think in terms of contour first and augmentation second [2].

Some of the most recognizable characteristics of gluteal aesthetics are:

- A lumbosacral depression that helps to distinguish the back from the buttock.

- Two sacral dimples, or supragluteal fossettes (one on either side).

- Two mild lateral depressions that correspond to the greater trochanter.

- Short infragluteal creases or folds.

- An absence of excess fat in the lumbosacral, medial thigh, and anterior thigh regions, and in areas commonly referred to as "love handles, saddlebags, and banana rolls".

- A point of maximum projection on the lateral view that corresponds to the level of the trochanter and the maximal mons pubis projection [2, 7, 8].

These superficial anatomic landmarks have clinical relevance to gluteal contouring surgery and have significant implications for the postoperative appearance of specific gluteal features that are judged to be appealing by our society [2]. The Posterior Superior Iliac Spines (PSIS) form two distinct depressions called the sacral dimples that result from the confluence of the PSIS, the multifidus muscles, the lumbosacral aponeurosis, and the insertion of the gluteus maximus [2]. These anatomic depressions are characteristic of attractive buttocks and serve as the superior corners of the sacral triangle, which is defined by the two PSIS with the coccyx as the inferior border of the triangle. Because it is esthetically pleasing, this triangle and its borders should be enhanced during surgery if possible. Another important topical landmark is the lateral trochanteric depression formed by the greater trochanter and insertions of thigh and buttocks muscles, including the gluteus medius, vastus lateralis, quadratus femoris, and gluteus maximus. This depression is important in the esthetics of an athletically toned buttock although some ethnic groups prefer that the trochanteric depressions not be emphasized or even filled if pronounced [2]. The iliac crest forms the superior border of the buttocks and is a palpable and often visible landmark for guiding incision placement [2]. The fixed, well-defined subgluteal sulcus and the infragluteal fold are formed by thick fascial insertions from the femur and pelvis through the intermuscular fascia to the skin corresponding to the ischial tuberosity, the insertions of the semitendinous muscle and long belly of the biceps femoris, and the lower border of the gluteus maximus, which serves as the inferior border of the buttock proper. The creases should not extend beyond the medial two thirds of the posterior thigh and should form in the midline an inverted V configuration defining the infragluteal "diamond" esthetic unit, which should have a downward-sloping 45° angle between the lower margin of the intergluteal crease and the top of the inner thigh. The infragluteal fold length and definition play important roles in esthetically pleasing buttocks. A longer infragluteal fold suggests an aged, ptotic, and deflated-looking buttock with skin and fascial excess. In contrast, a shorter infragluteal fold contributes to a full, taut, and youthful-looking buttock [2, 7, 8, 13].

Adipose tissue in the gluteal region is rich in dense connective tissue particularly in areas adjacent to the mean line, the pudendal region, and the gluteal crease. The connections made by this tissue to the gluteal region dermis and the osteo-ligamentous structures of the pelvis help to maintain and support the mass of buttocks soft tissue in position [3]. Moreover, a constant fascial layer can be found at the gluteal region, in continuation with Scarpa's fascia, and can be identified as part of the SFS though it is of lesser and unsteady thickness compared to Scarpa's fascia. The superficial and deep subcutaneous fat layers show differences related to sex and region. As in the abdominal region, lobular fatty tissue can be found superficial, whereas lamellar fatty tissue is deep to the superficial fascial system. The relation of thickness of the superficial fat layer compared to the deep is 1:3 in male, 1:2 in female for the cranial gluteal regions, and 1:1 for the caudal regions in both sexes. In the caudal regions the SFS fades to a more reticular collagen network [14]. In the aesthetically pleasing buttock, most of the fat is located directly posterior and has enough fullness superiorly to create a lifted, round look [13].

Analysis of gluteal aesthetics from the lateral view incorporating the buttocks, surrounding torso, and lower extremities reveals that in aesthetically pleasing buttocks, the ratio of the anterior superior iliac spine to the greater trochanter, and the greater trochanter to the lateral point of maximum projection of the buttock, should not exceed 1:2 and that the ideal proportion between trochanter-maximal gluteal projection and trochanter-maximal mons pubis projection is a 2:1 ratio [7, 8]. An additional feature that may contribute to beautiful buttocks is lumbar hyperlordosis. This hyperextension of the spine in the lumbosacral region is an ethnic feature, and sometimes is the result of a forced posture since childhood. In these cases, the sacrum is horizontalized an average of 5-7°, which in a side view gives the impression of a greater buttocks projection than what is normally produced by the muscles in this region [7].

The posterior-anterior view of the gluteal region is extremely important when planning gluteal contouring surgeries and assessing outcomes. From this view, the gluteal region appears to have 8 aesthetic units: 2 symmetric "flank" units, a "sacral triangle" unit, 2 symmetric gluteal units, 2 symmetric thigh units, and 1 "infragluteal diamond" unit. Clearly, each of these units can be further subdivided into subunits but this level of more detailed analysis seems to yield diminishing returns with respect to clinical outcomes. All 8 gluteal aesthetic units have a clinically relevant impact on outcome after gluteal body contouring, and their preservation, surgical enhancement, or reduction can independently enhance overall gluteal appearance. Consequently, these units should be considered during the surgical planning process, and the junctions between aesthetic units should guide incision placement during excisional procedures [2, 8].

Ethnicity and gender account for a wide variety in gluteal size and shape [3, 8]. Recently significant variations in aesthetic ideals among ethnic groups in the United States have been described. For example, Asian Americans prefer a shorter buttock with a higher point of maximum projection, providing the illusion of longer legs and a more balanced proportion between the torso and extremities. Hispanic and African Americans, on the other hand, seem to prefer more projection than either Asian or white Americans. A higher point of maximum projection, deeper lumbosacral depression, and an absence of lateral thigh depressions also appear to be favored by African Americans while white Americans prefer a more athletic ideal with greater definition of the muscular and bony anatomy and less anterior-posterior projection [8].

In the massive weight loss buttock, there is typically an excessive distance between the iliac crest and the perceived superior gluteal margin, translated as reduction of apparent gluteal height, and between the L5 dimple and the central crease. The medial upper and medial inferior quadrants severely lack volume and there is also comparatively more tissue in the lateral compared with medial quadrants [1]. Surgery for correction of the buttocks following MWL involves more than projection and volume which anyway must be in a balanced proportion with the rest of the body. To determine the appropriate type of surgical intervention required and whether additional liposculpturing (liposuction and/or lipofilling) may be required, the surgeon must thoroughly evaluate the buttocks regarding the frame in particular its type (round, square, A- or V-shape), the gluteus maximus muscle, the four key junction points of the muscle and frame (upper inner gluteal/sacral junction, intergluteal crease/leg junction, lower lateral gluteal/leg junction and lateral midgluteal/hip junction), and finally the degree of ptosis from lateral view [2]. He must also be familiar with the "signs" of beauty of this anatomical area that should be the aim of his surgical intervention [7].

3. AUTOLOGOUS GLUTEAL AUGMENTATION (AGA)

De-epithelialized flaps have been used in gluteal contouring during the last 3 decades [2]. Autologous gluteal augmentation in association with CBL has also been described to prevent deformities associated with classic CBL techniques. The natural evolution of what was learned, combined with experience in alloplastic augmentation, led to the concept of using available, well-vascularized autologous tissue in the buttock region to address the gluteal deformities of MWL patients and the flattening effects of a CBL. Instead of discarding all tissue removed during the posterior portion of a CBL, it can be molded into the shape of an implant and inset beneath the CBL skin flaps. Various autologous flap designs emerged over time [2, 5]. AGA can be accomplished using bilateral de-epithelialized island flaps, Le Louarn and Pascal Flaps, "moustache flaps", medially based transposition flaps, or superior gluteal artery perforator flaps [2, 4, 5, 10, 15]. In fact, all these flaps, except probably to a certain extent the medially based random transposition flap, are anatomically almost identical. They derive their blood supply from gluteal perforators. They differ, however, in their design, extent, degree of lateral undermining, advancement, rotation, and inset.

Island AGA flap (Figs. 2 and 3)

The early techniques of autologous gluteal augmentation, describe separate island flaps for autologous gluteal augmentation outlined one on each buttock within the upper and lower outlines of the CBL. The first flap design, called an island AGA flap, simulates the round, non-anatomic design of submuscular gluteal implants [2]. Flaps are de-epithelialized and the caudal incision is beveled down to the muscle fascia. Surrounding excess tissue is removed down to the level of the lumbosacral fascia. The island AGA flap is not undermined, only muscle fascia is released superiorly and laterally to increase its downward mobility [2, 5, 15]. After confirming the position of maximum projection, the islands are anchored to the gluteal fascia at the desired level and the inferior CBL skin flap, which is also very lax, is advanced superiorly over the de-epithelialized island after appropriate undermining [5]. It is sutured to the upper incision line while securing SFS suspension to achieve adequate wound closure and stable scar formation [16].

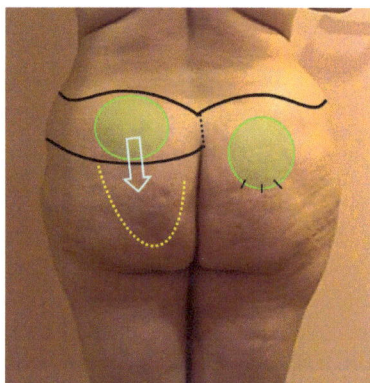

Fig. 2. Island AGA De-Epithelialized Flap. Flap Not Undermined. Undermining of the Gluteal Skin Marked by Dotted Yellow line Allowing Superior Advancement Over the Island Flap that is Shifted Inferiorly and Secured in Position.

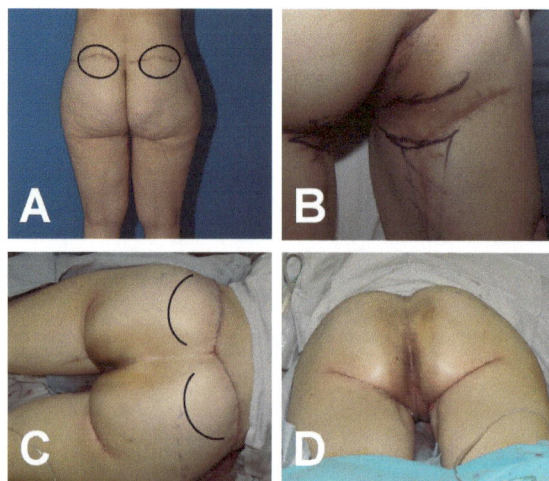

Fig. 3. (A) Patient with Previous Belt Lipectomy Scar Crossing Proposed Island AGA Flaps and Precluding Wide Undermining, Rotation, or Transposition. (B) Poor Infragluteal and Medial Thigh Scars. (C, D) Immediate Post-Operative Result. Relative High Placement of Island Flaps with No Augmentation Infero-Medially.

This technique allows one to fill the upper half of the buttocks, which is claimed to be the most important region aesthetically [4]. The supra-tochanteric depression may be similarly improved by advancing a superiorly or inferiorly based de-epithelialized flap or by rolling such a flap on itself for additional bulk if needed [16]. Although reasonable results are obtained, the amount of volume that is produced with such a flap is insufficient to overcome the gluteal flatness in most massive weight loss patients. Moreover, in most cases, the point of maximum projection achieved usually lies slightly above the transposed level of the mons pubis, which is higher than ideal. However, this location may be preferred by men, African American, and Asian women [2, 5, 15].

The "Moustache AGA Flap" (Fig. 4)

Experience gained with early flap designs led to the development of the "moustache AGA flap" [2] which is a modification regarding the placement of the central bridge of tissue as well as the lateral flap extensions or "handlebars" which are elevated from the underlying fascia and transposed infero-medially [5]. The moustache flap uses lower back and lateral flank tissue as a partial island and partial transposition flap based on perforators from the superior gluteal artery and lumbar perforators [2]. The anatomy of these perforators, as it relates to the moustache flap design, recently was documented as consistently being approximately 9 cm from the midline [2]. Moreover, the robust vascularization of an AGA flap and limitation of flap dissection to no more than two contiguous angiosomes seem to provide good flap perfusion and viability [2]. Inferomedial transposition of the "handlebar" portions of the moustache flap allows recruitment of additional tissue for augmentation purposes and lowers the point of maximum gluteal projection to the level of the mons pubis, which is considered the esthetic ideal. In addition, imbrication of the flap with sutures permits formation of a more anatomically shaped tissue mound that is reminiscent of anatomic gluteal implants. Because resection of tissue from the central area of the flap to allow easier in-setting likely would decrease projection, the tissue volume in this area is included in the flap and retained. This flap is indicated when significant long lasting esthetic augmentation is desired following MWL particularly in female patients [2, 5]. Unless there is inadequate postsacral tissue eliciting complaints of pain in the coccyx or sacral area when sitting, which is typical among patients who have experienced MWL [2], retention of the central portion of the flap may produce an undesirable sacral fullness in some patients and makes the definition of the sacral triangle difficult to attain.

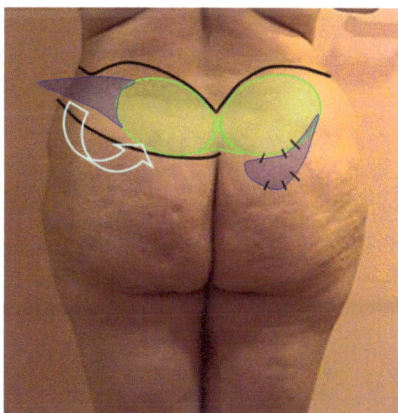

Fig. 4. Moustache Flap Design. Presacral Tissues Maintained and Handlebar Extension Undermined and Transposed Inferomedially.

Ovoid Dermal Fat Island Flap (Fig. 5)

The ovoid dermal fat flap designed as an autologous buttock implant is claimed to provide additional projection during belt lipectomies with lower body and buttocks lifts resulting in a more pleasing and natural body contour as well as more physician and patient satisfaction [9].

The flap originates in the medial half of the regularly excised supragluteal tissue and its size can be individualized according to the patients' buttocks contour. The ovoid flap is de-epithelialized and then dissected down to the fascia at an oblique angle with undermining the superior border of the island and the inferior gluteal CBL flap. An oblique column of tissues is thus created with almost equal flap dermal surface and base, with the base positioned more inferiorly (caudally) than the surface allowing greater mobility inferiorly with a longer, more mobile dermal fat flap. A pocket is then created for insertion of the flap by undermining the buttock in the plane above the fascia and extending it a sufficient length to reach the inferior gluteal crease. The ovoid flap is subsequently rotated caudally 180° over itself, inserted into the pocket and anchored to the fascia securing the de-epithelialized surface upside down which gives the flap a more rounded implant-like shape. The lower buttock skin is pulled in the reverse direction to cover the flap [9]. It has been argued that this design is superior to local flaps with a more cephalad base over the sacral area that have limited mobility and produce additional bulk to the sacral region after advancement or rotation which is usually not warranted [9]. However, this technique requires repeated repositioning in the lateral decubitus position, adding time, inconvenience, and difficulty in assessing symmetry [15].

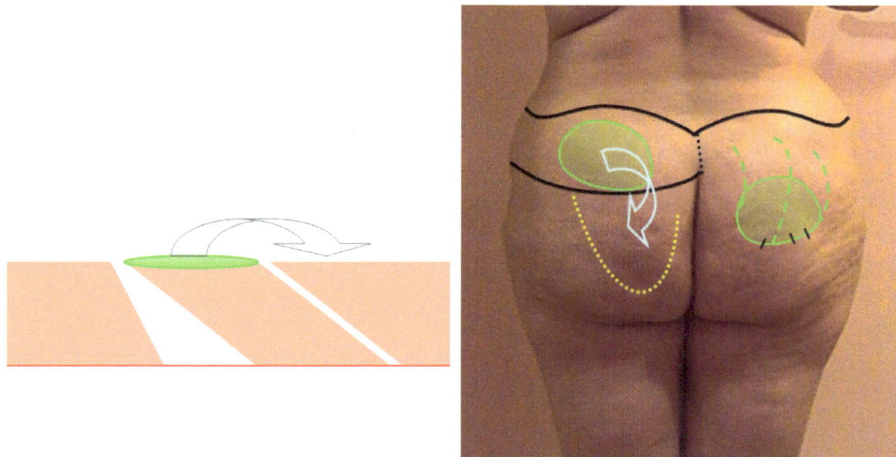

Fig. 5. Ovoid Dermal Fat Island Flap Design and Rotation. Limit of the Undermining, Marked by Dotted Yellow Line. Oblique Dissection of the Flap to be Rotated 180° Over Itself In-Setting the De-Epithelialized Surface Upside Down.

Medially Based Transposition Flap (Fig. 6)

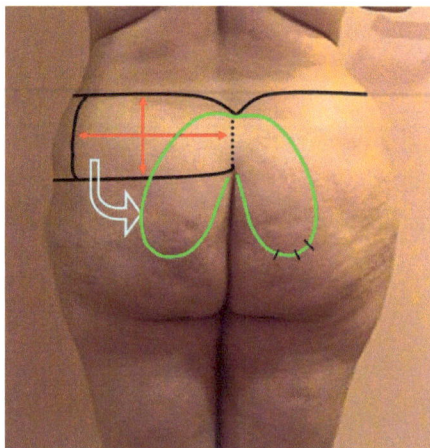

Fig. 6. Random Medially Based Transposition Flap. Superior Gull-Wing and Inferior Horizontal Incision Lines. Random Flap Not Exceeding 2:1 Length-To-Width Ratio Transposed Medially and Inferiorly for Autologous Buttock Augmentation.

A random medially based flap may be designed along the lower back, over the inferior aspect of the thoracolumbar fascia and the superomedial insertion of the gluteus maximus muscle with a superior gull-wing and inferior horizontal incision lines. The flap is composed of subcutaneous tissue, and dermis if desired, and based at the midline extending laterally to a distance based on the length needed to approximately reach the infragluteal fold when rotated, but limited by its width, not exceeding 2:1 length-to-width ratio. Its thickness is determined by the volume of tissue necessary to fill the buttock contour. The flap can be dissected from the underlying tissues, except at its medial base, and rotated inferiorly to augment the gluteal area. The buttock skin and subcutaneous tissue are then advanced superiorly and rotated medially to tighten the buttocks, lateral thigh, and posterior thigh. Despite the anatomic possibility of an axial property with perforators from the superior gluteal artery, lateral sacral arteries, and fourth lumbar artery that may provide blood supply to this flap, the flap is random and must not be considered more robust than a random flap, thus any attempt to extend its length would result in fat necrosis. In fact, it is recommended to limit its length to slightly less than a 2:1 length-width ratio [10].

Le Louarn and Pascal Flap (Figs. 7 and 8)

The flap is designed starting from the lower side of the excision and across the whole width of the buttocks. The external limit of the flap is the lateral end of the buttock fold. Thus, its main axis is horizontal and it may be up to 25cm in length and 10cm to 12cm in width [4]. In the original description of the technique, the flap designed within the CBL incisions is not mobilized; rather, the inferior gluteal tissue is extensively mobilized and brought over the stationary island flap [15]. For further augmentation and to give more volume, flap design can be elongated laterally. This requires, however, lateral undermining a third of the flap to be able to turn it downwards and inwards [4].

Depending on how much weight the patient has lost, and similar to the de-epithelialized island AGA flap, incision down to the muscular plane allows the creation of an island dermal fat flap consisting of very mobile tissues that can be considerably displaced. Often it is possible to advance the flap down to the infragluteal groove [4]. A pocket is created inferiorly by undermining the buttock superficial to the muscle fascia. The dissection plane extends a sufficient length downwards to accommodate the island flap and to allow suture suspension of the most caudal gluteal tissues to the advanced island flap [4]. Undermining the inferior skin flap beneath the SFS to the level of the trochanteric fat deposit releases the SFS zone of adherence and allows optimal transmission of the lifting forces to the distal lateral thigh [16]. The inferior skin flap is pulled superiorly to cover the de-epithelialized flap [4].

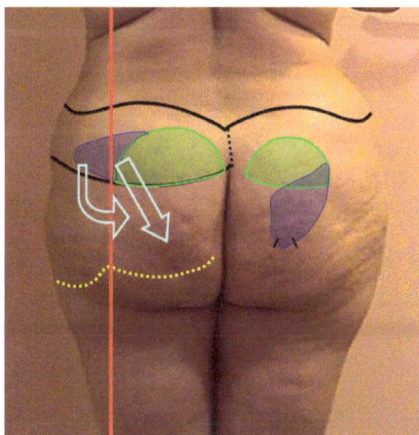

Fig. 7. Le Louarn and Pascal Flap Design with Inferior Advancement and Rotation. External Limit of the Flap Corresponds to the Lateral End of the Gluteal Fold Marked by a Vertical Red Line. Possible Lateral Extension with Undermining Marked in Blue. Inferior Limit of the Undermining, Above the Inferior Border of the Gluteus Maximus and Extending over the Trochanter Marked by Dotted Yellow Line. Lateral Extension Transposed Inferomedially.

Critics of the Le Louarn and Pascal flap claim that, similar to the simple island AGA flap, achieved gluteal fullness is too superior and often results in a "double-bubble" where the inferior ptotic gluteal tissue sags beneath the projecting superior portion. The Le Louarn and Pascal flap may not allow the surgeon to position the flap in the ideal location to fill the inferior medial quadrant of the buttocks, which is the most important area of volume loss in the MWL patient [15]. Moreover, there remains uncertainty about how far medially the dissection of the flap can safely proceed to obtain the desired degree of rotation [15].

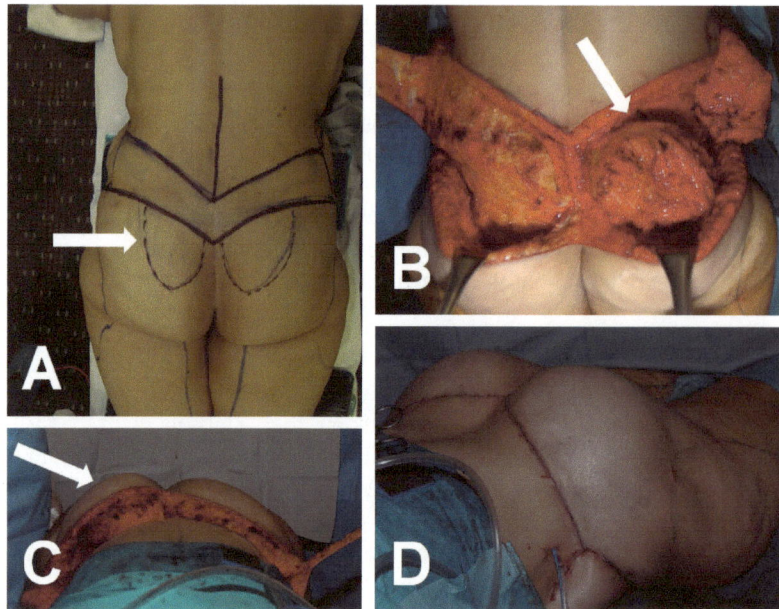

Fig. 8. (A) Planning of Belt Lipectomy. Area To Be Augmented (Arrow). (B) Le Louarn and Pascal Flap with Inferior Advancement and Rotation (Arrow) on the Left. Flap De-Epithelialized Still in Place on the Right. (C) Augmented Buttock (Arrow) as Compared to Un-Augmented Opposite Side. (D) Final Result of Autologous Buttock Augmentation Before Completing the Anterior Portion of the Belt Lipectomy. Highest Point of Augmentation Corresponds to the Level of the Mons Pubis.

Lumbar Hip Dermal Fat Rotation Flaps (Fig. 9)

Bilateral dermal fat rotation lumbar hip flaps, based on the region corresponding to the gluteal perforators vessels may be performed in selected patients to create a harmonic autologous augmentation of the buttocks region. This technique is applicable to patients with skin excess, skin flaccidity, and/or ptosis, and for redundant skin folds in the lower back region, that often occur following the surgical treatment of morbid obesity.

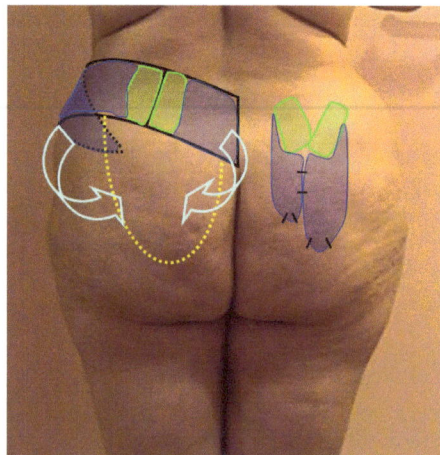

Fig. 9. Lumbar Hip Dermal Fat Rotation Flap. Fusiform-Shaped De-Epitelialized Flap Incised in the Midline Creating a Lateral and Medial Flap. 2 Flaps Undermined (Blue) with Pedicle Over Gluteal Perforators. Undermined Portions Rotated Inferiorly and Secured in Position.

Beginning approximately 4cm above the coccyx, a line is marked usually a little bit below the anterior iliac spine. This line joins the classic abdominoplasty line of incision when it curves anterolaterally. The amount of skin excess to be resected is determined by the pinch test and the superior line of resection is determined. This line is initially parallel to the first and then continues laterally above the anterior iliac spine to be joined ultimately with the first line delineating a long fusiform-shaped flap with its lateral limit slightly inferior to the anterior superior iliac spine. Occasionally the fusiform flap extends as far as the mid-clavicular line or even into the pubic region. The fusiform flap is then de-epithelialized and a sagittal incision is made in its midline delineating a lateral and a medial dermoglandular flaps which are undermined, maintaining a large pedicle in the region corresponding to the gluteal perforators. The two segments are then rotated, sutured together and fixed in the desired position. The inferior CBL adipocutaneous flaps are advanced over the dermal fat flaps in an upward direction and sutured to the upper lumbar hip skin border in a biplanar wound closure [17].

Superior Gluteal Artery Perforator Flap (Fig. 10)

Previous reports of lower body lift combined with autologous augmentation are limited by suboptimal positioning of the augmentation flaps and potential flap size restrictions. These reports do not provide a sound anatomical basis for vascular supply to the flaps, nor do they respect the natural anatomical boundaries between the buttocks and lower back [1].

The gluteal island flap is very safe thanks to its numerous perforators. Therefore, various types of undermining are possible, if not exceeding a third of the flap surface, to achieve a nice inferomedial mobilization [18]. However, to improve and extend the design of AGA flaps and optimize gluteal contour by medializing the rotation point, knowledge of vascular perforator anatomy of the gluteal area is essential [15]. With this knowledge, large flaps can reliably be raised, and better flap rotation and positioning may be achieved to augment the inferomedial gluteal quadrant and avoid buttock flattening associated with CBL procedures without augmentation [15]. By including perforators to provide adequate vascular supply, fat resorption and fat necrosis should be minimized [15].

The basis of the superior gluteal artery perforator flap in sound anatomy offers superior versatility in the design of AGA flaps to optimize gluteal aesthetics [15]. Markings are performed with the patient in the standing position and the L5 dimple and posterior superior iliac spine are identified serving as reference points for the upper transverse incision. The incision courses just one fingerbreadth above the posterior superior iliac spine and continues laterally approximately one fingerbreadth above the palpable origin of the gluteus medius muscle and the upper margin of the muscular component of the gluteus maximus. The upper V-shaped incision is therefore designed to yield a scar that lies, after 1 to 2cm inferior displacement, exactly within the natural anatomical boundary between the gluteal region and the lower back. The center point of the lower incision usually corresponds with the upper margin of the central crease and is checked by the pinch test. Coursing laterally, there is almost always a fairly distinct demarcation between smooth skin superiorly and wrinkled skin inferiorly. The inferior skin represents the true skin of the buttock and the lower incision is marked along this skin demarcation; the pinch test is used to make minor adjustments [1].

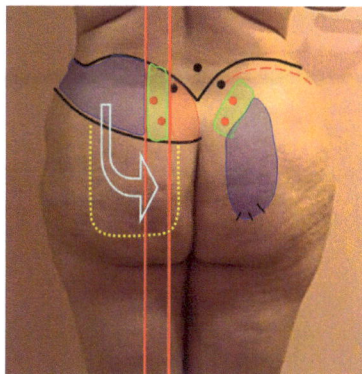

Fig. 10. Superior Gluteal Artery Perforator Flap. L5 Dimple and Posterior Superior Iliac Spines (Black Circles). Flap Perforators (Red Circles) Are Within 5-10 cm (Delineated by Vertical Red Lines) from the Midline. Superior Incision Made 1 Finger-Breadth Above The iliac Crest (Curved Interrupted Red Line). Lateral Half of the Flap (Blue) Is Undermined, and Medial 5cm Portion (Red) Excised. Area of Gluteal Undermining Delineated by Yellow Dotted Line. Undermined Flap Transposed Inferomedially.

Superior gluteal artery perforators lying along a line drawn from the posterior iliac spine to the greater trochanter may be identified by doppler ultrasound. Two distinct major perforators can be identified between 6 and 9 cm from the midline on each side. Oval flaps are then designed to include both perforators on their medial aspects. To minimize sacral fullness, the tissue between the midline and 5cm on either side, which does not contain vital perforators, is excised as in a standard CBL. The remaining large superior gluteal based flaps are de-epithelialized and dissected laterally to medially starting at the midaxillary line. The lateral half of the flap can be undermined until the lateral perforator is encountered approximately 9cm from the midline. The flaps are then analyzed for viability and trimmed to achieve the desired bulk [1].

A gluteal pocket is then created by undermining in a plane just superficial to the gluteal muscle from the CBL incision line inferiorly extending to within 5cm of the inferior gluteal crease to avoid injury of the cutaneous sensory nerves. Undermining of the lower limit of the pocket is typically performed with scissor and blunt dissection. This may help avoid any thermal injury to nerves in the vicinity. The gluteal pocket extends only over the medial half of the buttock which is the area of volume deficiency. The de-epithelialized flap is then rotated inferomedially and the inferior gluteal skin advanced superiorly over the perforator flap to meet the superior incision line [1, 15]. Though located higher than the bikini line, resultant scars remain between the anatomical subunits of the lower trunk and buttocks and are claimed to be hidden with most undergarments [15].

4. TECHNICAL CONSIDERATIONS OF AGA FLAPS

Despite numerous anatomical studies and reports, there is still a lot of controversy regarding blood supply and AGA flap rotation [18]. Perfusion to the skin overlying the gluteal region is supplied by perforating branches of the superior and inferior gluteal arteries, both of which branch from the internal iliac artery, with 13 to 20 vessels per gluteal region [2, 15]. The lumbosacral region also is supplied by lumbar perforators [2]. Some of these perforators must be sacrificed during the posterior portion of a CBL, an AGA with CBL, or a buttock-flankplasty; however, the abundant vascular supply of the gluteal region provides robust perfusion to surrounding tissue flaps [2].

Flaps based on medial perforators are very reliable and have been used extensively to cover sacral pressure sores and lumbosacral defects [15]. Two key superior gluteal perforators, at approximately 7 and 9 cm from the midline seem to be reliable and reproducible to specifically perfuse the flap [1]. Doppler ultrasound to locate the perforator blood supply of the flap is, in fact, unnecessary because the two key perforators are so reliably found between 5 and 10cm from the midline on each side [15]. Lateral perforators of the superior gluteal artery are not necessary for flap viability; if preserved, they may prevent adequate mobilization of the flap into the inferomedial quadrant of the buttocks [15]. However, converging anatomical studies demonstrate that, statistically, perforators are located more laterally (10 to 12cm) at a middle distance between the posterior superior iliac spine and the greater trochanter [11, 18]. This indicates that to ensure the preservation of most of the vital perforators only the lateral third of the flap can be undermined safely and not the lateral half [18].

On the other hand, it has been demonstrated that the superior gluteal zone combines 48.5 percent of perforators, whereas the central gluteal zone is the most poorly vascularized region [12]. Nevertheless, the ovoid dermal fat island flap in which the flap base originates more inferiorly than the surface, which is obviously not in the superior gluteal zone [9, 18] is a clear indication of the robust blood supply of the gluteal area which tolerates various types of flap design and undermining.

Various incision locations, however, have different effects on the overall appearance of the buttock. Even with an untrained eye, a proportional relationship and natural transition between the buttock and its surrounding area is perceived [8]. To achieve a more aesthetically pleasing postoperative result, the incision placement may vary superiorly or inferiorly with respect to the iliac crest; but choosing an incision location requires a trade-off between waist definition and buttock elongation [2]. A high incision better maintains waist-to-hip ratio laterally, however, if a classic CBL incision placement is too high it will deleteriously elongate the appearance of the buttock. It limits also flap placement so maximum projection is higher than ideal, violates sacral triangle esthetic unit, and resultant scar becomes visible with some clothes [2, 8]. Low placed incisions, on the other hand, shortens the buttocks, allows lower flap placement, maintains sacral triangle aesthetic unit, and results in well hidden scars, even in bikini. It diminishes, however, waist definition [2]. If an incision that runs straight across the back is either too high or too low, the buttock will appear, respectively, too rectangular or too square [2, 8]. Similar to treating the face, nose, and abdomen, preservation or enhancement of gluteal relationships can improve clinical outcomes [8].

Regardless of the flap design chosen, the volume of the flaps described can be adjusted as required by the clinical situation and the patient's wishes. Although larger flaps produce more augmentation, smaller ones are sometimes desired, especially if it becomes apparent that an increased margin of safety is needed to guarantee tissue perfusion or if a larger flap places excessive tension on the posterior CBL flaps [2]. All autologous flaps can be ''down-staged'' intraoperatively if needed or resected if an AGA flap cannot be positioned appropriately or accommodated when the posterior CBL or upper and lower buttock lift flaps are brought together. Should there be any concern about tissue perfusion, excessive tension on the CBL closure, or inability to close the flaps over the gluteal tissue mound, the auto-augmentation should be abandoned so as not to compromise the safety of the lower body lift [2].

5. ALLOPLASTIC IMPLANT GLUTEAL AUGMENTATION

Although implant insertion increases volume and projection, it introduces a foreign body into the buttocks and requires expertise limited by a technical learning curve for proper placement [15]. For MWL patients, no literature has yet appeared about the use of gluteal implants. Theoretically, in such patients gluteal implant designs have limitations, especially as a primary treatment for pronounced platypygia [2, 5]. Many patients who have experienced MWL, or patients with senile buttocks, lack sufficient fat or tissue volume to pad an implant, and their skin can be thin, which might make them more susceptible to less than optimal outcomes with implant migration, extrusion, palpability, or visibility [2, 5-7]. In fact such deformities cannot be corrected solely with the use of gluteal implants and liposuction. In most cases, it will be necessary to perform some type of wide dermo-cutaneous adjustment [7]. Despite this, augmentation with implants may still be applicable in certain subsets of carefully selected and well-informed patients [2].

6. AUTOLOGOUS FAT TRANSFER

Large-volume fat injections collect autologous fat from areas of excess and redistribute this fat in the area with a relative deficiency [15]. Autologous fat transfer for gluteal augmentation was reported first in 1990. Since then, multiple reports have verified the clinical efficacy of transferred fat for this application [2, 5]. Though fat transfer for gluteal augmentation may play a role in body contouring and patient satisfaction with the procedure is perceived to be high despite the lack of systematic documentation in most published reports, its efficacy as a primary modality in massive weight loss patients is still contested vigorously [2, 5, 6]. Lipografting has been shown to be an effective means of moderately increasing buttocks volume, but this procedure does not directly address ptosis [19]. Nevertheless, its application in MWL patients is increasing. For some, flap auto-augmentation should be reserved for special circumstances, and fat injections should be the preferred method of increasing buttocks projection in patients who have undergone MWL [20]. It is unclear though whether the full potential of autologous fat injection for gluteal augmentation in MWL patients will be realized. Several issues remain that may limit applicability in this population: the lack of enough available fat to overcome severe volume loss and skin laxity in MWL patients who have reached a low BMI; practical sequencing with excisional procedures because fat transfer is labor intensive; and the impact of compression from sitting on the resorption of fat if combined with excisional procedures that require supine postoperative positioning (eg. abdominoplasty and CBL) [2, 5, 6]. Presently, large volume autologous fat transfer can be used in select patients as an adjunctive mode to enhance the buttock shape [5, 6].

7. ADJUNCTIVE SURGICAL PROCEDURE FOR GLUTEAL CONTOURING AND ENHANCEMENT

For some patients, skin excess or laxity in the posterior thigh, medial thigh, and infragluteal fold cannot be corrected fully with AGA or CBL. Adjunctive excisional techniques at the time of autologous augmentation or CBL, or as a staged procedure in the most severe cases, often are necessary to obtain an optimal outcome [2].

An infragluteal diamond lift is useful for enhancing this esthetic feature and achieving the 45° downward-sloping angle at the junction between the infragluteal fold and the intergluteal crease [2].

The posterior thigh lift with dermo-tuberal anchoring is useful in contouring the posterior thigh and residual buttock ptosis at the time of CBL, provided that flap or implant augmentation is not performed. If autologous tissue or implant augmentation is incorporated into a CBL, it seems prudent to stage a posterior thigh lift as a secondary procedure, because it can involve significant undermining of the buttock skin. In contrast, the posterior thigh lift may be beneficial at the time of fat transfer, which requires no undermining [2].

The "inverted dart" modification of the posterior CBL incision is another useful maneuver for improving gluteal esthetics. It locates the incision between gluteal esthetic units and is a powerful tool for shortening and stabilizing the length of the inter-gluteal crease [2].

8. CONCLUSION

No single technique is applicable to all gluteal deformities. The wide spectrum of deformities demands an individualized approach to each patient [2]. Moreover, a better understanding of the anatomy and a new surgical armamentarium can improve the cosmetic results of gluteal contouring following MWL, and, thereby, enhance patient satisfaction in this challenging population [2].

In patients with MWL undergoing CBL, the most practical means of augmenting the inferomedial gluteal region is to use tissue that would otherwise be discarded [15]. Creation of an autologous buttock implant from a dermal fat flap provides additional projection during the buttock lift; together with an extensive subcutaneous undermining and preservation of the SFS, effective augmentation, suspension, and stable gluteal lift with an aesthetically acceptable outcome may be achieved [14, 19].

The use of a number of adipo-cutaneous flaps to address gluteal augmentation in different settings has been described. The ideal flap for gluteal enhancement should be versatile, not vascularly compromised, and give the maximum projection at the midlevel of the buttocks [19]. In a small subset of patients, the gluteal flap alone may not provide adequate projection. This might occur in patients with decreased lumbar lordosis or in those with decreased skin and soft tissue available for reconstruction. In this population, further augmentation may be subsequently provided at a second stage with large volume fat injection, if available, or implant reconstruction [15].

The results obtained with autologous tissue in combination with the CBL cannot match those typical of gluteal augmentation in patients who have not lost significant weight and have little skin excess. Nevertheless, the gluteal esthetics of patients who have sustained MWL can be enhanced greatly with an autologous tissue flap at the time of CBL [2]. Buttocks flap auto-augmentations with CBL, however, can be very difficult to master; they increase operating time by at least 45 minutes in the best of hands, and in inexperienced hands can lead to bizarre buttocks contour with serious complications of skin or fat necrosis, chronic seromas resistant to treatment, infections, and sepsis [20]. However, with experience and with the evolution of this procedure, it is possible to achieve lifting of the buttock and lateral thigh, reduction in the number and size of adipose cutaneous folds of the lower and middle back, improvement in the waist silhouette, and elimination of redundant flank tissue. The result is an improved body contour and tightening of the skin [19].

REFERENCES

[1] Colwell AS, Borud LJ. Autologous gluteal augmentation after massive weight loss. Aesthetic analysis and role of the superior gluteal artery perforator flap. Plast Reconstr Surg 2007; 119: 345-56.

[2] Centeno R, Mendieta C, Young V. Gluteal contouring surgery in the massive weight loss patient. Clin Plast Surg 2008; 35: 73-91.

[3] Gonzalez R. Etiology, definition, and classification of gluteal ptosis. Aesth Plast Surg 2006; 30: 320-26.

[4] Pascal JF. Le Louarn C. Remodeling body lift with high lateral tension. Aesth Plast Surg 2002; 26: 223-30.

[5]   Shrivastava P, Aggarwal A, Khazanchi RK. Body contouring surgery in a massive weight loss: An overview. Indian J Plast Surg 2008; 41: S114-S129.

[6]   Kenkel JM. Marking and operative technique. Plast Reconstr Surg 2006; 117 (1 Suppl.): 45S-73S.

[7]   Cuenca-Guerra R, Quezada J. What makes buttocks beautiful? A review and classification of the determinants of gluteal beauty and the surgical techniques to achieve them. Aesth Plast Surg 2004; 28: 340-47.

[8]   Centeno RF. Gluteal aesthetic unit classification: A tool to improve outcomes in body contouring. Aesth Surg J 2006; 26: 200-08.

[9]   Sozer SO, Agullo FJ, Wolf C. Autoprosthesis buttock augmentation during lower body lift. Aesth Plast Surg 2005; 29: 133-37.

[10]   Rohde C, Gerut ZE. Augmentation buttock-pexy using autologous tissue following massive weight loss. Aesth Surg J 2005; 25: 576-81.

[11]   Nojima K, Brown SA, Acikel C, *et al.* Defining vascular supply and territory of thinned perforator flaps: Part II. Superior gluteal artery perforator flap. Plast Reconstr Surg 2006; 118: 1338-48.

[12]   Kankaya Y, Ulusoy MG, Oruc M, *et al.* Perforating arteries of gluteal region: Anatomic study. Ann Plast Surg 2006; 56: 409-12.

[13]   Mendieta CG. Gluteoplasty. Aesth Surg J 2003; 23: 441-55.

[14]   Beck H, Kitzinger HB, Lumenta D, *et al.* The "gluteal smas" in the lower body lift (abstract). Plast Reconstr Surg 2009; 124(2S) Supplement: 678.

[15]   Colwell AS, Borud LJ. Autologous gluteal augmentation after massive weight loss. Reply. Plast Reconstr Surg 2008; 121: 1516-18.

[16]   Lockwood T. Transverse flank–thigh–buttock lift with superficial fascial suspension. Plast Reconstr Surg 1991; 87: 1019-27.

[17]   Raposo-Amaral CE, Cetrulo CL Jr, Guidi MCm *et al.* Bilateral lumbar hip dermal fat rotation flaps: a novel technique for autologous augmentation gluteoplasty. Plast Reconstr Surg 2006; 117: 1781-88.

[18]   Le Louarn C, Pascal JF. Autologous gluteal augmentation after massive weight loss. letter-to-the-editor Plast Reconstr Surg 2008; 121: 1515-16.

[19]   Sozer SO, Agullo FJ, Palladino H. Autologous augmentation gluteoplasty with a dermal fat flap. Aesth Surg J 2008; 28: 70-76.

[20]   Aly A. Body contouring after massive weight loss. Editorial Commentary. Clin Plastic Surg 2008; 35: 93.

CHAPTER 10

Breast Surgery in Patients after Massive Weight Loss

Moustapha Hamdi[*] and Serhan Tuncer

Professor and Chairman of Plastic Surgery Department, Brussels University Hospital, Laarbeeklaan 101, 1090 Brussels, Belgium

Abstract: Massive weight loss patients present characteristic breast deformities. Most of the techniques developed for correcting typically enlarged or ptotic breasts fail in this patient group, instead more complex and challenging manipulations are usually needed. Selection of the surgical technique is mainly based on the available breast volume. An algorithm for surgical correction of breast deformities following massive weight loss is presented.

Keywords: Breast reduction, Mastopexy, Implant, Auto-augmentation, dermal suspension, spiral flap, ICAP flap.

1. INTRODUCTION

The breast is undoubtedly the most crucial part of the female body symbolizing attractiveness and femininity. Frequently women seek for breast surgery to regain aesthetically pleasing and well-proportioned breasts with youthful curves. With the increasing demand, surgical techniques for breast reshaping evolved enormously within years and provide individualized correction for different types of breast deformities.

Despite the variety of deformities the fundamental goals in mammaplasty are constant and well defined. In any kind of breast surgery a symmetric, proportional and aesthetic appearance with adequate projection, medial and upper pole fullness and appropriately placed nipple areola complex is aimed. Besides, techniques developed for breast reduction and mastopexy focus on minimizing scars while providing a long-lasting result.

Tremendous worldwide growing of the obese population induced the development of modern bariatric surgery and this ended up with the rising of a new patient group in plastic surgery called post-bariatric or Massive-Weight-Loss (MWL) patient [1-3]. The number of individuals applying for body contouring after MWL is growing and reshaping of the complex breast deformity represents one of the considerable aspects of this patient group [4- 8].

2. BREAST DEFORMITY IN MASSIVE WEIGHT LOSS PATIENT

Massive weight loss patients present characteristic breast deformities. Most of the techniques developed for correcting typically enlarged or ptotic breasts fail in this patient group, necessitating more complex and challenging manipulations [9-12]. One should understand each distinctive deformity clearly to achieve an aesthetically pleasing and durable result (Table **1**). Massive weight loss in a certain period of time results with significant breast volume depletion and the breast mound loses its upper pole and medial fullness to form flattened parenchyma against the chest wall. This morphological alteration is known as "pancake" appearance [12, 13]. The breasts present grade III ptosis and are usually asymmetric because of unproportional volume loss. The breast mound is displaced laterally and shows continuity with the lateral chest rolls with significant medialization of the nipple areola complex. The quality of the skin envelope is also altered. Skin overlying the breast mound is always unstable and redundant with severe laxity and stretch marks. The severity of these deformities shows a great range of variety and the surgical technique should be individualized according to patient characteristics.

*Address correspondence to Moustapha Hamdi: Professor and Chairman of Plastic Surgery Department, Brussels University Hospital, Laarbeeklaan 101, 1090 Brussels, Belgium; Tel: +32-9-332-6040; Fax: +32-9-332-3899; E-mail: moustapha.hamdi@ugent.be, info@ibreast.be

Table 1. Breast Deformities in Massive Weight Loss Patient.

1. Volume loss
Loss of upper pole fullness
Loss of lower pole fullness
Flattened breast (pancake appearence)
Grade III ptosis
Assymetry
2. Changes in skin envelope
Redundancy
Laxity
Stretch marks
3. Miscellaneous
Laterally displaced breast mound
Side rolls
Medialization of nipple-areola complex

3. PATIENT SELECTION AND TREATMENT ALGORITHM

Selection of the surgical technique is mainly based on the available breast volume (Fig. **1**). The patients can be categorized into three groups. 1. Patients who have excessive breast volume, 2. Patients who have sufficient breast volume and 3. Patients who have insufficient breast volume.

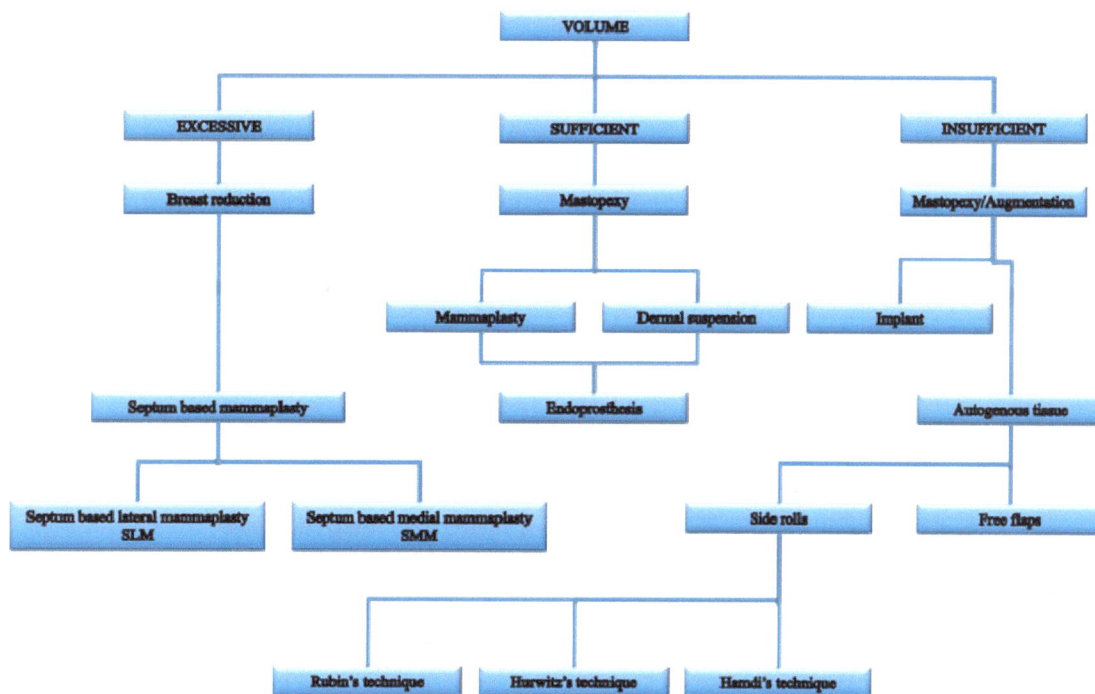

Fig. 1. Algorithm for Surgical Correction of Breast Deformities following Massive Weight Loss.

Unacceptable outcome can be expected when an insufficient preoperative approach is made (Fig. **2**).

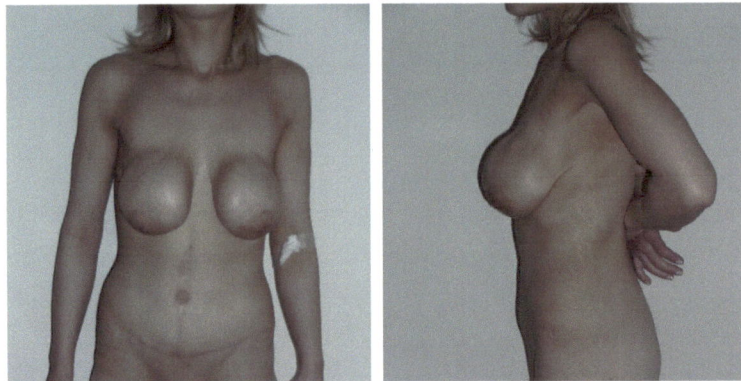

Fig. 2. Prepectoral-Implant Breast Augmentation in a Patient after MWL. The Outcome Was Extremely Poor Because the Surgeon Did Not Select the Right Technique in this Case.

Patients with Excessive Breast Volume

In fact, because of significant volume depletion, few of the MWL patients are included in this group. Women who have enlarged breasts with coexisting findings of a MWL patient should be treated with one of the appropriate breast reduction techniques. However, the majority of the standard reduction mammaplasty techniques disrupt the structural anatomy of the ligamentous suspension system of the breast and they are insufficient in MWL breast reshaping. In this patient group, the ligamentous system is attenuated and further disruption with surgery should be avoided. Instead preserving and re-tightening procedures should be preferred. We use a modified septum based mammaplasty technique for reshaping the breast in MWL patient with excessive breast volume [14]. This surgical method is the modification of the technique described by Wuringer in 1999 [15]. She demonstrated a horizontal septum that originates at the level of the fifth rib and heads toward the nipple. This horizontal septum guides the main nerve to the nipple, the cranial vascular layer, consisting of the thoracoacromial artery and a branch of the lateral thoracic artery, and the caudal vascular layer, consisting of the perforating branches of intercostal arteries 4 and 5. At its

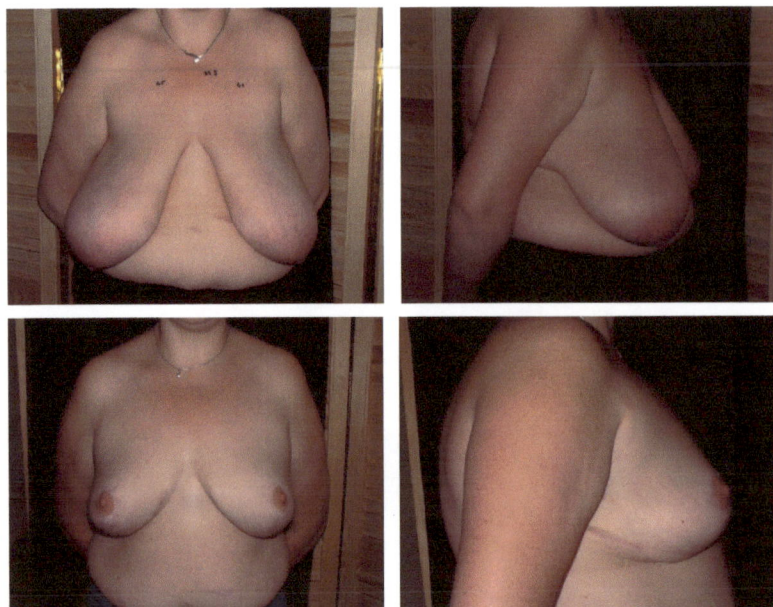

Fig. 3. A Septum-Based Lateral Mammaplasty (SLM) Technique for Reshaping the Breast in MWL Patient With Excessive Breast Volume.

borders, the septum curves upward into vertical ligaments, which attach the breast to the sternum and the axilla, thereby guiding vessels and nerves to the nipple-areola complex also. In her series she used a central pedicle with additional fixation of the pedicle by suturing the peri-areolar dermal ring circularly to the thoracic wall. She also used the medial and lateral ligaments for shaping the breast by their plication and fixation to the pectoral fascia. In our technique the Nipple Areola Complex (NAC) was based on either a medially or a laterally based pedicle depending on the breast shape and the location of the NAC. In both pedicles, the dermoglandular tissue bearing the NAC were designed on this septum (Fig. **3**).

If the NAC is significantly medialized and more resection from the lateral segment is desired, a medially based pedicle is preferred and the technique is called Septum-based Medial Mammaplasty (SMM). If the NAC is located laterally and resection from the medial pole is planned a laterally based pedicle is preferred and the technique is called Septum-based Lateral Mammaplasty (SLM). Either based on laterally or medially the technique allows sufficient amount of resection from the lower and upper pole of the breast with a safe NAC repositioning in terms of vascularity and sensation. Besides, preservation of the horizontal septum provides durability of the achieved result. The surgery ends with an inverted T scar and the amount of redundant skin and extension of the side rolls determines the length of the horizontal scar.

Patients with Sufficient Breast Volume

Most of the patients with MWL belong to this group. On patient evaluation there is significant loss of breast volume but still the remaining tissue after weight loss is enough to form a breast with adequate projection. Mastopexy techniques have been utilized to address the deformity of the MWL breast with sufficient volume.

Modifications of the standard vertical mammaplasty techniques are the most commonly applied approaches in this group [16]. Superior pedicle is usually preferred and central and inferior dermoglandular tissues are utilized for autoaugmentation instead of resection. In any kind of modification, the pedicle bearing NAC is anchored to the desired level, the remaining parenchymal tissues are reshaped to form an endoprosthesis, and the dermal suspension techniques are used to provide a durable result (Fig. **4**).

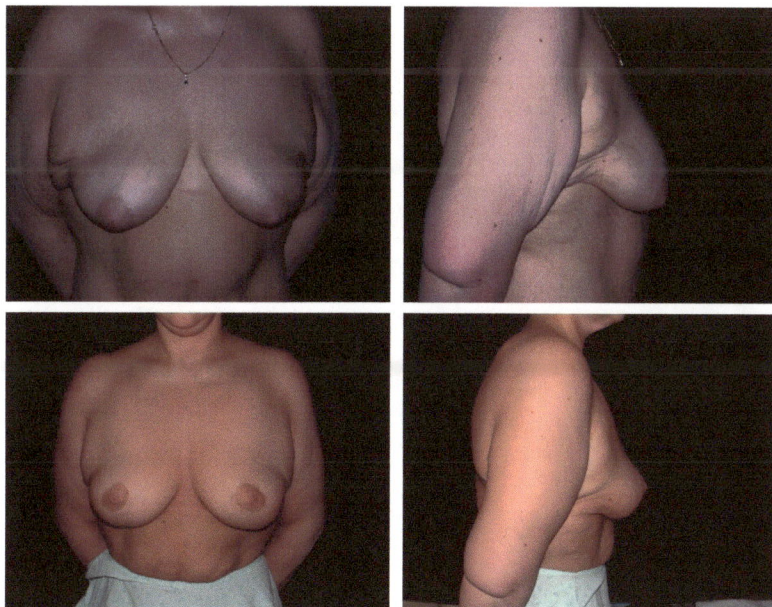

Fig. 4. Standard Vertical Scar Mammaplasty Technique.

As a modification of Lejour's [17, 18] classical vertical mammaplasty technique, the inferior dermoglandular extension can be lifted off the pectoral fascia and folded back superiorly beneath the superior pedicle and anchored with sutures to provide upper pole fullness (Fig. **4**). Medial and lateral breast pillars are then brought together to constitute a narrow based breast mound with projection. Several modifications of using inferior dermoglandular flap as an endoprosthesis have been described by different authors. Mobilization of the inferior based chest-wall flap to the upper pole was first described by Ribeiro [19]. Daniel used the same flap for upper pole fullness and he passed the flap under a pectoral muscle loop to obtain a long-lasting result [20]. Graf and Biggs reported their experience with this technique in different types of breast deformities [21, 22]. Goes advocated that in severe breast ptosis skin alone does not provide enough early support and suggested use of a polyglactine or a mixed mesh in addition to dermal suspension on a central pedicle [23].

Patients with Insufficient Breast Volume

If the patient has insufficient breast tissue, volume replacement is needed in addition to previously described mastopexy techniques. This can be provided by using silicone implants or by utilizing the "side rolls" for autoaugmentation.

1. Implant-Breast Augmentation in MWL Patient

Implant augmentation can be required for the patients who have completely deflated breasts after MWL. However there are some issues that should be considered for using implant with mastopexy in this patient group [24]. Skin is less able to support an implant because of severe laxity; besides applied mastopexy techniques further disrupt suspension structures of the breast. Therefore, small or moderate sized implants should be preferred to prevent recurrent ptosis. Patients with relatively good skin quality are better candidates for implant augmentation but still dermal suspension techniques or use of alloplastic materials such as acellular dermis should be considered for additional support [9, 25]. The combined approach is now used with caution in the massive weight loss patient, and it is crucial that the patients get informed preoperatively about the potential risks and high incidence of revisions. It is important to finalize mound adjustments through glandular shaping and implant augmentation before skin takeout and tightening of the skin envelope for safe closure. Some authors suggest staging the surgery in order to minimize the incidence of recurrent ptosis, prevent NAC viability problems and allow small adjustments during second stage [26]. In this way mastopexy and autoaugmentation techniques are applied in the first procedure and implant augmentation is performed safely as a second procedure to improve the outcome (Fig. **5**).

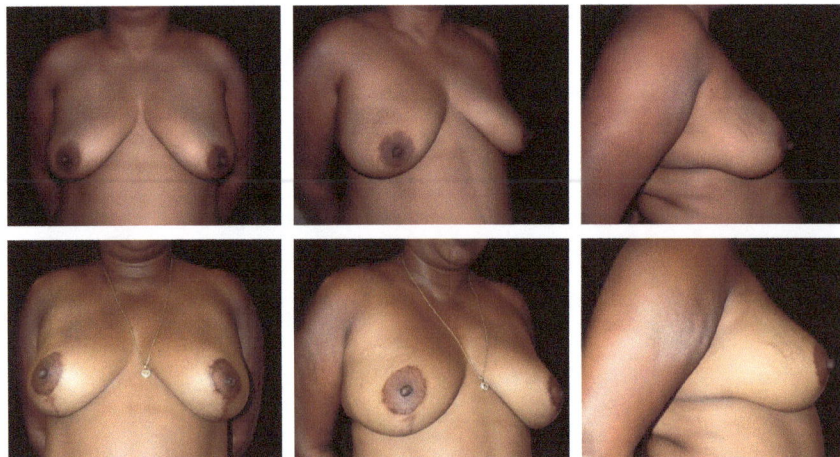

Fig. 5. A 34-Year Old Patient Who Lost 45 kg. Due to the Insufficient Volume of the Breast, a Combined Vertical Scar Mastopexy with a Retro-Pectoral (245 gr.) Gel-Implant Was Performed.

2. Mastopexy and Augmentation with Autologous Tissue in MWL Patient

In most of the MWL patients, breast deformity is complicated by nearby excess tissue. Correction of the lateral skin redundancy or 'side rolls' in the lateral thoracic area can be challenging during breast reshaping. However, this axillary extension of the breast can be used to provide autoaugmentation in patients with insufficient breast volume. Rubin described a total parenchymal breast reshaping with dermal suspension mastopexy using the excess tissues around the breast mound [10, 13]. He used a modified Wise pattern, extending laterally to encompass any significant lateral rolls (Fig. **6**).

In his technique, after de-epithelialization of the entire region, lateral and medial dermoglandular flaps based on intercostal perforators were elevated. The central pedicle is secured to the second or third rib periosteum with permanent sutures, followed by securing the lateral extension flap to the third rib periosteum close to the central suspension suture creating a discrete lateral curvature. The medial breast flap is suspended and secured to the chest wall. Next, the lateral and medial flap dermis are plicated to the central dermal extension. Finally the inferior pole of the breast is plicated to shorten the nipple to inframammary fold distance and provide extra projection. Rubin's technique allows correction of the lateral chest wall and the entire aesthetic unit of the upper chest. By extending the lateral incision onto the back, mild back rolls can be corrected or brachioplasty procedure can be conjoined by axillary and medial arm extension. Dermal suspension and parenchymal fixation to the rib periosteum can prevent early recurrent ptosis. Long incisions and relatively higher complication rates are the disadvantages of the technique but can be overcome by appropriate patient selection.

Hurwitz *et al.* described the use of lateral side rolls as a spiral flap to augment and reshape the breast in MWL patients [27]. In his technique he de-epithelialized fasciocutaneous flap extensions of the Wise pattern mastopexy. Upward flip of the inferior portion and a twist, rotation, and advancement of the lateral portion of the de-epithelialized dermoglandular tissue is performed and because of this mobilization pattern, the technique is called spiral flap. They suggested that their technique provided four intertwined upper body lift effect including correction of epigastric looseness through a reverse abdominoplasty, superior positioning of the inframammary folds, elliptic excision of the lateral chest and mid-back skin rolls and reshaping the breasts by a mastopexy using the spiral flap technique.

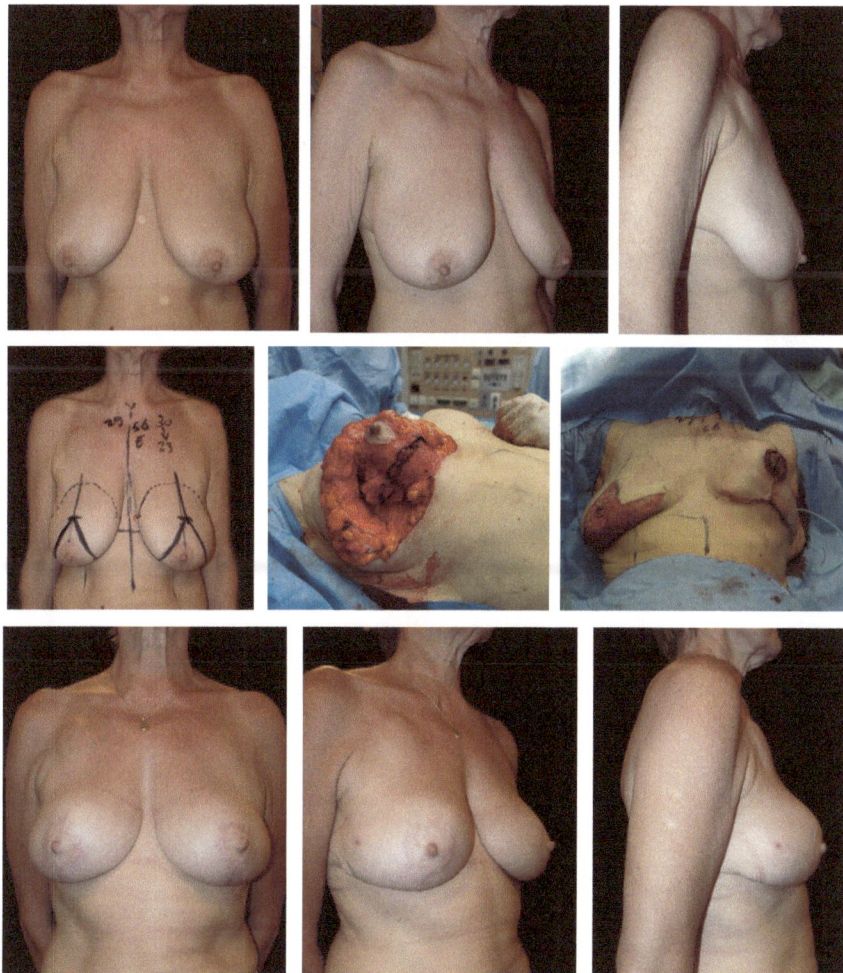

Fig. 6. Modified Wise Pattern with Lateral Extension Pour Auto-Augmentation Technique.

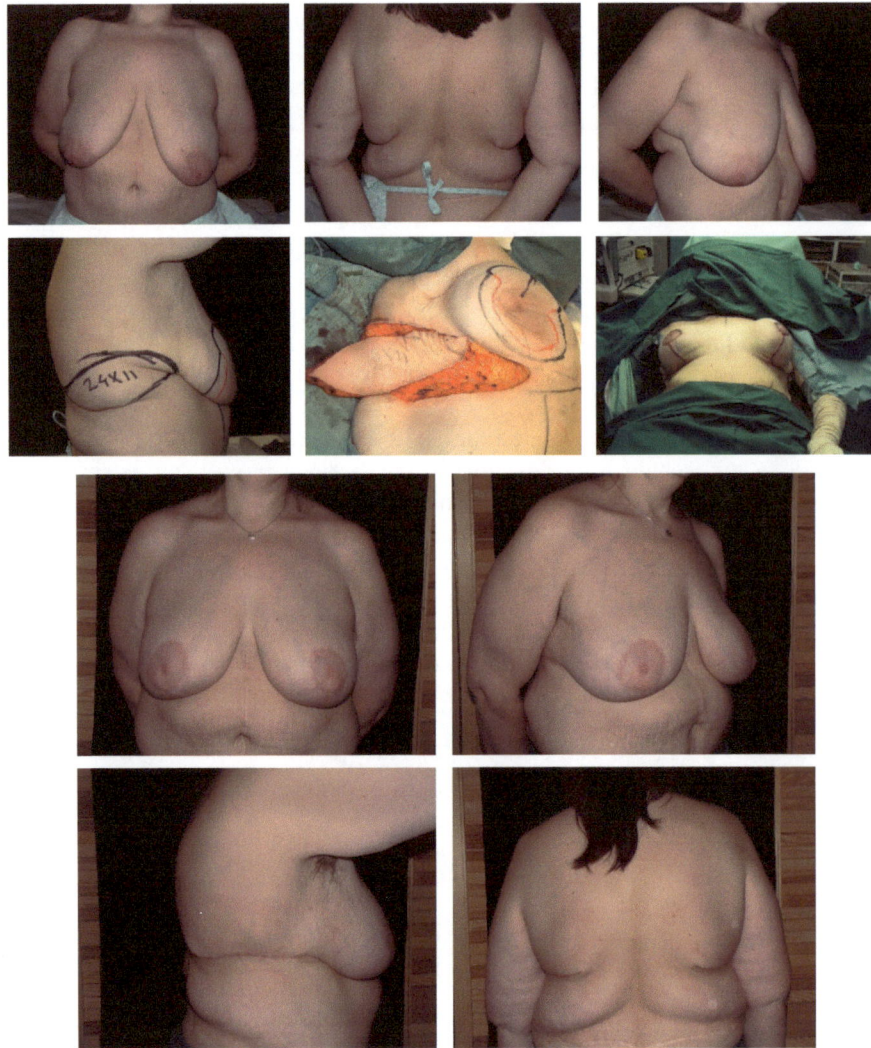

Fig. 7. Mastopexy and Autoaugmentation with Lateral Redundant Dermoglandular Tissue Based on Lateral Intercostal Artery Perforators.

We suggested the use of lateral redundant dermoglandular tissue based on Lateral Intercostal Artery Perforators (LICAP) with some personal modifications [11]. The LICAP was first reported in partial breast reconstruction within a clinical algorithm based on the location of the defect and the availability of these perforators [28]. A case report of pedicled perforator flaps for breast augmentation was subsequently published [29]. We also reported our clinical experience with ICAP flaps and addressed the use of pedicled LICAP flaps in massive weight loss patients [11]. In a recent study, we described the anatomical details of the perforators and the flap design [30]. The flap is designed lateral to the breast over the axilla and lateral thoracic area. The anterior border of the flap should include the junction of the inframammary fold (IMF) with the anterior axillary line to allow primary donor site closure. The width of the flap depends on skin redundancy and varies between 9 and 13cm. The perforators are located with a Doppler and the closest and the most anterior perforator to the breast is chosen for adequate arc of rotation of the flap. Latissimus Dorsi (LD) muscle is exposed after incision and the flap is elevated above the muscle fascia. Once the largest perforator closer to the pectoralis major muscle is found it is freed off the surrounding tissue. The serratus anterior muscle is split and the perforator is dissected until its exit above the rib. Once dissection of the perforator is complete, the rest of the flap is elevated easily above the muscle fascia. The inferior incision of the flap is then extended into the IMF. The breast gland is dissected and a retroglandular pocket is prepared for the flap. The mastopexy is first marked in a vertical scar mammaplasty pattern. The horizontal extent of the reduction pattern is determined during surgery, after harvesting and insetting the flap (Fig. **7**).

An alternative method for breast reshaping in a MWL patient with insufficient volume is transferring autogenous tissue as a free flap. In majority of the MWL patients an abdominoplasty procedure is needed. Lifting-excision procedures can be combined as a total body lift or can be staged to minimize operation time and complications [31, 32]. In carefully selected patients, who will also receive an abdominoplasty, the whole abdominal tissue can be raised on Deep Inferior Epigastric Artery Perforators (DIEAP) on each side of the abdomen to be transferred to the breast as a free flap [33]. Preoperative mapping of the perforator anatomy with Multidetector Computerized Tomography (MDCT) facilitates identification of the dominant perforator and intramuscular course of the vessels [34]. This complex procedure is long and prone to well-known complications of a long free flap procedure [35, 37], but it provides effective and satisfactory result in patients who need both breast and abdominal reshaping together and have limited volume at the breasts and lateral side rolls for autoaugmentation.

4. OUTCOMES AND COMPLICATIONS

Depending on the experiences and on the literature, it is evident that massive weight loss patients are amenable to various complications after body contouring procedures [38-40]. In a previous study of 449 post–bariatric surgery patients, 42 percent of the subjects developed complications and the rates ranged from 8 to 66 percent in different series [41]. Complications in the postoperative period include wound healing problems, seromas, thromboembolic complications, infection, hematoma and necrosis [38-41]. Patients' body mass index, overall medical condition and co-morbid factors play an important role on developing postoperative problems [38, 40]. A very important determinant is combining procedures or more aggressively applying total body lifting procedures. The medical condition of the patient, the operative team, the surgeon's experience with major body contouring procedures, the operative facility and resources, and the anesthesia team should all be seriously taken into consideration when deciding to combine procedures [31].

The most common complications in the early postoperative period are wound healing problems and seroma formation [39, 41]. Although it was suggested that molecular abnormalities in the wound healing cascade may play an important role in MWL patients, the main cause seems more likely multifactorial [41]. However both seroma formation and wound healing problems are seen at the abdominal and thigh regions and the incidence of such complications is lower at the breasts. Tension free closure during breast reshaping techniques along with the use of suction drains prevents most of the early postoperative complications.

5. CONCLUSION

Breast reshaping after massive weight loss is an effective procedure and provides high patient satisfaction. Although characteristic breast abnormalities in this patient group is well described, deformities may vary and surgical techniques should be individualized depending on the breast volume, quality of the skin envelope and coexisting nearby excess tissues. The algorithm presented here guides the surgeon to choose the ideal surgical approach to establish a satisfactory and long lasting result with minimum complications. Correction of the breast deformity can be performed as part of a combined procedure together with reshaping of other regions such as abdomen, arms and thigh or staging the procedures can be preferred according to the medical condition of the patient and surgeon's experience with major body contouring procedures.

REFERENCES

[1]   Sarwer DB, Thompson JK, Mitchell JE, *et al*. Psychological considerations of the bariatric surgery patient undergoing body contouring surgery. Plast Reconstr Surg 2008; 121: 423e-434e.

[2]   Song AY, Jean RD, Hurwitz DJ, *et al*. A classification of contour deformities after bariatric weight loss: The Pittsburgh Rating Scale. Plast Reconstr Surg 2005; 116: 1535-44.

[3]   Gusenoff JA, Rubin JP. Plastic surgery after weight loss: current concepts in massive weight loss surgery. Aesthet Surg J 2008; 28: 452-55.

[4]   Mitchell JE, Crosby RD, Ertelt TW, *et al*. The desire for body contouring surgery after bariatric surgery. Obes Surg 2008; 18: 1308-12.

[5]   Cintra W Jr, Modolin ML, Gemperli R, *et al.* Quality of life after abdominoplasty in women after bariatric surgery. Obes Surg 2008; 18: 728-32.

[6]   Taylor J, Shermak M. Body contouring following massive weight loss. Obes Surg 2004; 14: 1080-85.

[7]   Matarasso A, Aly A, Hurwitz DJ, *et al.* Body contouring after massive weight loss. Aesthet Surg J 2004; 24: 452-63.

[8]   Santry HP, Gillen DL, Lauderdale DS. Trends in bariatric surgical procedures. JAMA 2005; 294: 1909-17.

[9]   Colwell AS, Driscoll D, Breuing KH. Mastopexy techniques after massive weight loss: An algorithmic approach and review of the literature. Ann Plast Surg 2009; 63: 28-33.

[10]   Rubin JP, Gusenoff JA, Coon D. Dermal suspension and parenchymal reshaping mastopexy after massive weight loss: statistical analysis with concomitant procedures from a prospective registry. Plast Reconstr Surg 2009; 123: 782-89.

[11]   Hamdi M, Van Landuyt K, Blondeel P, *et al.* Autologous breast augmentation with the lateral intercostal artery perforator flap in massive weight loss patients. J Plast Reconstr Aesthet Surg 2009; 62: 65-70.

[12]   Kwei S, Borud LJ, Lee BT. Mastopexy with autologous augmentation after massive weight loss: the intercostal artery perforator (ICAP) flap. Ann Plast Sur 2006; 57: 361-65.

[13]   Rubin JP, Agha-Mohammadi S, O'Toole JP. Breast reshaping after massive weight loss. In: Aly A, ed. Body Contouring After Massive Weight Loss. St. Louis: Quality Medical Publishing; 2006; 361-78.

[14]   Hamdi M, Van Landuyt K, Tonnard P, *et al.* Septum-based mammaplasty: a surgical technique based on Würinger's septum for breast reduction. Plast Reconstr Surg 2009; 123: 443-54.

[15]   Würinger E. Refinement of the central pedicle breast reduction by application of the ligamentous suspension. Plast Reconstr Surg 1999; 103: 1400-10.

[16]   Hall-Findlay E J. A simplified vertical reduction mammaplasty: Shortening the learning curve. Plast Reconstr Surg 1999; 104: 748-59.

[17]   Lejour, M. Vertical mammaplasty and liposuction of the breast. Plast Reconstr Surg 1994; 94: 100-14.

[18]   Lejour M. Vertical mammaplasty: update and appraisal of late results. Plast Reconstr Surg 1999; 104: 771-81; discussion 782-84.

[19]   Ribeiro L. A new technique for reduction mammaplasty. Plast Reconstr Surg 1975; 55: 330-34.

[20]   Daniel M. Mammaplasty with pectoral muscle flap. Paper Presented at: The 64th American Annual Scientific Meeting in Montreal, Quebec; October 7-11, 1995.

[21]   Graf RM, Mansur AE, Tenius FP, *et al.* Mastopexy after massive weight loss: extended chest wall-based flap associated with a loop of pectoralis muscle. Aesthet Plast Surg 2008; 32: 371-74.

[22]   Graf R, Biggs TM. In search of better shape in mastopexy and reduction mammoplasty. Plast Reconstr Surg 2002; 110: 309-317; discussion 318-22.

[23]   Góes JC. Periareolar mammaplasty: double skin technique with application of polyglactine or mixed mesh. Plast Reconstr Surg 1996; 97: 959-68.

[24]   Spear SL, Boehmler JH, Clemens MW. Augmentation/mastopexy: a 3-year review of a single surgeon's practice. Plast Reconstr Surg 2006; 118: 136S-147S; discussion 148S-149S, 150S-151S.

[25]   Colwell AS, Breuing KH. Improving shape in mastopexy with cadaveric or autologous dermal slings. Ann Plast Surg 2008; 61: 138-42.

[26]   Coon D, Michaels J 5th, Gusenoff JA, *et al.* Multiple procedures and staging in the massive weight loss population. Plast Reconstr Surg 2010; 125: 691-98.

[27]   Hurwitz DJ, Agha-Mohammadi S. Postbariatric surgery breast reshaping: the spiral flap. Ann Plast Surg 2006; 56: 481-486; discussion 486.

[28]   Hamdi M, Van Landuyt K, Monstrey S, *et al.* Pedicled perforator flaps in breast reconstruction: a new concept. Br J Plast Surg 2004; 57: 531-39.

[29]   Van Landuyt K, Hamdi M, Blondeel P, *et al.* Autologous breast augmentation by pedicled perforator flaps. Ann Plast Surg 2004; 53: 322-27.

[30]   Hamdi M, Spano A, Van Landuyt K, *et al.* The lateral intercostal artery perforators: anatomical study and clinical application in breast surgery. Plast Reconstr Surg 2008; 121: 389-96.

[31]   Hurwitz DJ. Single-staged total body lift after massive weight loss. Ann Plast Surg 2004; 52: 435-441; discussion 441.

[32]   Gusenoff JA, Coon D, Rubin JP. Brachioplasty and concomitant procedures after massive weight loss: A statistical analysis from a prospective registry. Plast Reconstr Surg 2008; 122: 595-603.

[33] Hamdi M, Weiler-Mithoff EM, Webster MH. Deep inferior epigastric perforator flap in breast reconstruction: experience with the first 50 flaps. Plast Reconstr Surg 1999; 103: 86-95.

[34] Hamdi M, Van Landuyt K, Van Hedent E, *et al.* Advances in autogenous breast reconstruction: the role of preoperative perforator mapping. Ann Plast Surg 2007; 58: 18-26.

[35] Hofer SO, Damen TH, Mureau MA, *et al.* A critical review of perioperative complications in 175 free deep inferior epigastric perforator flap breast reconstructions. Ann Plast Surg 2007; 59: 137-42.

[36] Lundberg J, Mark H. Avoidance of complications after the use of deep inferior epigastric perforator flaps for reconstruction of the breast. Scand J Plast Reconstr Surg Hand Surg 2006; 40: 79-81.

[37] Garvey PB, Buchel EW, Pockaj BA, *et al.* The deep inferior epigastric perforator flap for breast reconstruction in overweight and obese patients. Plast Reconstr Surg 2005; 115: 447-57.

[38] Greco JA 3rd, Castaldo ET, Nanney LB, *et al.* The effect of weight loss surgery and body mass index on wound complications after abdominal contouring operations. Ann Plast Surg 2008; 61: 235-42.

[39] Shermak MA, Rotellini-Coltvet LA, Chang D. Seroma development following body contouring surgery for massive weight loss: patient risk factors and treatment strategies. Plast Reconstr Surg 2008; 122: 280-88.

[40] Shermak MA, Chang DC, Heller J. Factors impacting thromboembolism after bariatric body contouring surgery. Plast Reconstr Surg 2007; 119: 1590-1596; discussion 1597-98.

[41] Albino FP, Koltz PF, Gusenoff JA. A comparative analysis and systematic review of the wound-healing milieu: implications for body contouring after massive weight loss. Plast Reconstr Surg 2009; 124: 1675-82, discussion 1683-84.

CHAPTER 11

Gynecomastia and Male Chest Contouring

Alaa Gheita[1*] and Bishara S. Atiyeh[2]

[1]Faculty of Medicine Cairo University, 12 Hassan Sabry Street, Zamalek, Cairo, Egypt [2]American University of Beirut Medical Center, Beirut, Lebanon

Abstract: Gynecomastia is an extremely disturbing deformity especially when it occurs in young male adolescents. It is of frequent occurrence in obese persons and even more so after severe weight loss. Following massive weight loss, there is excessive redundancy of the skin breast mound as well as the chest skin around it. This deformity needs to be corrected to regain a male chest contour or if possible, even better a masculine or an ideal male's chest wall appearance. Gynecomastia following massive weight loss can be classified in only two categories based on the required correction. Type 1 gynecomastia characterized by mild skin redundancy or breast ptosis that can be corrected by concentric circumareolar excision of the excess skin and Type 2 gynecomastia characterized by major skin redundancy and breast ptosis that necessitates excision of chest wall skin with shifting of the nipple position for correction.

Keywords: Gynecomastia, inframammary fold, nipple-areola complex.

1. INTRODUCTION

Gynecomastia is a word derived from Greek with two parts "Gyne" meaning woman and "mastos" meaning breast, in another word the woman breast in a male. Development of abnormally large mammary glands in males with a feminin look can be emotionally devastating. Feelings of shame, embarrassment and humiliation are common. There is a multitude of pathological reasons for this deformity; some are due to hormonal imbalance between the stimulatory effect of estrogen and the inhibitory effect of androgen [1]. However, in the majority of cases no abnormalities can be detected. Gynecomastia is common among healthy men, and is often incidentally noted on routine physical examination. In an asymptomatic healthy man with long standing stable gynecomastia with a benign history and detailed physical examination, no further hormonal evaluation is necessary [2].

2. CLINICAL PRESENTATION

Gynecomastia is an extremely disturbing deformity especially when it occurs in young male adolescents (Fig. **1**). It is of frequent occurrence in obese persons (Fig. **2**) and even more so after Massive Weight Loss (MWL) (Fig. **3**). With loss of weight the fat decreases in volume while the breast skin cover does not retract and becomes loose and redundant producing a breast ptosis appearance as in a female. In those male patients, withdrawal occurs and a feeling of self pity sets in as a result of the severe psychological trauma caused by the deformity. Patients tend to refrain from certain social activities, and unfortunately, those are precisely the patients who tend to delay surgical correction.

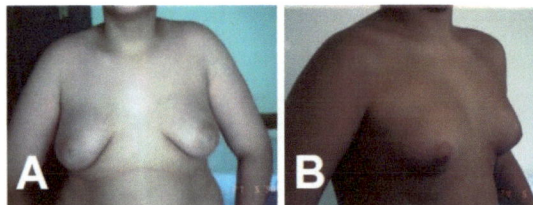

Fig. 1. (A) 14 Years Old Obese Adolescent with Gynecomastia. (B) Average Weight Adolescent with Gynecomastia.

***Address correspondence to Alaa Gheita:** Faculty of Medicine Cairo University, 12 Hassan Sabry Street, Zamalek, Cairo, Egypt; Tel: +202-2-7367734; E-mail: gheita@link.net

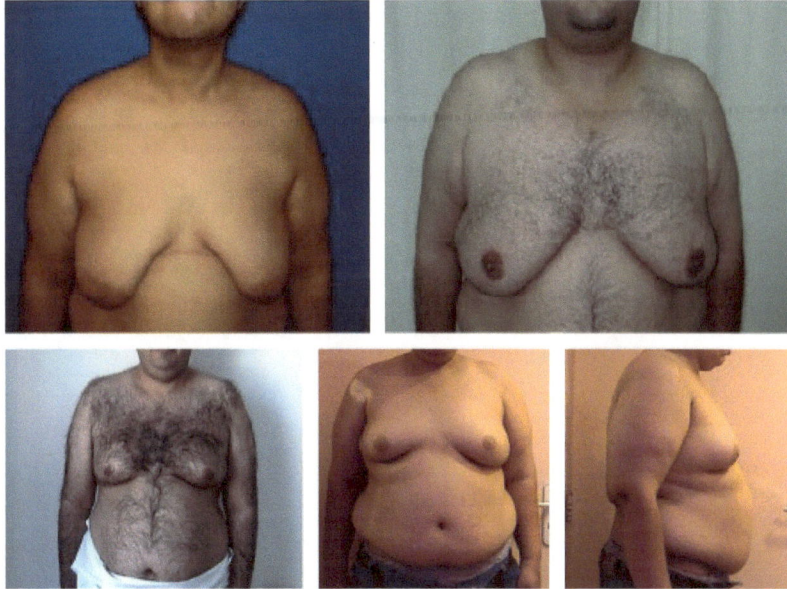

Fig. 2. Gynecomastia in Obese Adult Males Showing Breast Hypertrophy and Ptosis.

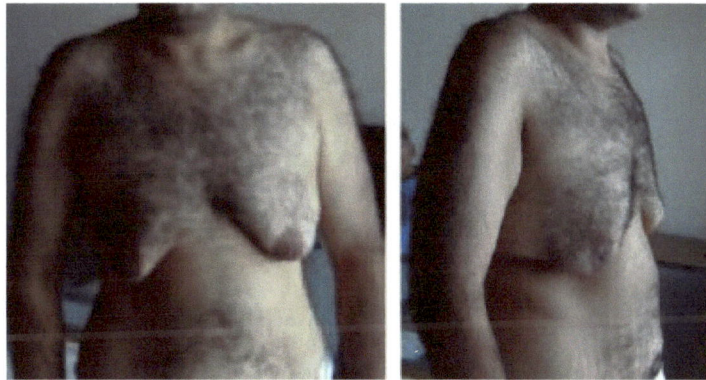

Fig. 3. 32 Years Old Male Patient with Gynecomastia and Breast Ptosis after Massive Weight Loss.

Men represent a small portion (15 to 20 percent) of the total number of patients currently presenting for gastric bypass surgery. However, this number is increasing and, even though at present the number of male patients desiring body contouring after MWL remains small (3 to 14 percent), theoretically, this number should grow [3]. Following MWL, male breasts are one of the most disturbing body regions and can be a difficult area to treat because of varying degrees of ptosis, nipple malposition, excessive parenchyma/fat, and loss of the inframammary fold, with a general loss of definition or shape [4]. Excessive redundancy of the skin breast mound as well as the chest skin around it needs to be corrected to regain a male chest contour or, if possible, even better a masculine or an ideal male's chest wall appearance.

There are a multitude of classifications for gynecomastia suggested since the term was introduced by Galen in the second century. Galen defined it as an unnatural increase in the breast fat of males [5, 6]. Currently, however, there is no classification system for pseudogynecomastia after MWL, with breast tissue being mostly fat and skin [3]. Several classification schemes with various treatment modalities have been devised for the treatment of gynecomastia. MWL patients are often grouped into the most severe categories of gynecomastia classifications. These classification schemes do not address the deformities that extend onto the chest wall and abdomen. Failure to address the entire aesthetic unit of the chest in continuity with the lateral chest and abdomen can lead to suboptimal results. As more male patients present for body contouring, a formal classification and treatment algorithm for this population group is increasingly justified [3].

Deformities of the male chest after MWL vary significantly and are challenging to correct without a major skin excision [3]. For reasons of simplicity and clarity, the deformity could be classified in only two categories based on the required correction. Type 1 gynecomastia characterized by mild skin redundancy or breast ptosis that can be corrected by concentric circumareolar excision of the excess skin, and Type 2 gynecomastia with major skin redundancy and breast ptosis that necessitates excision of chest wall skin with shifting of the nipple position for correction (Figs. **4** and **5**).

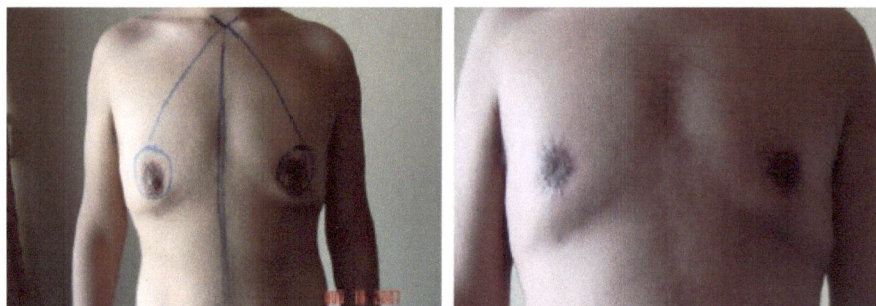

Fig. 4. Type 1 Gynecomastia.

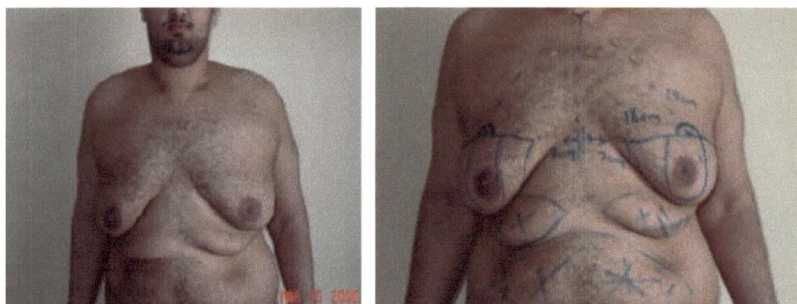

Fig. 5. Type 2 Gynecomastia.

3. AESTHETICS OF MALE BREAST

Men in general have a very consistent body habitus, with truncal fat deposition predominantly affecting the chest, abdomen, and flanks. The male breast is a region that symbolizes manhood and strength, and the male chest is supposed to be flat and muscular. For an aesthetically pleasing or ideal male chest, the fat or breast tissue should be minimal in front of the pectoral muscle. Male breast is more or less oval in shape going slanting up to the axilla and fitting in with the shape of the pectoralis muscle underneath it from origin to insertion. All attempts for correcting or reducing the feminine looking male breast or gynecomastia should excise the excessive fat and loose skin in front of the pectoralis muscle and end up, as close as possible, with a male muscular looking chest wall. Ideal male chest is clearly seen in Leonardo Da Vinci's painting *"The Vitruvian Man"*. Realistically only very few will have this kind of ideal chest, however, a non professional athlete patient with gynecomastia may achieve a "muscular athletic-looking chest" after surgical correction with exercises and physical training [7].

4. SURGICAL CORRECTION OF GYNECOMASTIA AFTER MASSIVE WEIGHT LOSS

The ideal method of gynecomastia correction must provide a technical means of removing the excess breast tissue without compromising the blood and nerve supplies to the nipple-areola complex. At the same time, it should also provide a method of recontouring the breast mound and handling the problem of skin excess without leaving unsightly or long scars [8]. The first reported surgical treatment of gynecomastia was by Paulus Aeginata in the seventh century AD who used a lunate incision below the breast or, for larger breasts, two converging lunate incisions to enable the excision of excess skin [9]. Since then, a multitude of methods have been suggested in the literature [10, 11]. A wide range of excisional and lipoplasty procedures have been described and adopted by different authors for the treatment of gynecomastia and are also applicable for correction of post-MWL breast deformities (Fig. **6**). Various incisions on and under the breast including transareolar, periareolar, and inframammary incisions have been used. The Nipple-Areola Complex (NAC) can be relocated as a full-thickness graft or preserved on a de-epithelialized flap [4].

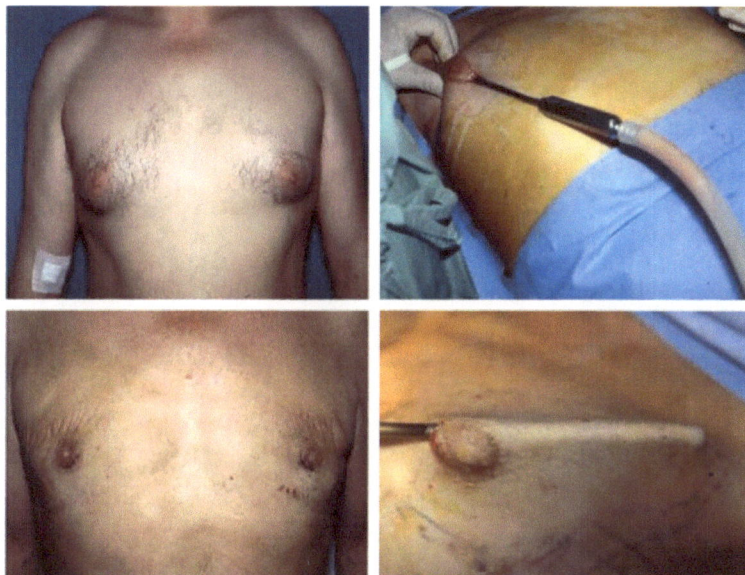

Fig. 6. Gynecomastia in a Young Patient Following Loss of 30 kg Corrected by Liposuction Only (Not by the Authors).

Many of the procedures suggested for correction of the male breast are usually derived from the methods of reduction mammaplasty applied in females [12]. These available surgical techniques present some disadvantages represented mainly by the multiplicity of scars which remain apparent in the male with unusual shape, and by the lack of symmetry in the size of both breasts and / or the nipple position [13]. They usually result in scars that may not be aesthetically acceptable and may migrate or shift over time [4] (Fig.**7**).

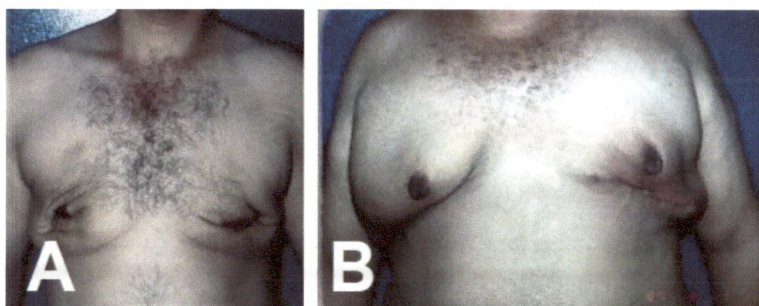

Fig. 7. (A) Post Subcutaneous Mastectomy for Gynecomastia with Ptosis (Not by the Authors). (B) Gynecomastia Corrected by Inverted-T Reduction Mammaplasty (Not by the Authors).

In a male patient with skin redundancy, an inverted-T mastopexy placing scars along the anterior chest can have a positive effect by eliminating the excess skin and providing more control over the position of the NAC. In some cases, the inframammary fold is fairly well defined, but the NAC complex lies well below it. In these situations, an inverted-T mastopexy with an inferior glandular nipple-areolar pedicle may not be adequate, because it may result in bulkiness of the central chest. This type of patient may be best managed by direct excision of the excess tissue at the level of the inframammary fold, with a free graft of the NAC. Even though the vertical scar is eliminated, scars can be more prominent with this technique. In some cases, excess skin can be resected by a circumareolar approach either with or without a pursestring suture. Redundant inelastic skin can also be removed through long obliquely-oriented anterolateral chest wall incisions in the form of a boomerang above the nipple [4].

We believe that most of the deformities can be corrected by only two methods as stated previously: (i) the "circumareolar" method for minor skin redundancy and (ii) the "horizontal ellipse" technique whenever there is excessive skin redundancy of the breast and the chest around it, which is frequently the case in patients after severe weight loss.

i. The "Concentric Circular" Method [14]

This is probably a most convenient approach in patient with mild skin redundancy and breast ptosis following weight loss (Fig. **5**) as it leaves minimal scarring, good blood supply to the areola and nipple and allows nipple relocation to a higher level if necessary. The upper border of the new areolar site is located. The amount of skin to be excised is estimated. A circular epidermal incision around the areola is traced followed by de-epithelialization of demarcated excess skin. In case where it is judged necessary to remove fat to diminish breast volume, liposuction may be performed or the lower half of the circle is deeply incised, the NAC is elevated on a superior flap to maintain its vascularity and excess fat or fibro-glandular tissue is excised [14]. This technique, however, has the disadvantage of possible outward stretching of the circular scar, wrinkling of the skin at the circumareolar closure particularly in cases with excessive skin redundancy, and frequent residual skin redundancy [15].

ii. The "Horizontal Ellipse" Method [13]

This newly suggested method consists of a horizontal elliptical excision of excessive redundant skin of the breast and deep tissues while keeping a superior pedicle flap carrying the areola nipple complex to its new position on the chest wall.

Preoperative Markings and Planning

The ideal site for the neo-nipple location on the chest wall should fall on the breast axis, which begins at the sternal notch; this location varies individually according to the built of each patient. Recently, an easy and reliable method to determine the horizontal and vertical coordinates of the male nipples has been proposed. These coordinates are in golden proportion with two easily measurable distances, umbilicus-anterior axillary fold apex and umbilicus-suprasternal notch, the inter-nipple distance and the position of the horizontal nipple plane relative to the suprasternal notch can be calculated with great reliability and accuracy [16].

Regardless of the guidelines utilized to determine the new NAC position, the distance from the suprasternal notch and from the midline to the new areola should be identical on both sides to achieve symmetrical positioning of the areola on the chest wall. A horizontal ellipse of breast resection is then planned beginning at the flat border of the breast medially and ending at the flat border laterally. The upper limb of the ellipse is on the horizontal line passing at the suggested site of the new nipple. The midline of the lower limb of the ellipse falls at around 6 cm from the infra-mammary line (Figs. **8 and 9**).

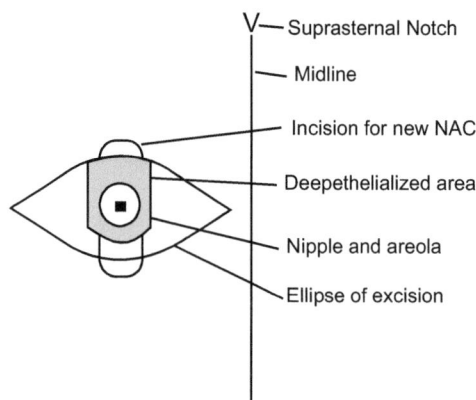

Fig. 8. Preoperative Markings. The Ellipse Covers the Area of Excessive Breast Tissue. The Grey Area Represents the Superiorly Based Flap to be De-Epithelialized and Constitutes a Dermo-Glandular Pedicle with a Width of Around 8cm and Variable Length Depending on the Extent of Ptosis.

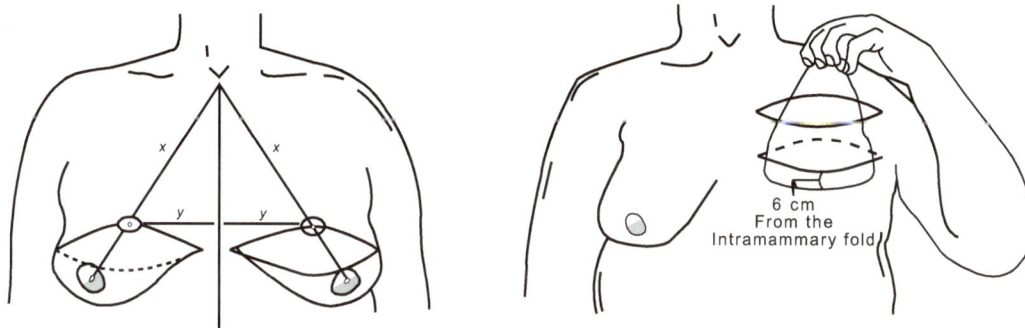

Fig. 9. The Distances from the Suprasternal Notch to the Nipple *(x)* Should be Equal on Both Sides, as well as from the Midline to the Nipple and the Lower Limb of the Horizontal Ellipse is at 6cm from the Infra-Mammary Fold.

Operative Technique

The pre-marked ellipse of skin is incised and the superior pedicle flap maintaining blood supply to the areola nipple complex is de-epithelialized and elevated. The excess skin glandular and fatty tissues in front of the pectoralis muscle are excised. The two edges of the ellipse are closed and the areola is buried. The new circular site of the areola is de-epithelialized in its upper half while full thickness skin excision is performed in its lower half. The buried areola is then delivered and sutured to the skin. No drainage is necessary. Compressive dressings is used in the form of a figure of 8 and left for 10 days (Fig. **10**).

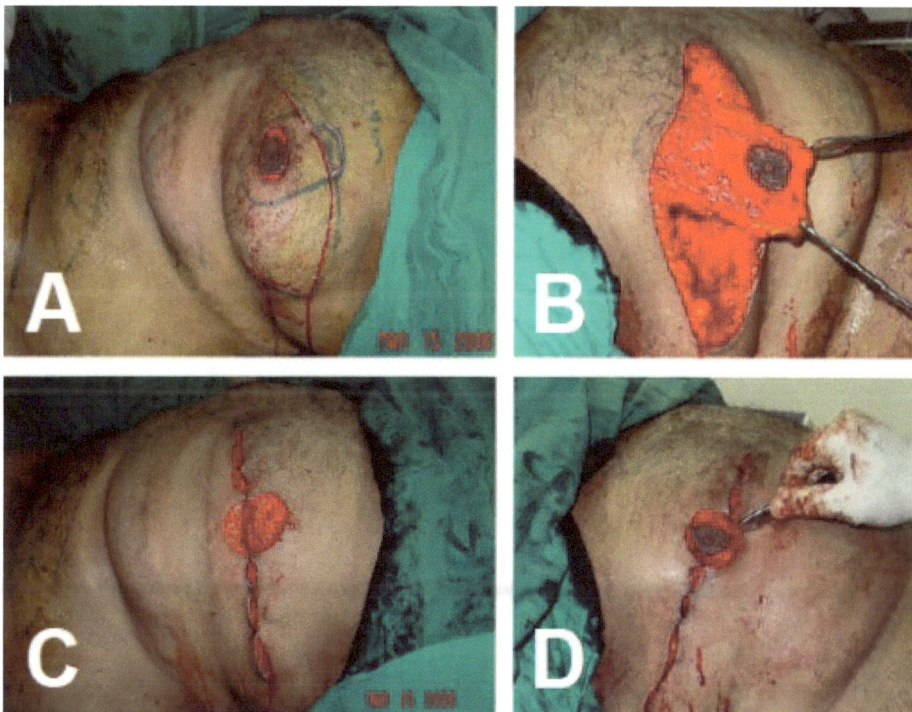

Fig. 10. (A) Incisions for the Ellipse of Excision and Circular around the Areola Preparing for the Dermo-Glandular Flap Carrying the Nreola-Nipple Complex. (B) Resection of the Ellipse of Ptotic Skin and Breast Tissues in Excess Completed. Superior Pedicle Dermo-Glandular Flap Maintaining the Blood Supply to the Areola and Nipple Complex Elevated. (C) Ellipse Closed and the Circular Defect in the Skin to Receive the Areola at its New Site on the Male Chest Wall is De-Epithelialized. (D) Nipple-Areola Complex Delivered on the Surface and fixed at its New Site. There is a Single Horizontal Scar Interrupted in its Middle by the Areola.

This simple procedure gives excellent results with no or minimal complications such as seroma, hematoma, or infection. The single resulting scar was hardly conspicuous falling into Langer's line and is interrupted at its middle by the areola (Figs. **11-14**).

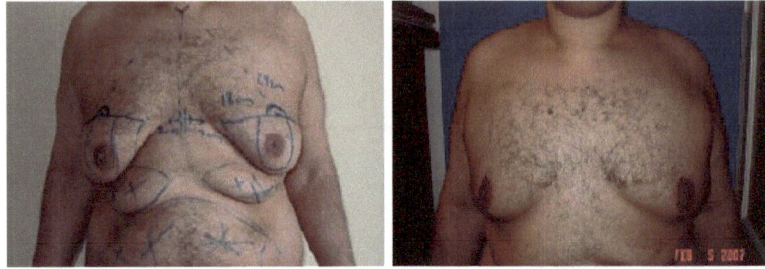

Fig. 11. Gynecomastia Corrected with the Ellipse Technique and Final Result One Year after Surgery. The Scar Falling into Langers Line is Hardly Noticeable.

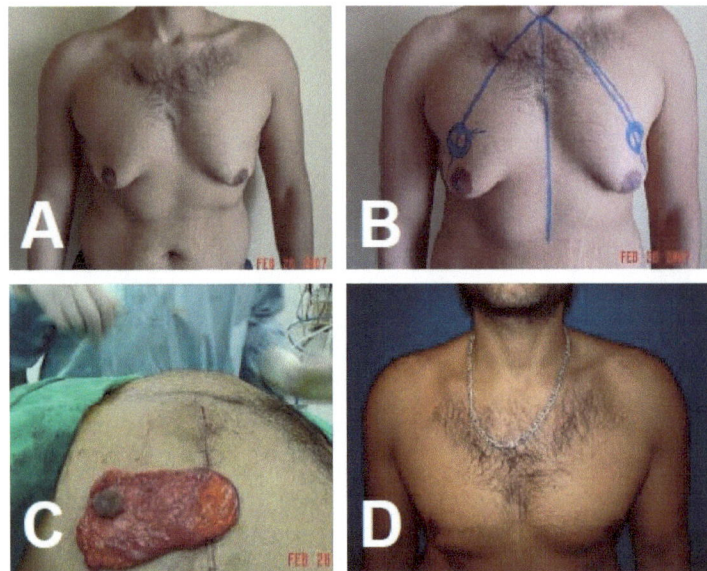

Fig. 12. (A) Gynecomastia following Severe Weight Loss. (B) Planning of new Nipple-Areola Position. (C) Superior Pedicle Flap After Closure of the Horizontal Elliptical Excision Medially and Laterally. (D) One Year after Correction by the "Horizontal Ellipse" Method.

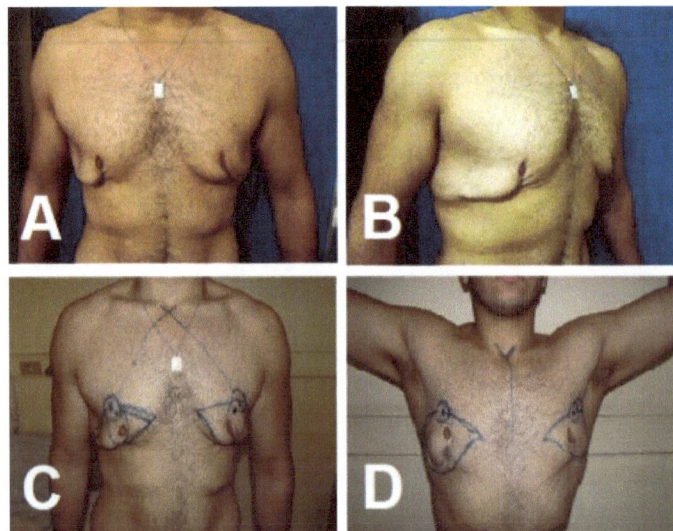

Fig. 13. (A and B) Patient with Previous Surgical Correction by an Inverted-T Method Showing Marked Disfiguring Result (Not by the Authors). (C and D) Planning for Correction by an "Ellipse Horizontal Method" Both in Relaxed Position and with Elevated Arms Showing Previous Scars, Residual Excessive Skin Redundancy and Result of Ischemic Changes and Areola Necrosis.

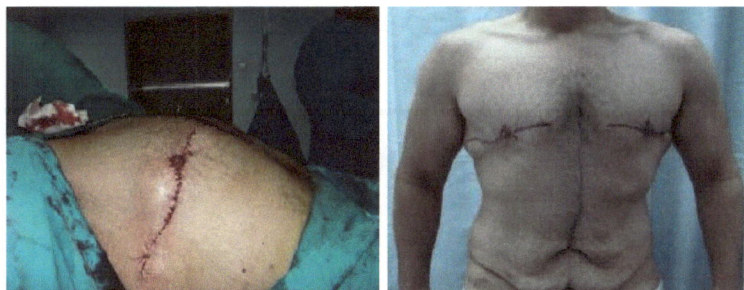

Fig. 14. Immediate Result of the Ellipse Method and Result at 3 months with Slight Scar Hypertrophy which is Exceptional as the Incision is in Langers Lines.

It is worth noting that lower body lift and brachioplasty procedures can affect the anterior chest improving greatly skin redundancy and gynecomastia. In selected cases, contouring of male anterior chest wall may also be possible with lateral torsoplasty and circumareolar nipple-areola transposition as recently described [14] (Fig. **15**).

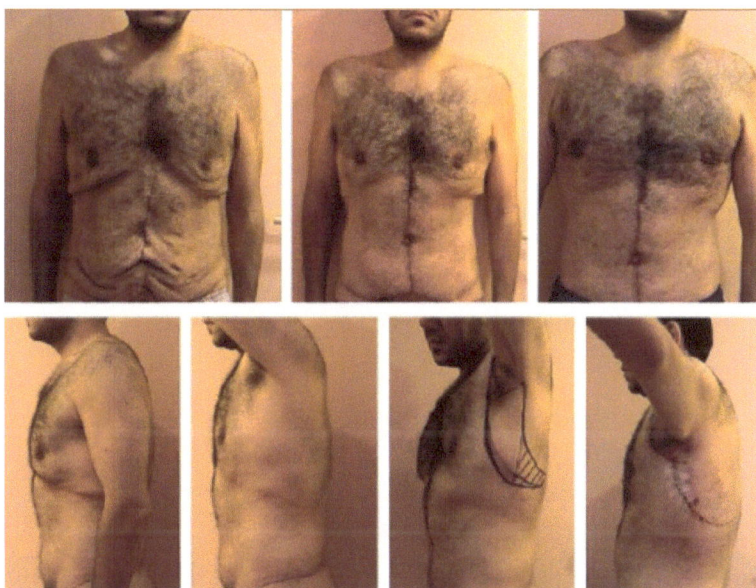

Fig. 15. Gynecomastia following Massive Weight Loss Greatly Improved after Lower Body Lift. Residual Chest Deformity Corrected by Lateral Torsoplasty and Circumareolar Nipple-Areola Transposition.

5. CONCLUSION

Following MWL, male breasts are one of the most disturbing body regions and can be a difficult area to treat because of varying degrees of ptosis, nipple malposition, excessive parenchyma/fat, and loss of the inframammary fold, with a general loss of definition or shape. Various surgical designs have been described to contour the male chest following MWL, however, all have four main goals to tighten, flatten, and harmonize the surface over the frontal thoracic wall with minimally noticeable scars in the frontal view. Regardless of the technique, contouring male chest after MWL has a high patient satisfaction rate. The critical components include proper placement of the NAC and the adaptation of the surgical technique to the type of deformity and existing anterior chest wall asymmetries.

REFERENCES

[1]   Leibovitch I. Incidence and management of gynecomastia in men treated for prostate. J Urol 2006; 175: 1961-63.

[2]   Hanavadi S, Monypenny IJ, Mansel RE. Is mammography overused in male patients? Breast 2006; 15: 123-26.

[3]   Gusenoff JA, Coon D, Rubin JP. Pseudogynecomastia after massive weight loss: detectability of technique, patient satisfaction, and classification. Plast Reconstr Surg 2008; 122: 1301-11.

[4]   Atiyeh BS, Hayek S, Dibo S. Contouring of male breast after bariatric surgery and massive weight loss – a case report. Aesthet Surg J 2008; 28: 688-96.

[5]   Simon BE, Hoffman S, Hahn S. Classification and surgical correction of gynecomastia. Plast Reconst Surg 1973; 51: 48-52.

[6]   Abramo AC. Axillary approach in suction-assisted lipectomy of gynecomastia. In: Shiffman MA and Di Giuseppe A, Eds. Liposuction Principles and Practice. Springer, 2006; First edition; pp. 465-66.

[7]   Blau M. American Society of Plastic Surgeons. 2005 Gender quick facts: Cosmetic Plastic Surgery. Available at: www.plasticsurgery.org/public.

[8]   Courtiss EH. Gynecomastia: Current recommendations for treatment. Plast Reconst Surg 1987; 79: 740-50.

[9]   Fruhstorfer BH, Malata CM. A systematic approach to the surgical treatment of gynaecomastia. Br J Plast Surg 2003; 56: 237-46.

[10]   Rohrich RJ, Ha RY, Kenkel JM, *et al.* Classification and management of gynecomastia: Defining the role of ultrasound-assisted liposuction. Plast Reconstr Surg 2003; 111: 909-23.

[11]   Handschin AE, Bietry D, Husler R, *et al.* Surgical management of gynecomastia—a 10-year analysis. World J Surg 2008; 32: 38-44.

[12]   Goldwyn RM, Cohen MN. The unfavorable result in plastic surgery: Avoidance and treatment. 3rd ed. Baltimore, Md: Lippincott Williams & Wilkins 2001; pp. 663-73.

[13]   Gheita A. Gynecomastia: The Horizontal ellipse method for its correction. Aesth Plast Surg 2008; 32: 795-801.

[14]   Davidson BA. Concentric circle operation for massive gynecomastia. Plast Reconstr Surg 1979; 63: 350-75.

[15]   Tashkandi M, Al-Qattan M, Hassanain JM, *et al.* Gynecomastia. Ann Plast Surg 2004; 53: 17-20.

[16]   Atiyeh B. Dibo S, el Chafic AH. Vertical and horizontal coordinates of the nipple-areola complex position in males. Ann Plast Surg 2009; 63: 499-502.

CHAPTER 12

Torsoplasty

Joachim Graf von Finckenstein[*]

Department of Plastic and Aesthetic Surgery, Klinikum Starnberg, Germany

Abstract: The male chest appears that of females in patients with gynecomastia and in patients presenting after huge weight loss. The disadvantages of reduction mammaplasty in men are the visible scars on the chest wall. The aim of the chest lifting is to reposition the breast in a male appearance by thinning the amount of glandular and fatty tissue and avoiding noticeable scars in the chest wall, the wound being hidden in the anterior axillary line.

Keywords: Adolescent surgery, bariatric surgery, body lift, body lifting, body contouring, chest lifting, chest lift, female-like breasts, feminization, gynecomastia/surgery, lipectomy, liposuction-assisted mastectomy, mammaplasty, male breast enlargement, male feminization, mastectomy, reduction mammaplasty, male breast lesion, male breast abnormality, feminized male breast, male breast mass, reconstructive surgical procedures, skin laxity, skin redundancy, surgical flaps, weight loss.

1. INTRODUCTION

When male patients with gynecomastia, (from a variety of causes, be it hereditary, age, drugs, weight loss...), undergo a "classic" mammaplasty to achieve skin tightness over the breast by skin resection a technique which cannot be performed by a simple periareolar approach, the chest wall becomes scared. In reduction mammaplasty for women, scars limiting the aesthetic unit of the breast are hidden in the shadow of the organ, which is lacking in male breasts.

The region of skin redundancy mostly treated, especially after weight loss, other than the breasts is the lower abdomen. The satisfying results after abdominoplasty and lower body lifting are due to scars limiting the aesthetic units as well. As the only aesthetical limit in male breasts is the areola, scar positioning is very limited and remains a problem if more skin resection is needed. The presently described surgical treatment achieves skin tightness in the frontal thoracical area while avoiding at the same time noticeable scars; redundant skin is shifted and resected in the axilla placing wound closure in the mid-axillary line by a so called Chest Lifting. Male patients are by far more concerned than females.

2. MATERIAL AND METHODS

Patients concerned are male patients with shape disharmony of the breast and the thoracic wall, following the Grade III and IV classification of Simon and Mc Kinney [1]:

- Cases of gynecomastia

- Cases of pseudo- gynecomastia (especially in elder men)

- Cases of weight loss

Female patients with indication for chest lifting are cases of skin redundancy in the area of the axilla and lateral thoracic wall. Although we also had 6 female patients in our series, men are by far more affected than women; in women, the problem of chest disharmony can mostly be solved by mammaplasty, where scars can be hidden in the shadow of the organ much easier than in male breasts.

***Address correspondence to Joachim Graf von Finckenstein:** Department of Plastic and Aesthetic Surgery, Klinikum Starnberg, Germany; Tel: +49-8-151-29968; Fax: +49-8-151-89149; E-mail: dr.med@finckenstein.de

Skin, fat or breast gland surplus of the upper torso with subsequent development of mammary creases lead to a feminine appearance of the breast shape in male patients. If the nipple-areola complex is positioned too medially, the indication for Chest Lifting is ideal. The aim of Chest Lifting is to restore the breast shape preventing noticeable scars in the chest wall and to place the scars in the shadow of the upper arm in the middle axillary line.

3. SURGICAL PROCEDURE

Skin redundancy of the chest wall leads to an unaesthetic breast appearance in male patients. With lifting up the arms, the skin is pulled upwards and the infra mammary crease disappears; at the same time the nipples are lifted up in a more natural lateral position on the chest wall. It is possible to imitate this upwards elevation of the shoulder by pulling the skin under the axilla in the direction of an 80° vector; pulling means stretching and stretching -in surgical terms- means tightening and cutting (Fig.1). Breast shape becomes more harmonious.

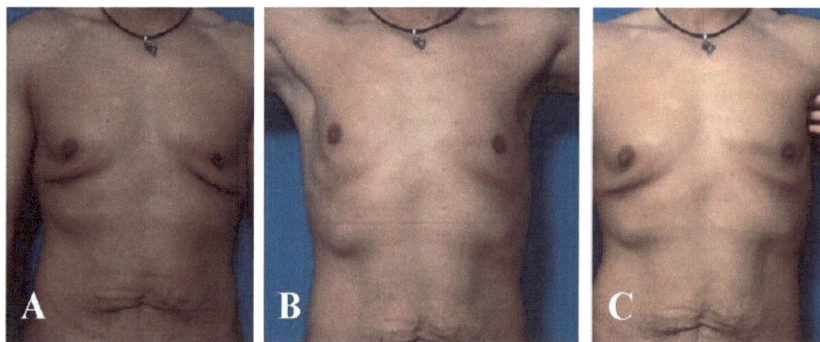

Fig. 1. (A) Unaesthetic Chest Appearance with Skin Redundancy. (B) Lifting up the Arms Improves Chest Appearance. (C) Pulling up the Skin Towards the Axilla, Achieves Adequate Correction.

4. MARKINGS

The marking of the skin to be removed corresponds to an elliptic resection whose cranial part ends in the crease of the axilla and whose distal part ends at the area of the 7th or 8th rib in the middle line of the axilla. A vector of approximately 80° must be marked in the standing position to define the line for tissue transposition (Fig.2). The resulting scar will correspond to the mid-axillary line.

It is impossible to define the markings as the patient is lying on the operation table. Even the control of aesthetic success in tissue transposition is nearly impossible during the operation; that is why it is indispensable to perform a meticulous marking preoperatively in a standing position.

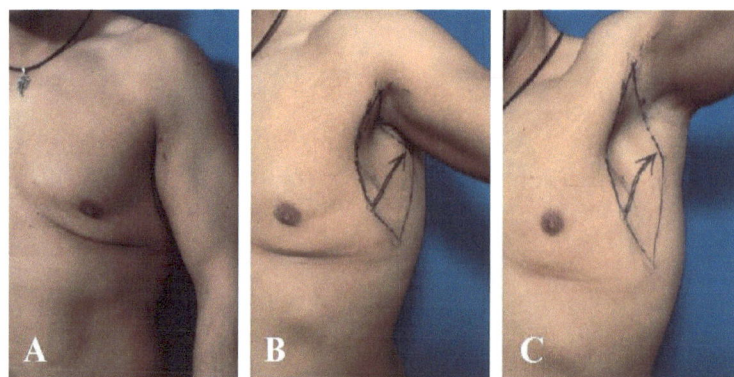

Fig. 2. Chest Lifting Markings. (A) Markings Completely Hidden by Arm. (B and C) Skin Ellipse to be Excised with Vector of Pull Indicated by Arrow.

5. OPERATIVE PROCEDURE

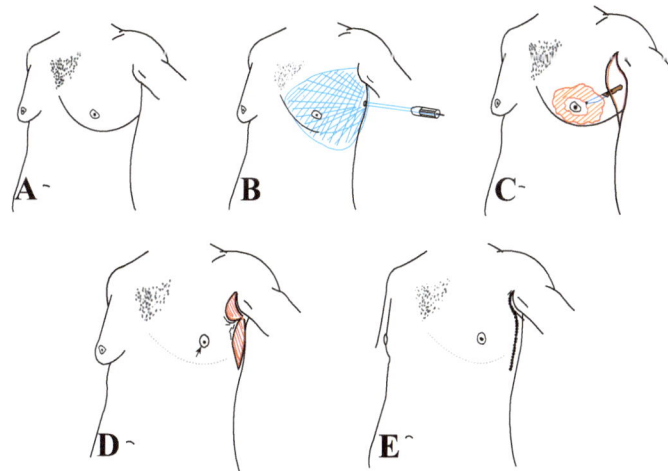

Fig. 3. Surgical Steps.

1) After conventional infiltration with the solution of Klein, intensive initial liposuction is performed to aspirate excess fatty tissue, to soften and release skin tension, and to achieve a pre-undermining of the area to be treated. It is important to include in the liposuction the pectoral area that is to be tightened (Fig. **3A** and **B**).

2) The marked skin is excised with its underlined attached fat. Wide sharp undermining of the frontal thoracic wall is then performed reaching the anterior part of the mammary gland. Parts of the gland can be removed without causing torsion of the nipple making sure to leave enough glandular tissue attached to the nipple-areola complex (Fig. **3C**).

3) Rotation-advancement of the skin flap along the 80° upward vector. Some tension reducing sutures are placed between the detached subdermal attachments of the Superficial Fascial System (SFS) and the pectoralis muscle fascia to prevent seroma formation (Fig. **3D**).

4) Wound closure is performed with quilting sutures of the superficial fascial system to reduce skin tension and intra-dermal suturing (Fig. **3E**).

Drainage and compression bandage are needed in the post-operative period. In selected cases women may profit from a Chest Lifting procedure particularly when tissue surplus of skin and fat in the area of the frontal axillary crease cause a disharmonious shape of the lateral appearance of the thoracic wall.

6. RESULTS

We have performed chest lifting since 2002 in 51 cases; 7 women and 44 men were treated.

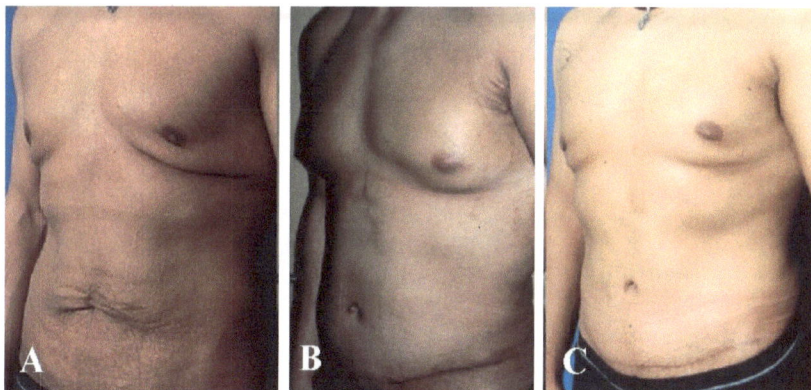

Fig. 4. (A) 28 Years Old Patient after Weight Loss of 30 kg. (B) Result of Chest Lifting (and Abdominoplasty) at 6 days and (C) at 3 Months Post-Operatively.

In 44 males, 9 patients presented with isolated gynecomastia and 35 patients had massive weight loss with subsequent disharmonious breast shape (Figs. **4** and **5**). In 28 of these cases abdominoplasty was performed prior to chest lifting or simultaneously.

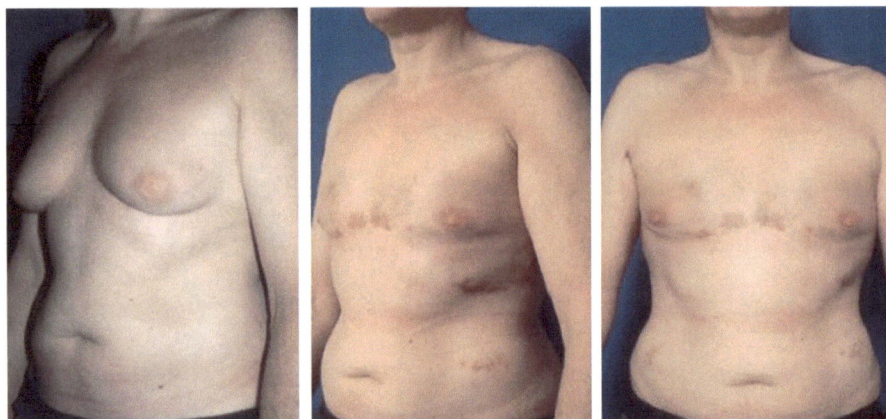

Fig. 5. 52 Years Old Patient with Gynecomastia. After Chest Lifting no Visible Scars are Left on the Frontal View of the Chest Wall 5 Days Post-Operatively; Glandular Tissue is Excised from an Axillary Approach.

Of the 7 female patients, 2 patients presented with a shape disharmony in the transition between the upper breast and the anterior axillary fold (**Fig. 6**).

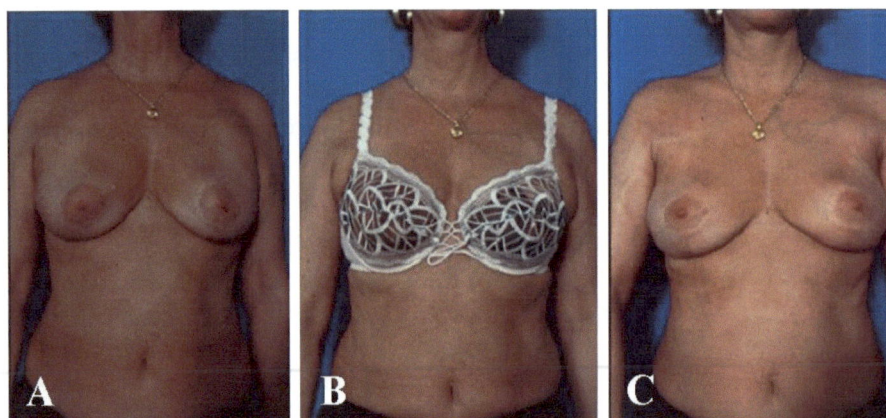

Fig. 6. (A and B) 48 Years Old Woman after Mammoplasty Performed 5 Years Earlier. She Disliked the Medial Positioning of her Nipple-Areola Complex and a Lateral Tissue Surplus Especially Visible when Wearing a Bra. (C) Result after Chest Lifting 3 Months Post-Operatively.

The aim of chest lifting is to tighten, to flatten and /or to harmonize body form over the frontal thoracic wall without leaving noticeable scars in the frontal view. The scars are hidden by the arm shadow. In all cases performed, the set goals were achieved. Results were aesthetically pleasing and easily reproducible. Early secondary surgery was necessary in 3 cases due to hematomas and in 2 cases due to infection with abscess formation in one of them. Delayed scar revision was performed in 2 cases due to dog ears in one case and scar hypertrophy in another. All secondary interventions were performed on male patients.

7. DISCUSSION

Surgical correction of gynecomastia through an axillary approach is not new [2]. The new idea in this presentation is an additional tightening of the chest wall area, by excising skin and hiding the scar in the mid-axillary line. From the plastic surgical view, gynecomastia and pseudo-gynecomastia are treated with two main aims: the resection of excrescent tissue and a minimum of visible scaring. In cases of simple fat excess liposuction is sufficient provided skin retraction ensues (Simon and Mc Kinney Grade I and II) [1, 3]; glandular surplus without soft skin redundancy can be managed by a periareolar approach to remove hypertrophied glandular tissue. As gynecomastia is often a combination of fat and gland excess, both liposuction and gland removal through the periareolar approach, are usually performed (Simon and Mc Kinney Grade II and III). Some authors even suggested a 2 stage procedure. The periareolar scar, however, might distort the nipple-areola complex if the gland tissue underneath the mammilla is not wide enough, or when tension on the suture line is too excessive. In cases of skin redundancy (Simon III) a "classic" mammaplasty to achieve skin tightness beside gland excision, regardless of the specific technique (inverted T, vertical or horizontal skin resection), invariably scars the frontal chest wall (Fig. **7**).

Chest Lifting avoids noticeable scaring of the frontal chest wall. This surgical technique concerns patients with gynecomastia Simon Grade III and IV for whom a periareolar approach would not lead to a satisfying result because wider skin resection is required. In the Chest lifting approach the redundant skin is pulled laterally towards the axilla and wound closure is placed in the mid-axillary line, the scars are thus hidden under the arm (Fig. **8**). Patients requiring this approach are mainly males with fat, glandular and/or skin surplus over the pectoral thoracic wall due to general body constitution or after massive weight loss.

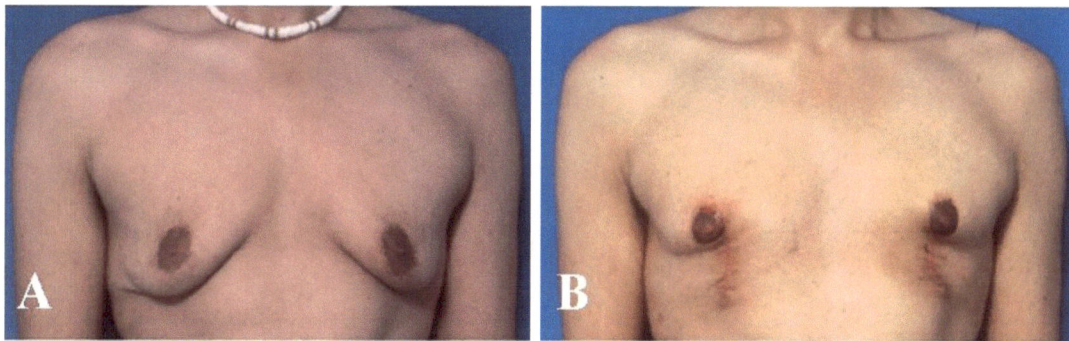

Fig. 7. (A) 23 Years Old Man after Weight Loss of 25 kg. A Periareolar Approach Would not Resolve the Problem of Skin Redundancy. (B) A Mammoplasty Was Needed Leaving Visible and Unaesthetic Scars on the Frontal Chest Wall.

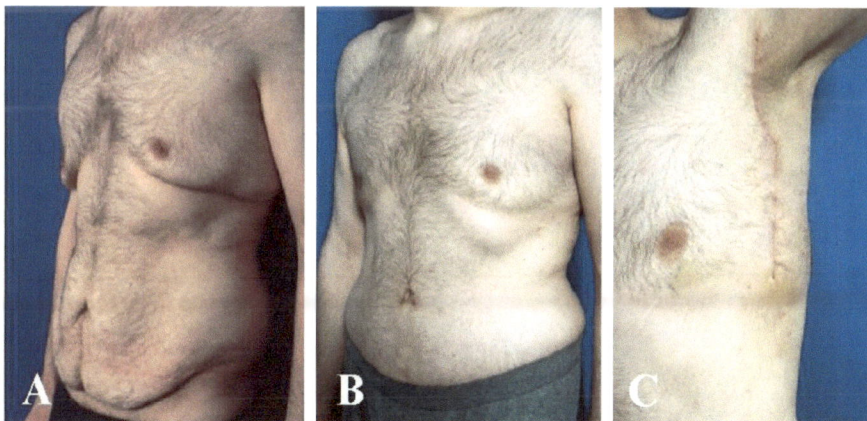

Fig. 8. (A) 36 Years Old Man after Weight Loss of 46 kg. (B) 3 Months after Chest Lifting and Abdominoplasty. (C) Scar Falling in the Mid-Axillary Line.

8. CONCLUSION

Male patients are by far more served by this surgical technique than females. In some selected cases, however, this method may be used in female patients to correct a disharmonious transition between the lateral breast shape and the axilla. The technique has demonstrated clearly that anterior chest wall contouring may be adequately achieved by lateral pulling and excision of skin.

The question to be answered in the future is whether adequate gynecomastia correction can be achieved endoscopicaly. For Simon Grade I to II cases, endoscopic glandular resection might become a common procedure in combination with liposuction which amounts to a minimal invasive procedure. For Simon Grade III and IV gynecomastia, however, skin resection is definitely needed. Lateral chest lifting might prove to be the ideal procedure.

REFERENCES

[1]   Simon BE, Kahn S. Classification and surgical correction of gynecomastia. Plast Reconstr Surg 1973; 51: 48-52.
[2]   Balch CR. A transaxillary incision for gynecomastia. Plast Reconstr Surg 1978; 61: 13-16.
[3]   American Society of Plastic and Reconstructive Surgeons Board of Directors. Gynecomastia / recommended criteria Position Paper of the ASPS, Oct. 1995.

CHAPTER 13

Back Contouring after Massive Weight Loss

Bishara S. Atiyeh[*] and Saad A. Dibo

American University of Beirut Medical Center, Beirut, Lebanon

Abstract: Management of excessive laxity of the abdomen, breasts, arms, thighs and buttocks following massive weight loss has received much attention, however little has been written about the surgical approach to the back rolls and folds. "Back rolls" can range in a number from one to four on either sides of the midline. In post-bariatric patients, the excess tissue must be excised. Liposuction is rarely indicated because the skin and subcutaneous tissue remains in excess in these patients. In addition, back tissue is dense and fibrous, making it less amenable to liposuction. Post-bariatric contouring of back-roll deformities requires an approach that allows for direct excision of the two lower folds (lower thoracic and hip rolls) during the circumferential lower body lift procedure and of the two upper folds.

Keywords: Back rolls, festoon curtain deformity, back contouring.

1. INTRODUCTION

After dramatic weight loss, patients are usually left with redundant skin and subcutaneous tissue that is unwieldy and many patients present with multiple folds or rolls of skin and subcutaneous fat laterally and posteriorly in addition to a large abdominal redundancy. The excess tissue is most apparent in the abdomen, thighs, buttocks, arms, back, and breasts. In addition to the poor unattractive cosmetic appearance, the overhanging redundant tissue can lead to skin irritation, pain, intertrigo, continual problems of hygiene and cleansing, and decreased activity [1, 2].

Management of excessive laxity of the abdomen, breasts, arms, thighs, and buttocks following massive weight loss has received much attention, however, little has been written about the surgical approach to the back rolls and folds [3].

2. ANATOMY OF THE BACK AND "FESTOON CURTAIN DEFORMITY"

In patients with general body relaxation, the maximum vertical relaxation of the trunk/thigh aesthetic unit occurs along the lateral body contour, not along the abdominal or medial-thigh area as is commonly believed [4]. A strong superficial fascial system adherence to the linea alba in the epigastrium limits vertical descent, and observed laxity frequently results from progressive horizontal loosening secondary to relaxation of the tissues along the lateral trunk [5]. Likewise, strong midline fascial adherences in the back prevent excessive midline tissue laxity and vertical descent.

"Back rolls" can range in number from one to four on either side of the midline. The highest fold is a posterolateral extension of the breast fold. In descending order, the next fold is the scapular roll. Next is the lower thoracic roll, and the lowest is the hip roll depending on the patient's pre-bariatric anatomy and the amount of weight loss achieved, one or more of the usual four folds may be absent [3].

In fact, in obese patients, fat accumulation occurs in the back primarily lateral to the spine and, due to gravity, folds form in a manner similar to festoon curtains (Figs. **1** and **2**).

These folds become more obvious after massive weight loss and deflation, accentuating horizontal tissue loosening. When severe, the cascading skin rolls, sweep like a Viennese festoon curtain from midline to lateral chest [6] (Fig. **3**).

*Address correspondence to Bishara S. Atiyeh:** American University of Beirut Medical Center, Beirut, Lebanon; Tel: +961-3-340032; E-mail: batiyeh@terra.net.lb

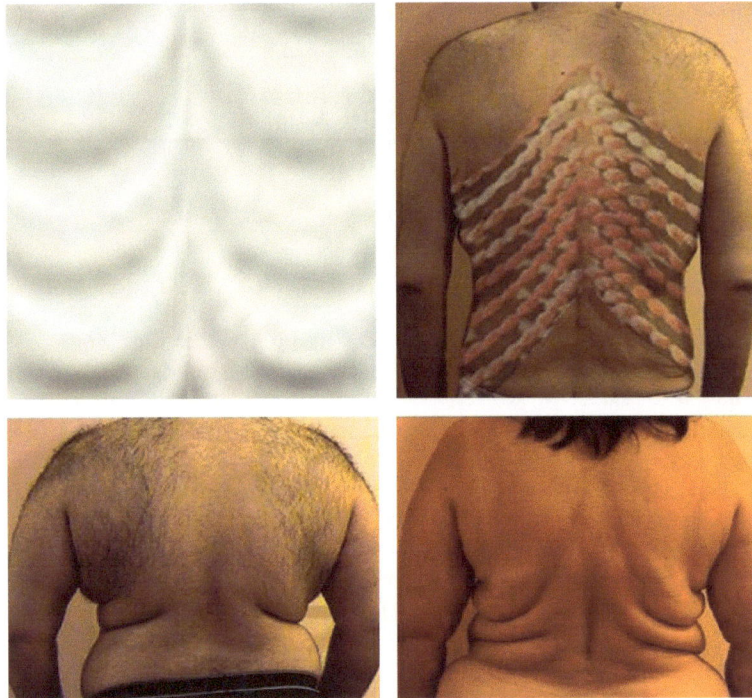

Fig. 1. Viennese Festoon Curtain and "Festoon Curtain Deformity". Evident Back Folds (Rolls) of Obese Patients: Upper Breast, Middle Scapular, and Lower Thoracic Folds.

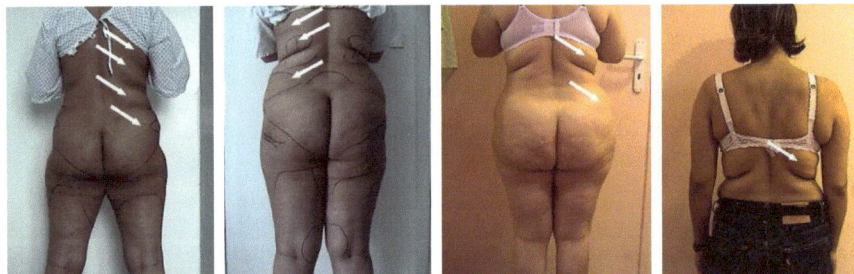

Fig. 2. Variable Distribution of "Back Rolls" in 4 Overweight and Moderately Obese Patients.

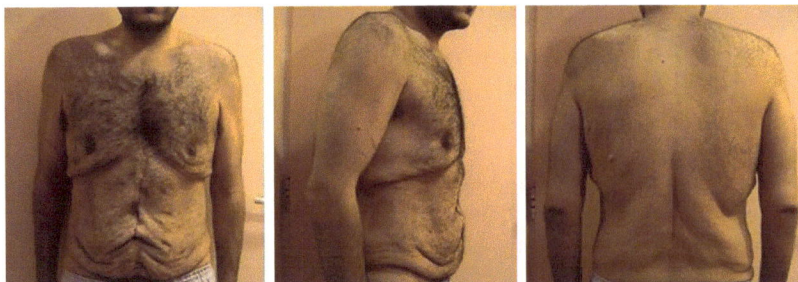

Fig. 3. Patient after Bariatric Surgery and Massive Weight Loss Illustrating Horizontal Tissue Laxity and "Festoon Curtain" Deformity.

3. BACK CONTOURING SURGERY

In post-bariatric patients, the excess tissue must be excised. Liposuction is rarely indicated because the skin and subcutaneous tissue remain in excess in these patients. In addition, back tissue is dense and fibrous, making it less amenable to liposuction [3]. Anterior and posterior truncal excess in the form of redundant skin is best treated with excision [7].

Patients with massive weight loss and their surgical needs are quite complex, and the surgical approach for these patients is not, and should not, be standardized [8]. Post-bariatric surgery patients can be successfully treated with a staged body contouring approach, improving their quality of life in a safe and effective manner [1]. The degree and type of skin excision is determined by other areas to be addressed, as well as prior surgical scars, hernias, degree of pannus, and medical and smoking history [8]. The area of most concern about redundant skin to post-bariatric surgery patients is the abdomen and mid body. Therefore, this area is addressed first [1]. The back tissue is contoured during several different body contouring operations, depending on the needs of each individual patient [3].

Traditional abdominoplasty is woefully inadequate. It poorly addresses the redundant lateral flank and hip rolls deformities. The lateral excesses may even be emphasized by the classical operation increasing lateral fullness or leaving dog-ears [1, 2]. With circular lipectomy and lateral thigh and buttock lift it is currently possible to achieve lifting of the buttock and lateral thigh, reduction in the number and size of adipose cutaneous folds of the lower and middle back, improvement in the waist silhouette, elimination of redundant flank tissue and abdominal skin, and plication of the rectus abdominis muscle [9]. After a circumferential body lift, many patients return to have their breasts, arms, and thighs contoured. However, troublesome areas of excess back tissue are usually not sufficiently addressed with standard procedures [3].

The lower thoracic and hip rolls serve as accurate markers for the upper and lower lines of resection in the circumferential abdominoplasty [3]. The upper back roll (*i.e.*, the breast fold) extends posteriorly from the inframammary crease. This may be excised at the time of the mammaplasty. The lateral breast fold is incorporated into the breast operation by extending the Wise-type pattern laterally to the posterior axillary fold, removing the lateral fold with the breast excision. It is better, however, to incorporate only that part of the breast fold to the posteroaxillary line during the mammaplasty. The balance of the upper breast roll and scapular rolls are best excised together in a separate procedure with the patient supine [3]. The spiral flap breast reshaping as part of an upper body lift has also been described to correct redundant lateral and back folds with simultaneous autologous breast augmentation. This lift is a reverse abdominoplasty that ends along the inframammary fold incision of the Wise pattern mastopexy and continues laterally along the back roll to the inferior tip of the scapula to correct the midback rolls of skin. This results in a scar along the anticipated brassiere line [6]. Excision of these rolls may be incorporated also in the lateral torsoplasty design, particularly in males [10] (Figs. **4 and 5**).

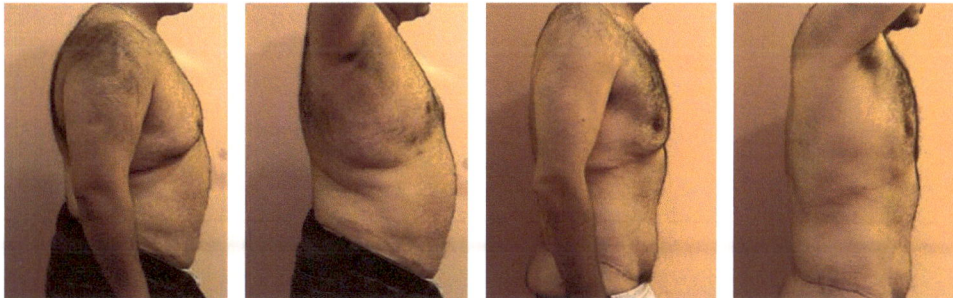

Fig. 4. Lateral Trunk and Upper Back Folds Can Be Markedly Improved by Arm Elevation Indicating the Possibility of Surgically Correcting These Folds in Selected Patients by Lateral Torsoplasty.

A consideration of prior scars is important in designing the circumferential lower body lift. Some authors have advocated a vertical midline incision to facilitate excision of excess tissue [1]. Use of a Fleur-de-Lis pattern in abdominal body contouring is a safe and effective technique for properly selected massive weight-loss patients. It is particularly appropriate for those patients with excessive horizontal laxity and significant upper midline abdominal fullness [11]. Described Fleur-de-Lis patterns can improve both horizontal and vertical laxity by removing a circumferential component in the lower abdomen combined with an inverted "V" pattern excision from the upper abdomen. Incorporating the Fleur-de-Lis pattern for the abdomen with techniques described for the flank and buttock regions can treat the circumferential component successfully and greatly improve back as well as chest contour [11] (Fig. **6**).

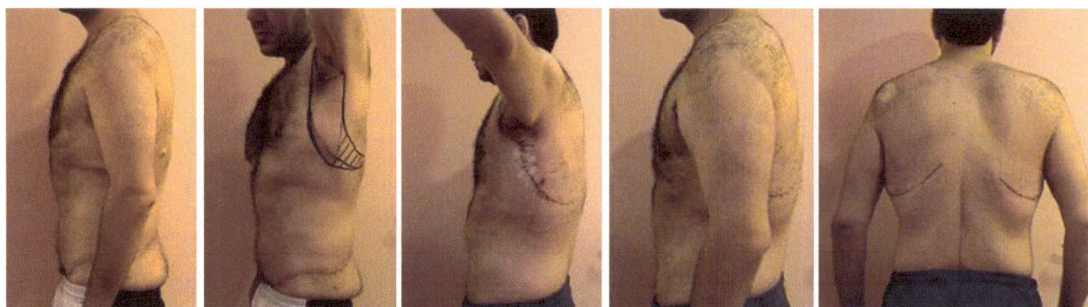

Fig. 5. Lower Back Folds Excised with Circular Lower Body Lift Procedure. Upper Fold Excised as Part of a Lateral Torsoplasty Procedure. Lateral Torsoplasty Immediate Post-Operative Result.

In open gastric bypass procedures, the patients are often left with midline scars that would have seemed to make a similar scar from a Fleur-de-Lis abdominoplasty acceptable. However, because of the advent of closed procedures, scarring now has been limited to small upper midline scars or laparoscopic port-site scars resulting from laparoscopic gastric bypass. Patients at present often are not willing to accept a long vertical component to their abdominal scar [1].

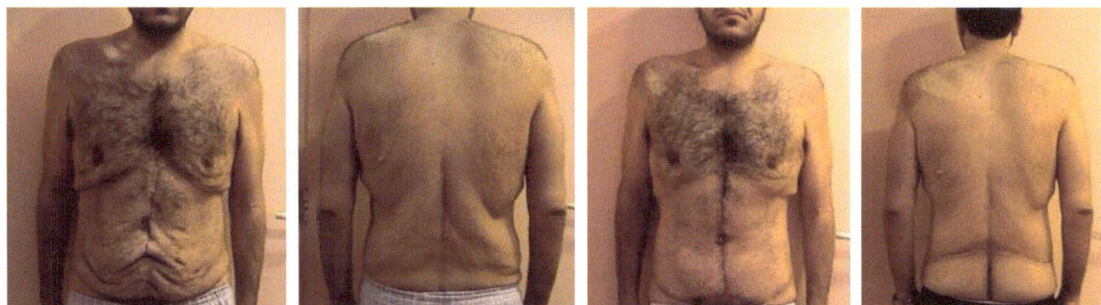

Fig. 6. Markedly Improved Back Contour after Fleur-De-Lys Circular Lower Body Lift.

However, for those patients who have undergone "open" bariatric techniques with midline vertical scar, Fleur-de-Lis abdominal contouring may be ideal. The midline abdominal scar that already exists, in fact, may hinder the redraping of the soft tissues by tethering the superior flap as it is pulled inferiorly during final tailoring. This technique can also be applied to older patients in whom scars are less of a problem, and who may have greater abdominal tissue laxity [11].

4. CONCLUSION

Post-bariatric contouring of back-roll deformities requires an approach that allows for direct excision of the two lower folds (lower thoracic and hip rolls) during the circumferential lower body lift procedure, and of the two upper folds (breast and scapular rolls) at a separate procedure. Direct excision of each of these pairs at separate sittings produces a well-contoured back with the lowest morbidity [3].

REFERENCES

[1]   Strauch B, Herman C, Rohde C, *et al.* Mid-body contouring in the post-bariatric surgery patient. Plast Reconstr Surg 2006; 117: 2200-11.

[2]   Van Geertruyden JP, Vandeweyer E, de Fontaine S, *et al.* Circumferential torsoplasty. Br J Plast Surg 1999; 52: 623-28.

[3]   Strauch B, Rohde C, Patel MK, *et al.* Back contouring in weight loss patients. Plast Reconstr Surg 2007; 120: 1692-96.

[4]   Lockwood TE. Lower-body Lift. Aesthet Surg J 2001; 21: 355-70.

[5]   Lockwood T. High-lateral-tension abdominoplasty with superficial fascial system suspension. Plast Reconstr Surg 1995; 96: 603-15.

[6]   Hurwitz DJ, Agha-Mohammadi S. Postbariatric surgery breast reshaping: The spiral flap. Ann Plast Surg 2006; 56: 481-86.

[7]   Rohrich RJ, Gosman AA, Conrad MH, *et al.* Simplifying circumferential body contouring: The central body lift evolution. Plast Reconstr Surg 2006; 118: 525-35.

[8]   Taylor J, Shermak M. Body contouring following massive weight loss. Obes Surg 2004; 14: 1080-85.

[9]   Sozer SO, Agullo FJ, Wolf C. Autoprosthesis buttock augmentation during lower body lift. Aesthetic Plast Surg 2005; 29: 133-37.

[10]   Atiyeh BS, Hayek SN, Dibo SA.Contouring Of Male Breast After Bariatric Surgery And Massive Weight Loss – A Case Report. Aesth Surg J 2008; 28: 688-96.

[11]   Wallach SG. Abdominal contour surgery for the massive weight loss patient: the fleur-de-lis approach. Aesth Surg J 2005; 25: 454-65.

CHAPTER 14

Brachioplasty

Sadri Ozan Sozer[1*] and Francisco J. Agullo[2]

[1]Texas Tech University Health Sciences Center, El Paso, Texas, El Paso Cosmetic Surgery, USA and [2]Texas Tech University Health Sciences Center, El Paso, Texas, El Paso Cosmetic Surgery, USA

Abstract: The arms represent a challenge in body contouring surgery. Deformities of the upper arms vary in presentation from minor defects to extensive skin excesses. When skin elasticity is good and fatty deposits limited, traditional forms of liposuction may be utilized. In the massive weight loss patient, the deformity often encountered is that of extensive excess skin, which will be the subject of discussion in this chapter.

Keywords: Brachioplasty, liposuction, bicipital groove.

1. INTRODUCTION

Brachioplasty was described by Correa-Iturraspe and Fernandez in 1954 [1]. Years later, Pitanguy [2] and Baroudi [3] popularized the procedure. There have been many variations of the procedure in regards to location and length of the incision as in management of the axillary scar [4-8]. Juri [7] advocated a quadrangular flap with T-closure in the axilla, Guerrero-Santos described a Z-plasty [8], and Hurwitz [9] popularized the L-plasty for the massive weight loss patients. Other modifications included securing of the superficial fascial system as described by Lockwood [10] and aggressive liposuction of the skin to be excised to preserve lymphatics and avoid lymphoceles or edema as proposed by Pascal [11].

2. EVALUATION

The patient is evaluated in regards to the degree of excess adipose tissue and excess skin on the arm, axilla, and lateral chest wall. Liposuction alone is effective in patients with excess adipose tissue and good skin elasticity. Although short scar and other limited forms of brachioplasty have been described and are indicated in a select group of patients, almost all patients desiring improvement of the contour of the arms after massive weight loss require a brachioplasty extending into the lateral chest wall.

A small subset of patients present with full or over-inflated arms despite stabilization of weight loss. Treatment of the arms in these patients is staged with an initial liposuction followed by a subsequent excisional procedure six months later. A brachioplasty can be performed as an isolated procedure, although most weight loss patients choose to combine other procedures. The combination of mastopexy with brachioplasty is common and provides improved three dimensional contour as the tissue from the lateral chest wall can be used to augment and increase projection of the breasts [11].

Brachioplasty exemplifies a tradeoff of improved arm contour for scar. The resulting scar is very prominent and visible, often thickened and red for many months before maturing. It is very important to counsel patients in this regard prior to surgery

Upper extremity treatment zones [12] and the classification of brachial ptosis [13] aid in documenting the defects and operative planning (Tables **1** and **2**). Appelt *et al.* proposed a new classification system and described an algorithmic approach to upper arm contouring based on the presence on excess fat, skin laxity, and the location of excess skin [14] (Table **3**). Under this protocol, patients with excess adipose tissue are

---

*Address correspondence to Sadri Ozan Sozer:** Texas Tech University Health Sciences Center, El Paso, Texas, El Paso Cosmetic Surgery, USA; Tel: +915-351-1116; Fax: +915-351-8790; E-mail: doctor@elpasoplasticsurgery.com

Bishara S. Atiyeh and M. Costagliola (Eds)

treated with liposuction. Excess skin limited to the proximal one third of the arm are treated with limited medial brachioplasty, those limited to the entire upper arm are treated with traditional brachioplasty, and those with extension to the forearm and chest wall are treated with extended brachioplasty.

Table 1: Four Treatment Zones of the Upper Extremity Proposed by Strauch *et al.*

UPPER EXTREMITY TREATMENT ZONES	
ZONE I	**Forearm**
ZONE II	**Region between the olecranon and the anterior axillary fold**
ZONE III	**Defined by the borders of the axilla**
ZONE IV	**Subaxillary lateral chest wall**

Table 2: Classification of Arm Ptosis Based on the Vertical Distance Between the Brachial Sulcus and the Inferior Curve of the Arm Proposed by El Khatib.

CLASSIFICATION OF BRACHIAL PTOSIS	
GRADE 1	**Distance is less than 5 cm**
GRADE 2	**Distance is 5 to 10 cm**
GRADE 3	**Distance is more than 10 cm**

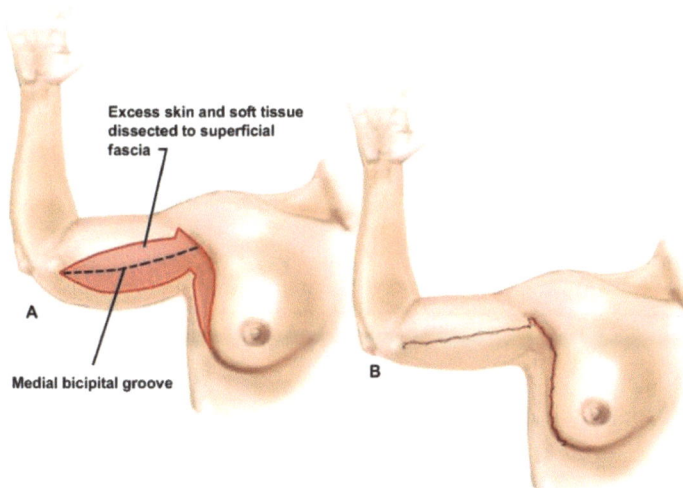

Fig. 1. Preoperative Markings.

3. MARKINGS

The markings are of utmost importance in brachioplasty (Fig. **1**). Although there is controversy over the ideal location of the scar placement, the authors prefer for the scar to be placed along the bicipital groove, as the scar is more difficult to be seen in this location and does not restrict range of motion at the elbow. Others prefer to place the scar posterior to the brachial groove at the inferior margin of the arm. The reasoning is that there is less gravitational tension and thus the scar is less prone to hypertrophy or widening. In this location the scar is not visible to the patient when looking in the mirror, but might be visible to someone looking at the patient from behind [12, 15, 16]. Regardless of the surgeon's preference, it is of utmost importance to discuss the scar location with the patient.

Table 3: Classification System Based on that Proposed by Rohrich *et al.* adapted from Applet *et al.*

OLD CLASSIFICATION SYSTEM		NEW CLASSIFICATION SYSTEM	
TYPE I	Minimal skin excess Moderate fat excess	I	Minimal skin excess Moderate fat excess
TYPE II	Moderate skin excess Minimal fat excess	IIa	Moderate skin excess Minimal fat excess Proximal arm
		IIb	Moderate skin excess Minimal fat excess Entire arm
		IIc	Moderate skin excess Minimal fat excess Arm and chest
TYPE III	Moderate skin excess Moderate fat excess	IIIa	Moderate skin excess Moderate fat excess Proximal arm
		IIIb	Moderate skin excess Moderate fat excess Entire arm
		IIIc	Moderate skin excess Moderate fat excess Arm and chest

The patient can be sitting or standing with the arms and forearms abducted at 90 degrees with the palms facing forward. The bicipital groove is marked first, extending from the medial elbow to the deltopectoral groove across the dome of the axilla, representing the intended scar placement location. The superior line of the incision is marked by pulling the skin downward to estimate the skin excursion under stretch. In general the superior line is 2-3cm above the bicipital groove line and tapers towards the ends. The estimated inferior line of incision is determined by pinch and is usually marked conservatively, as more tissue can be excised intraoperatively.

The distal extent of the area to be excised is proximal to the elbow joint. Extending the excision into the forearm results in a very visible scar, but is sometimes necessary in order to avoid a dog ear deformity. The proximal extension of the superior line is placed high in the dome of the axilla, creating a Z-plasty to avoid linear contracture of the axilla and continued inferiorly along the chest wall following the lateral contour of the breast. The amount of area to be excised in the lateral chest is also approximated using the pinch test. Symmetry in markings on each side are evaluated and modified as necessary.

The lateral chest wall marking can be continued with the lateral Wise pattern markings if performing a mastopexy at the same time (Fig. **2**). The anterior line marking of the chest wall excision is continued with the superolateral Wise pattern marking, and the posterior chest wall marking is continued with the inferolateral Wise pattern marking which correlates with the inferior mammary fold. The excess lateral wall tissue can be deepithelialized and used as a dermal fat flap to increase upper pole fullness and projection of the breast. Excision of excess back tissue can also be extended from the same markings (Figs. 2 and 3). Combining all of these procedures when indicated produces an improved three-dimensional contour when compared to performing each of these procedures separately.

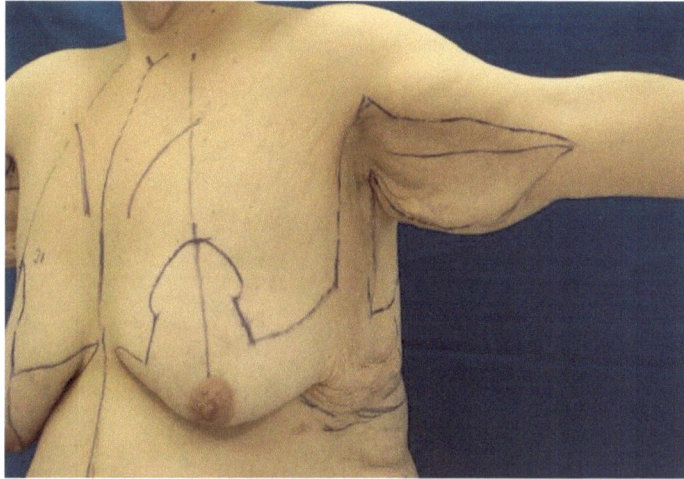

Fig. 2. Preoperative Markings Combining Brachioplasty with Mastopexy and Excision of Excess Back Tissue. Bicipital Groove is Marked with a Straight Line for Reference.

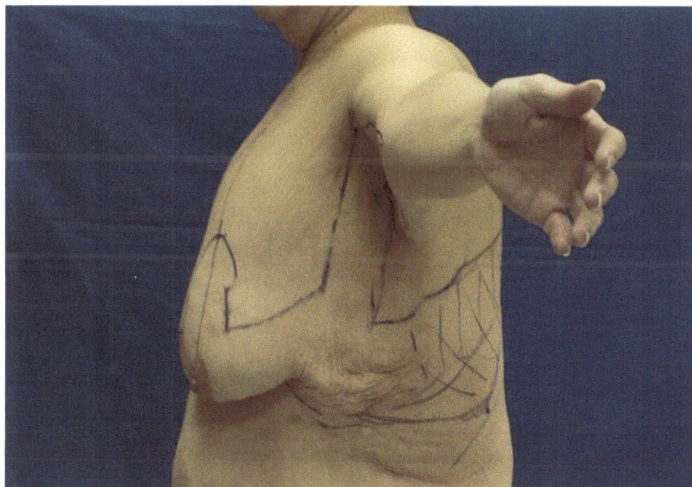

Fig. 3. Lateral View of Preoperative Markings Combining Brachioplasty with Mastopexy and Excision of Excess Back Tissue.

4. PROCEDURE

In the operating room, the arm is prepped and draped circumferentially in the standard surgical fashion giving exposure to the elbow, shoulder and lateral chest wall. The arms are abducted 80 degrees on the arm boards. Intravenous catheters can be placed distal to the elbow or on the lower extremities. Blood pressure cuff may be placed in the forearm or lower extremity. The area to be excised is infiltrated with tumescent solution through an incision at the level of the medial elbow. Extensive liposuction is performed in the area to be excised until most fat is removed. This delineates the dissection plane leaving behind a web of connective tissue and vessels between the skin and fascia, facilitating the excision of the tissue. Limited lipoplasty can be performed as necessary in other areas of the arm.

The superior incision with its extension into the lateral chest wall is made and it is carried down to a level above the brachial fascia. A flap consisting of skin and subcutaneous tissue is then elevated at this level. This dissection is simplified by the aggressive liposuction, which defines the plane and the tissue elevated is mainly skin. The depth of excision in the lateral chest wall is to the superficial fascial system and at the level of the axilla to the clavipectoral fascia without entering it.

After dissecting down to the previously marked inferior incision, the amount of tissue to be excised can be precisely defined by draping and cutting. This is done in a stepwise fashion, from distal to proximal. The amount of tissue excised is often less than the initially marked due to intraoperative edema of the arm. Excessive tension is to be avoided.

The incisions are then closed in layered fashion by approximating the superficial fascial system with an absorbable suture. The subdermal layer is then approximated with an absorbable suture and the skin closed with a subcuticular running absorbable suture.

We do not routinely place drains as there is not a large amount of dead space. The wounds are dressed and a pressure garment is applied. Efficiency in surgical time is critical when performing brachioplasty resections, especially if combined with other body contouring procedures. Examples of pre and postoperative photos are shown depicting typical improvements in arm contour achieved with this technique (Figs. **4-6**).

5. POSTOPERATIVE CARE

Use of the pressure garment is recommended for a period of four weeks. The patient is allowed to shower on the first postoperative day. The patient is advised to avoid heavy lifting for at least 2 weeks. Range of motion is not limited postoperatively although strenuous activity is discouraged.

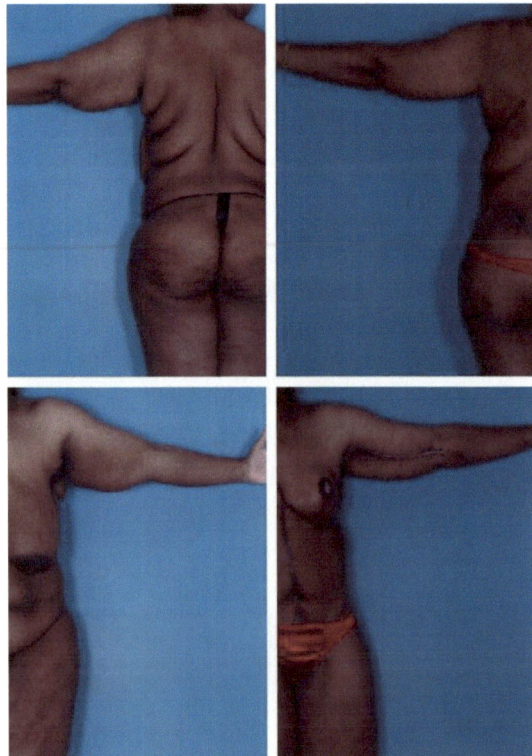

Fig. 4. Preoperative and Postoperative Photo Comparison Demonstrating Improved Arm Contour after Extended Brachioplasty Combined with Mastopexy in a Massive Weight Loss Patient. Note that Because of the Redundant Skin the Incision Extends Past the Elbow.

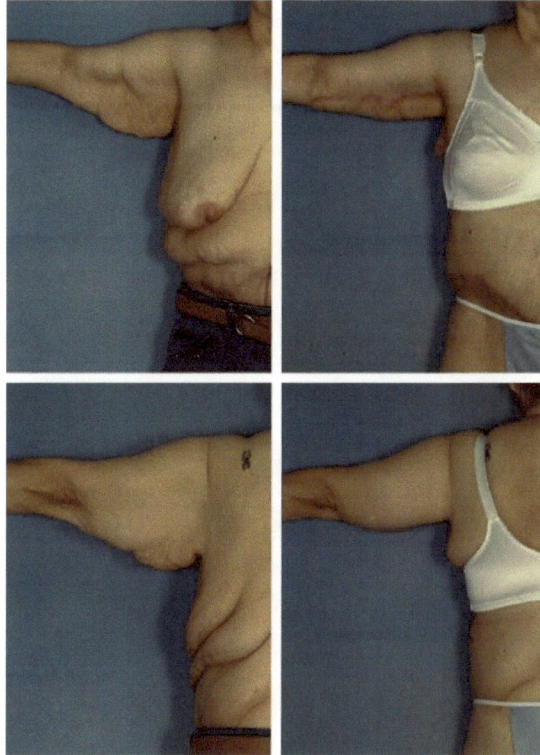

Fig. 5. Preoperative and Postoperative Photo Comparison Demonstrating Improved Arm Contour after Extended Brachioplasty Combined with Mastopexy in a Massive Weight Loss Patient.

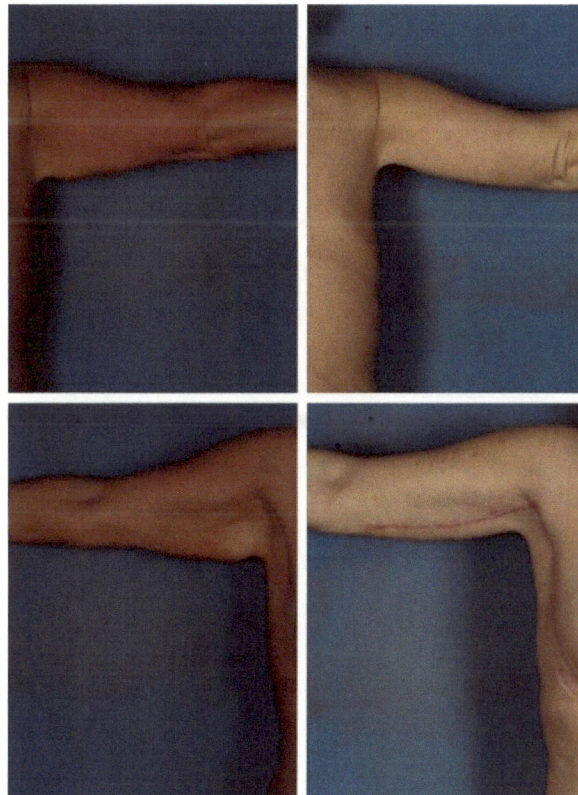

Fig. 6. Preoperative and Postoperative Photo Comparison Demonstrating Improved Arm Contour After Extended Brachioplasty Combined with Mastopexy in a Massive Weight Loss Patient.

6. COMPLICATIONS

Complications that can occur after brachioplasty include delayed wound healing, seromas, dehiscences, lymphoceles, bad scarring, infection, bleeding, nerve compression, neuromas, and sensory loss. Dehiscence can be avoided with good technique and tension free closure. This is usually a result of overresection of tissues. Seromas and lymphoceles in the arms present 4 to 6 weeks after surgery and typically resolve with aspiration and localized pressure. The use of liposuction of the area to be excised has dramatically decreased the incidence of lymphoceles in the literature and in our experience [10]. Sensory loss is traditionally limited to the area around the scar and resolves over a period of 6 months. Complete maturation of scars can take longer than one to two years in most patients. Some patients may require an additional more limited reduction in the future due to recurrence of excess skin due to the poor elasticity in massive weight loss patients.

Despite efforts to classify arm ptosis and formulate treatment algorithms based on zones and degree of ptosis and lipodystrophy [12, 13], unresolved problems of current brachioplasty techniques continue to include postoperative residual contour deformities resulting from skin resection errors with tissue over correction centrally and under resection proximally and distally, transverse cutaneous folds, hypertrophic scars, widened scars, and patient dissatisfaction with scar location [12, 17] (**Fig. 7**).

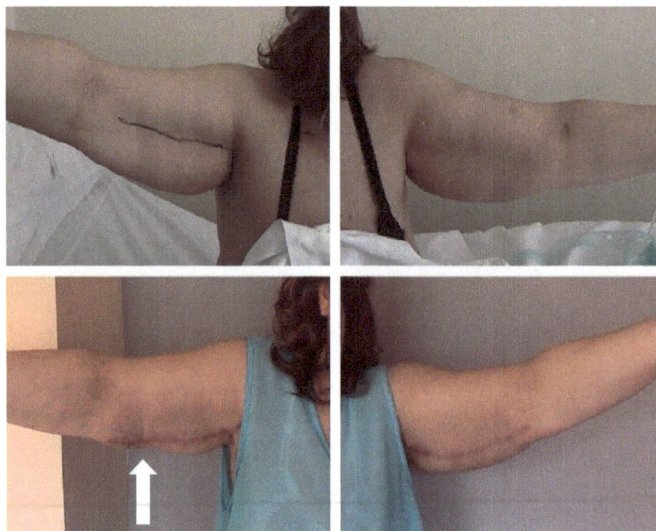

Fig. 7. Asymmetrical Improvement in Arms Contour Following L-Plasty with Tissue over Correction of the Right Arm Centrally and Under Resection Proximally and Distally and Overall Under Correction of the Left Arm.

7. CONCLUSION

With the current safety and popularity of weight-reduction surgery for morbid obesity, the number of patients presenting with severely ptotic skin and relatively little excess lipomatous tissue is increasing significantly. A growing number of patients are presenting for brachioplasty after MWL, however it is rarely performed alone and is frequently combined with other procedures [12, 18].

Contour deformity similar to a bat's wing that extends from the olecranon across the axilla to the chest wall is particularly challenging to the surgeon because of the long incision length needed to adequately resect the ptotic skin of the arm. Very often it requires crossing the axilla with the excision to correct skin laxity within the axilla and chest-wall region [12, 15]. Multiple techniques for upper arm rejuvenation have been described. Some may be somewhat unpredictable and commonly associated with significant untoward results, and none is completely satisfactory for all deformities [12, 17]. Brachioplasty is not a simple technique. It requires a long learning curve, and is strongly operator dependant; thus, a good experience in body-contouring plastic surgery is needed [19].

REFERENCES

[1]   Correa-Inturraspe M, Fernandez JC. Dermolipectomia braquial. Prensa Med Argent 1954; 34: 24.

[2]   Pitanguy I. Correction of lipodystrophy of the lateral thoracic aspect and inner side of the arm and elbow. Clin Plast Surg 1975; 2: 477-83.

[3]   Baroudi R. Dermolipectomy of the upper arm. Clin Plast Surg 1975; 2: 485-91.

[4]   Aly A. Brachioplasty in the patient with massive weight loss. Aesthet Surg J 2006; 26: 76-84.

[5]   Strauch B, Linetskaya D, Baum T, *et al.* Brachioplasty and axillary restoration. Aesthet Surg J 2004; 24: 486-8.

[6]   Juri J, Juri C, Elias J. Arm dermolipectomy with a quadrangular flap and "T" closure. Plast Reconstr Surg 1979; 64: 521-5.

[7]   Guerrerosantos J. Brachioplasty. Aesthet Surg J 2004; 24: 161-9.

[8]   Hurwitz DJ, Holland SW: The L brachioplasty: An innovative approach to correct excess tissue of the upper arm, axilla and lateral chest. Plast Reconstr Surg 2006; 117: 403-11.

[9]   Lockwood T. Brachioplasty with superficial fascial system suspension. Plast Reconstr Surg 1995; 96: 912-920.

[10]   Pascal JF: Brachioplasty. Aesthet Plast Surg 2005; 29: 423-29.

[11]   Hurwitz DJ, Agha-Mohammadi S. Post bariatric surgery breast reshaping: The spiral flap. Ann Plast Surg 2006; 56: 481-86.

[12]   Strauch B, Greenspun D, Levine J, *et al.* A technique of brachioplasty. Plast Reconstr Surg 2004; 113: 1044-48.

[13]   El Khatib HA. Classification of brachial ptosis: strategy for treatment. Plast Reconstr Surg 2007; 119: 1337-42.

[14]   Appelt EA, Janis JE, Rohrich RJ. An algorithmic approach to upper arm contouring. Plast Reconstr Surg 2006; 118: 237-46.

[15]   Aly A, Soliman S, Cram A. Brachioplasty in the massive weight loss patient. Clin Plast Surg 2008; 35: 141-7.

[16]   Selinger R. The posteromedial brachioplasty. Ann Chir Plast Esthet 2008; 53: 480-6.

[17]   Cannistra C, Rodrigo V, Benelli C, *et al.* Brachioplasty after massive weight loss: a simple algorithm for surgical plane. Aesth Plast Surg 2007; 31: 6-9.

[18]   Gusenoff JA, Coon D, Rubin JP. Brachioplasty and concomitant procedures after massive weight loss: a statistical analysis from a prospective registry. Plast Reconstr Surg 2008; 112: 595-603.

[19]   Migliori FC, Ghiglione M, Alessandro GD, *et al.* Brachioplasty after bariatric surgery: personal technique. Obes Surg 2008; 18: 1165–69.

CHAPTER 15

Correction of Facial and Neck Excessive Laxity

Mutaz B. Habal[*]

Tampa Bay Craniofacial & Plastic Surgery Center/USF, Tampa, Florida 33647-1164, USA

Abstract: The face is the window of the person to the surrounding world and the eyes are the avenue to his soul. Having started with this concept that we have in craniofacial surgery, we can look then at the patients in our daily practice and see how this part of the body presents the most important aesthetic unit that needs to be corrected especially after bariatric surgery and massive weight loss. As more patients become aware of the obesity issues, they will seek more help to put them. Their psyche and their bodies on track by cutting off the fat, they have accumulated over the years some from teen years. After the loss of adipose tissue, patients focus on the remaining excessive skin they have developed over the obese years. The part that people look at especially in a conservative society is the face and secondarily the neck when they are well dressed and mingling with friends and acquaintances. The face is divided into the upper components and the lower components engulfing the neck and jaws. The face is the first to show the reflection of obesity and the last to show the effect of changes in body fat composition. Contouring the face is part of the plastic surgery principles to complete the changes after the patient departs from the obesity stigma. Complications are minimal when the surgeon abides by the basic principles and rules of plastic surgery.

Keywords: Face and neck in morbid obesity, face deformity, counter plasty, neck plasty.

1. INTRODUCTION

It is imperative in a discussion about bariatric surgery and contour abnormalities to have a part on the correction of the facial contour deformities. The patients with morbid obesity come consulting the plastic surgeon. Common encounters seen in practice are the patient referral for suction lipoplasty in the hope of reducing weigh, and correct the abnormal adipose deposits. In fact, patients who have tried in vain everything from dieting on will need to be referred to a bariatric center where the appropriate operative procedures are done. They usually re-consult when the weight loss stabilizes asking for the help. The abdomen, in the majority of patients, requires attention because the panniculus interferes with daily functions and personal self esteem. We proceed with all other skin problems and then resort to the correction of the face and neck skin lastly. Such patients are usually in the older age group, however, it is hard to pin point a specific age group because it varies on how much skin care and solar radiation have produced its benefit or toll on the skin.

2. ANATOMICAL CONSIDERATIONS

Considering all the facts, the anatomical units of the face are essential to be considered before any approach to these sites is done surgically. The face and the neck are separate aesthetic units so that the approach can be well considered separately. To start with, we will look at the forehead with the brow ptosis, part of which is skin laxity the other part being gravitational sliding of the forehead over the patient's galea aponeurotica and the frontalis muscle. These are nevertheless lumped together in the corrective maneuvers used. Such changes following bariatric surgery are minimal since the adipose tissue deposits in that anatomical part are minimal compared to the abdomen and other parts of the body. Secondly we look at the eyelids; as noted by many, eyelids changes following weight loss are different from those due to starvation. The peri-orbital fat is significantly reduced in the starvation weight loss, while after bariatric surgery its effect may be to a less drastic extent. The eyelid skin folds to a lesser extent than what is seen in patients with blepharochalasis and blepharospasm, primarily a skin fold problem.

Address correspondence to Mutaz B. Habal: Tampa Bay Craniofacial & Plastic Surgery Center/USF, Tampa, Florida 33647-1164, USA; Tel: +1-813-238-0409; Fax: +1-813-238-1119; E-mail: mbhabal@verizon.net

Bishara S. Atiyeh and M. Costagliola (Eds)

The cheek fat changes are mainly gravitational and the pouch fat pad with a buccinators hollows as dimples follow the fat pad loss. These anatomical changes are minimal in the central components of the face and maximal in the periphery near the jaw lines where they become more redundant and gravitational. In the central part of the face, the bony prominences are always attached to the skin making it difficult to have any adipose tissue deposits in the central facial anatomic units. The neck as a unit will come last and the problem with the neck is merely gravitational. Part of the massive weight loss issue is the loss of adipose tissue in the sub-mental area. However, due to obesity, fibrous attachments between the neck muscles and the skin in most patients are lost, so the skin by itself is already loose before the massive weight loss starts taking effect on the neck.

Putting all those anatomical issues together will make the discussion of the surgical procedures, which are the traditional procedures, easy to communicate in a discussion on massive weight loss.

3. SURGICAL CONSIDERATIONS

The surgical corrections of the face and neck will be complete if we consider each one as an anatomic site by itself. For that we have divided the different operations that will be needed in such a fashion that we will call them surgical units. Each unit can be done alone or in combinations, as a matter of fact they can all be done together. It is preferable in our experience if the units under considerations are all to be done as a coverable items by the third part payers, then we will do the two procedures together. However, sometimes it is the preference of the patient or the preference based on realistic expectation or monetary reason that we perform what is desired by the patient and leave the rest to a later date [1].

Forehead

The initial surgical approach was to go ahead with the full scale surgical treatment that will include a hairline incision that extends from the temple to the temple. The incision line usually heals very well leaving in the area a faint white line. The dissection is done in a supra periosteal level freeing the supra orbital attachment so that the skin can be pulled back and the wrinkles flattened as well as the brow ptosis corrected. However this traditional approach is now replaced with a simpler and less invasive method where as the forehead is dissected well through small incisions in the hair bearing area and suspension sutures are placed. The suspension suture may be anchored with a screw placed in the skull adequately correcting brows ptosis and forehead wrinkles. Since we do so many of these procedures, we found in our practice that this is a very helpful technique if not the technique of choice. Some still feel that the use of endoscope is necessary to complete the supra-orbital separation and the freeing of the frown muscles. Our technique is less complicated. The choice of the proper technique will always be a personal preference as we look at the neck and face in conversational distance Fig. **1** and laterally Fig. **2**.

Upper Eyelids

The upper eyelids blepheroplasty is the classical and traditional way in which the excessive skin is removed in a lid fold incision. Due to the thin nature of the skin and its pliability, this incision is hardly noticeable in few months. Fat removal usually is not needed since massive weight loss patients have lost a lot of fat and to remove further fat will produce a hollow orbit and will not be aesthetically appropriate.

Lower Eyelids

The lower eyelids follow the same traditional removal of the excessive skin and the avoidance of removing any further fat pads as with the massive loss of weight these fat pads in and around the orbits will be the first to go. It is recommended that in massive weight loss patients the surgeon always avoids any further removal of the peri-orbital fat. This will help avoid the production hollow orbits in these patients.

Cheeks

There is always some improvement to be done to the lower part of the cheeks where the skin is redundant. A short scar face lift will be indicated as it will be the only way to remove the extra skin. If the patient is having a simultaneous neck lift, this will be a combination maneuver to correct both issues. Combing the face lift with a neck lift will be one approach that will result in a happy patient at the end.

Jawels

The use of the Hammock maneuver which consists primarily in placing a suspension suture of PDS from the strap muscles to the tip of the mastoid periosteum for great anchorage, thus recreating the 90 degrees triangle as seen in Fig. **2** in combination with neck lift as described by the group from Mexico will be one way to correct the problem by removing the excessive skin from the lower face particularly the jawels and cheeks. The other way to correct the described problem will be with the neck lift. Either way, the jawels presence is not aesthetic and removing its redundant skin will help the patient aesthetic configuration of the face.

Neck

The neck is considered the center of beauty and attraction by most cultures particularly so in the female patient. The beauty triangle between the two nipples to the chin is considered as the center piece of the female sexuality; it is almost always exposed and it is always complemented by the attire in daily fashion (Fig. **1**). Aesthetic contour deformity of this triangle following massive weight loss can be categorized into three different neck types.

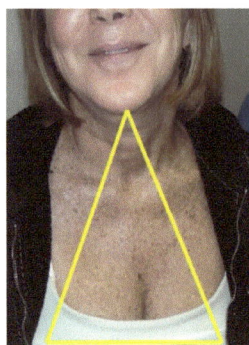

Fig. 1. Beauty Triangle Depicted in Yellow, This Area is Usually Exposed by Attire in Most Cultures.

Type I Neck: Is that of the slim neck that has lost all the supporting fat and what is left is primarily redundant skin that is well depicted in front of the ears and in the lower part of the neck. Gravitational forces change the position of the skin as well seen in this patient (Fig.**2**). The yellow triangle is to demonstrate that the cervical-mandibular angle is still within normal configuration; the problem is specifically the skin. Removing the redundant skin will alleviate the problem in these patients. We do not need to remove sub-fascial fat or sub-muscular fat that does not require attention or carving. Patients are concerned about the skin redundancy. We do not do or recommend placing incisions on the anterior part of the neck or violating the beauty triangle of the patient. The skin tightening is done through a pre-auricular classical incision; after excising excess skin, the neck skin is lifted in a vertical vector with two anchoring suspension sutures in the traditional fashion.

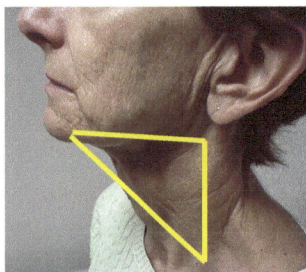

Fig. 2. Type I Neck – Slim Neck.

Type II Neck: Is the fat neck; even though the patient has lost a massive amount of adipose tissue the neck remains fat (Fig. 3). The fat in these patients is sub-muscular in two compartments. Under the platysma muscle, sub-platysma and under the strap muscles, sub-strap muscles. The treatment is the same as in Type I neck as far as the skin is removed but a supplemental incision is done under the chin in the crease for better access to remove the excess fat in the neck and tighten the muscles in a traditional fashion (midline plication).

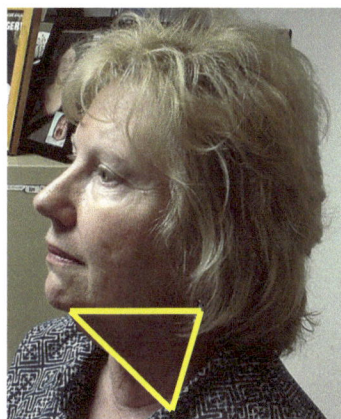

Fig. 3. Type II Neck – Fat Neck.

Type III Neck: Is the fat big and inherent neck, as a complex problem; there is redundant skin but minimal, seen mostly in men (Fig. **4**). The treatment requires removing the adipose tissue which is mostly sub-muscular and tightening the strap muscle first then the interdigitation of the platysma muscles, for complete rejuvenation of the neck that will allow the 90 degree angle to be reformed in the center of the neck avoiding a vertical scar. This can all be done through a sub-mental incision with good lighting equipment for adequate visualization from the small incision. Minimal skin is removed through that incision. The midline muscles that are separated in the center of the neck are approximated in the midline. Moreover, the patient wears a neck brace in the post operative period as a soft collar at night to help the skin adhere to the underlying structures avoiding bow stringing of the neck skin.

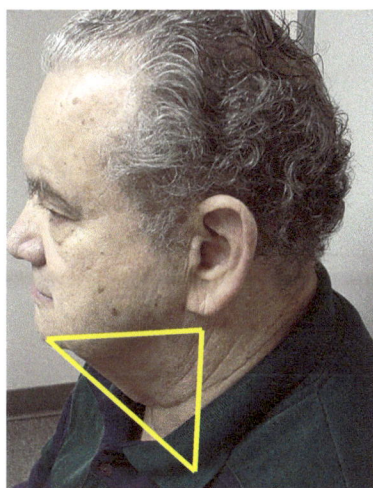

Fig. 4. Type III Neck – Complex.

The adipose tissue that is removed is in such a large volume that will make it necessary to place a suction drain to avoid the dead space effect and prevent as well the bow stringing of the neck skin due to gravitational forces. Rejuvenation of patients with weight loss is well seen in the example of the patients in Fig. **5** illustrating adipose tissue removal. This fat was residual adipose tissue and was all sub-muscular. It has necessitated removal via a small sub mental incision.

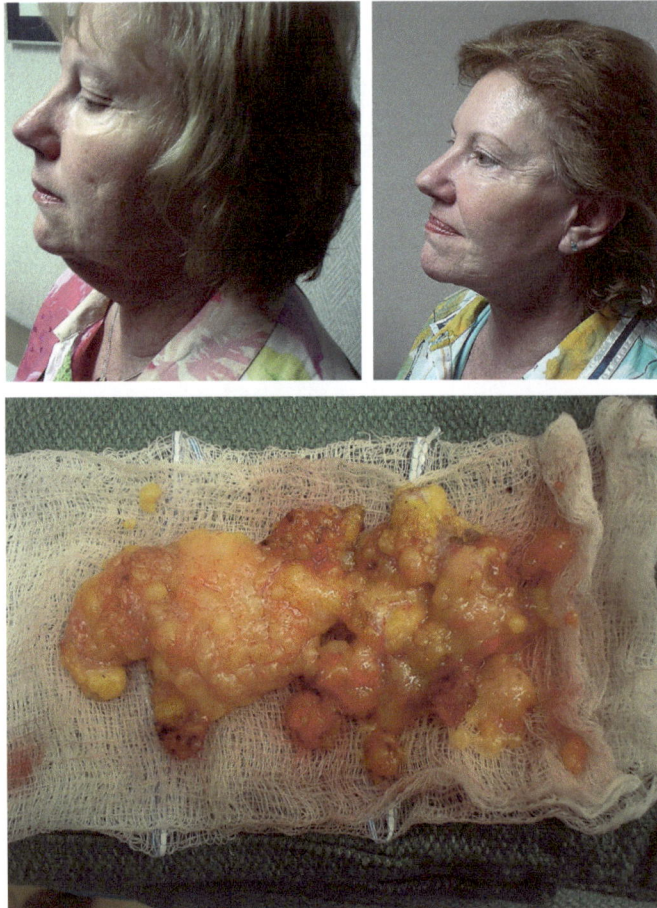

Fig. 5. An Example of Adipose Tissue Removed from a Patient.

Total rejuvenation is also illustrated in this patient with neck and face tightening of the skin (Fig. **6**).

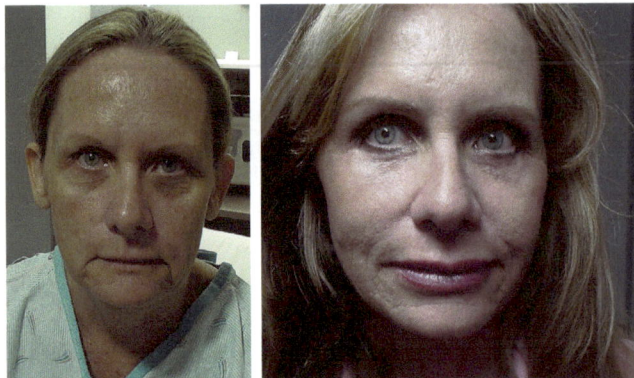

Fig. 6. Before and After Achieving a Youthful Look.

Results of the reparative procedures last for a long time since. We see patients several years later with total stability and maintenance of the corrected configuration (Figs. **7** and **8**).

4. SURGICAL MISHAPS

The most common surgical mishaps seen in patients with contour correction in the head and neck area are similar to the rest of the other procedures that is infection and bleeding. Good homeostasis is paramount in any procedure with morbid obesity correction. The next step is adequate placement of drains that are kept a short period postoperatively. There are also two measures that we have learned in the last ten years; the first is the use in any procedure in that region of Platlet-Rich Plasma (PRP) in the form of autologous blood. The second is that of controlled hypotensive anesthesia during the procedure, maintained up to the application of the pressure dressing. With these measures we do not need to use the massive blood replacement that was common in the early seventies when these procedures became a common surgical treatment in general hospitals. Today, due to these advances and measures that are used routinely, very rarely any patient undergoing massive surgery requires blood replacement. Patients can go home with a low hematocrit and will heal in an expedited well controlled way.

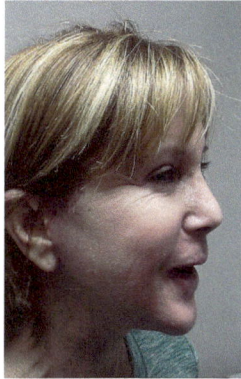

Fig. 7. Patient 3 Years after Surgery. Stable Neck.

Fig. 8. Patient 10 Years After Surgery on the Fat Neck.

Prophylactic antibiotics are the major reason for the total reduction in post operative wound infection, to reduce SSI Surgical Site Infection (SSI) or deep wound infection. These infections may happen when a large segment of tissue gets devitalized, then and only then more wound care measures are instituted in a traditional fashion.

5. DISCUSSION

In the last decade, body contouring has reached a most important stand in the surgical arena as a sophisticated procedure for patients who are suffering from morbid obesity. After bariatric surgery, patients consult the plastic surgeon seeking help in getting rid of the excessive skin they have grown which drapes body parts and causes loss of normal and aesthetic anatomical landmarks. Men and women want to look at their best when undressed. We have declared in the health system that obesity is a disease in our country so its treatment becomes a coverable item for third part payers. In the USA, disregarding the age and general condition of the particular patient, men and women can get the proper obesity treatment needed. Such treatment is now a covered procedure by all the health systems and patient thus receives all the benefits [2].

The morbid obesity procedures end result and consequences of massive weight loss with sudden disappearance of the large volume of the cushioned fat between the skin and the underling tissues result in loss of the normal skin elastic recoil thus the skin becomes lax and redundant and looses all its configurations. Usually, all approaches of skin surgery are dependent on skin elasticity. However, following bariatric surgery and massive weight loss, we have to deal with two parameters: redundancy and loss of elasticity. These will dictate the main basic principles for the surgical correction of the patient's condition after bariatric surgery and massive weight loss. Few other principles that will need to be addressed by the operating surgeon need to be adhered to so as the procedures are performed adequately to produce an acceptable outcome.

A first such principle that must be noted is the fact that the face is the first part of the body to puff out with morbid obesity and weight gain and the last to show the effect of weight loss due to the greater elasticity of the facial skin. That specific thin and elastic skin is different than the skin in other parts of the body. This is to protect it from the daily gravitational forces and environmental factors during the regular activities especially to face and neck. A look at an example of these biological and anatomical differences, we should look at the abdomen panniculus that showed up first after weight loss, wrinkly accordion skin, gravity pulls it down and have no form or substance. The second principle is that the face has two components, the static anatomic unit of the upper face and the dynamic unit in the jaws which are in constant motion with mastication, deglutition and speech during wakeful hours of the individual. The neck on the other hand comes second in its three dimensional movements which, without the support of the adipose tissue, tends to be pulled down by the gravitational forces. Neck skin is less elastic and thinner than the other facial skin with less pilocebacious and adnexal units. The third principle is that with weight loss, effect of gravity on the neck becomes most apparent on the outside uncovered skin and the extra skin needs to be addressed separately only for biological and anatomical needs, to regain the natural configuration in the face and neck as part of the facial contour surgery.

In the description of the surgical approach, we have discussed facial units separately considering the fact, however, that we must always look at the face as the window of the patient to the outside world as we must always consider that fact of the eyes are the avenue to the soul. The patient seeks attention of the surgeon to address these issues and correct the deformity that ensues. They will come also looking for a solution to the problems ensuing after their massive weight loss and point to the eyelids and the neck as a source of unhappiness. These are the apparent anatomical components that need some help. Refusal to cover these items by the USA health system is not always encountered [3].

It must be noted also that what is seen following massive weight loss is not similar to starvation noted and discussed during the great wars where individuals were subjected to the atrocities of war.

Having said all that, anatomical sites needing surgical corrective maneuvers are addressed according to priorities. As noted before, in an obese patient, patients' desire for correction of the neck and face is in the forefront. Following massive weight loss when the patient reaches the stage of surgical contour correction and is left helpless, the neck and the face take the toll while they were the first priority when bariatric surgery and weight loss were initiated. Most patients elect to correct abdominal contour deformity first since the abdomen is the part that interferes with function, drapes the genitals, subject to foul smell and skin maceration making the need before the appearance a must.

6. SUMMARY

In summary facial contouring and neck lift are the essential parts of the total picture of body contouring surgery. These components of the face and neck represent the most important body image of the patient and should be addressed very early as part of the treatment plan. Even if coverage by the health care system is denied, correction of face and neck contour deformities following massive weight loss may be considered first since excessive upper lid skin, particularly in the elderly, may interfere though not so frequently with the neuro-visual transmission representing both functional and aesthetic concerns. The fact to be considered as well is that these elective operations are surgical in nature and may result sometimes in mishaps requiring attention such as infection hematomas, and lastly unhappy patients with unrealistic expectations. Due to the nature of the condition and sometimes ensuing diabetes, prevalence of infective complications and hematoma formation may be high.

REFERENCES

[1]   Hurwitz DJ, Rubin JP, Risin M, *et al.* Correcting the saddlebag deformity in the massive weight loss patient. Plast Reconstr Surg 2004; 114: 1313-2.

[2]   Song AY, Jean RD, Hurwitz DJ, *et al.* A classification of contour deformities after bariatric weight loss: the Pittsburgh Rating Scale. Plast Reconstr Surg 2005; 116: 1535-44; discussion 1545-46.

[3]   Bassetto F, Vindigni V, Scarpa C, *et al.* Use of oxidized regenerated cellulose to stop bleeding after a facelift procedure. Aesthet Plast Surg 2008 32: 807-9. Epub 2008 Jul 1.PMID: 18592302.

CHAPTER 16

Anesthesia for Contouring Body in Post-Bariatric Surgery Patient

Marwan S. Rizk*, Cynthia J. Karam and Chakib M. Ayoub

Department of Anesthesiology, American University of Beirut Medical Center, Beirut, Lebanon

Abstract: Obesity is a major health problem throughout the world. Its management by bariatric surgical interventions is associated with a good success rate, leading to a necessary removal of excess skin after weight loss. The popularity of bariatric surgery has created this new patient population characterized by major anatomical and physiological changes with Massive Weight Loss (MWL) who still maintains many of the co-morbidities of obesity. As a part of a multidisciplinary management, anesthetic management of these patients should take the specific problems into consideration associated with obesity and optimize them before surgery. All these factors such as physical changes, co-morbid medical conditions (pulmonary, cardiovascular and metabolic changes such as vitamin deficiencies *etc.*), airway difficulty, propensity for venous thrombosis, embolism, patient positioning, avoidance of hypothermia, pharmacodynamics and pharmacokinetics of anesthetics and changes related to bariatric surgery as well are all addressed in this chapter.

Another important aspect that is addressed is the role of regional anesthesia and its advantages over general anesthesia for these patients, taking into consideration the difficulty in palpating bony landmarks or even identifying the midline because of skin laxity and residual excess adipose tissue and therefore neuroaxial blockade could be challenging.

Finally the postoperative management is optimized by development of clinical pathways involving the surgeon, anesthesiologist, patient and support staff. Although obvious physical and physiological improvement may be attained by such surgical procedures, the patient must be fully informed of the multiple risks and complications inherent in such procedures.

Keywords: Anesthesia in obese patients, body contouring, post massive weight loss surgeries, regional anesthesia and landmarks, physiology in post MWL patients, phramacodynamics in post MWL patients.

1. INTRODUCTION

Obesity is a major health problem throughout the world. Its management by dietary restriction alone is associated with a poor success rate [1]. Laparoscopic adjustable gastric banding, vertical banded gastroplasty, as well as other techniques (e.g., Roux-en-Y or biliopancreatic gastric bypass) have been shown to decrease excess weight by more than 50%, leading to a necessary removal of excess skin after the weight loss [2]. While the anesthetic techniques for bariatric surgery are well described, to the best of our knowledge, there are no guidelines outlining the anesthetic management of patients with major weight loss following bariatric surgery [3, 4].

A new type of patient population characterized by major anatomical and physiological changes has emerged and their anesthetic care has become a challenge for the anesthesiologist.

This special group of patients with exaggerated weight loss in relatively short periods of time seeks the help of plastic reconstructive surgeons to improve their deteriorated and at times grotesque body image. In fact, 84.5% of patients losing weight through gastric bypass request body contour surgery. Gusenoff *et al.* showed that of the 11% of patients who underwent body contouring following gastric bypass, 47% had multiple operations [5, 6]. According to the authors, the lower incidence of body contour surgery recorded in this study is attributed to a low socioeconomic factor.

*Address correspondence to Marwan S. Rizk:** Department of Anesthesiology, American University of Beirut Medical Center, Beirut, Lebanon; Tel: +961-3-224940; E-mail: mr04@aub.edu.lb

Bishara S. Atiyeh and M. Costagliola (Eds)

The body dysmorphisms suffered by these patients are related to excess skin as well as subdermal fat accumulation in such areas as the face, neck, breasts, torso, abdomen, pelvis, gluteal region and extremities. The folds of excess skin become damaged by intertrigo, hyperkeratosis, mycosis, acanthosis nigricans and cellulitis.

These body changes interfere with daily activities, provoke isolating behaviors and frequently cause awkwardness in the sexual relationships of these patients. These changes negatively impact the patient's quality of life and, if appropriate management is not instituted, they may fall into old behaviors which will lead them back to extreme or morbid obesity, a vicious cycle characteristic in such patients. In contrast, when these patients undergo body contour surgery, their quality of life has been shown to change significantly: better personal, social and cultural adaptation, increased productivity and changes in their state of health [7].

2. CO-MORBIDITIES OF THE PREVIOUSLY MORBIDLY OBESE PATIENT

The evaluation and management of the previously obese should be integral, involving various specialists. When evaluating these patients it is important to remember that, while one group of patients will experience improvement or even resolution of the comorbid conditions that accompany obesity [8], a variable percentage of the patient population will continue to suffer the consequences of these same co morbidities, as well as the side effects associated with bariatric surgery and, moreover, must face long term aesthetic reconstructive procedures. One study of 152 French patients subjected to gastric banding, in whom the BMI dropped from 44.3 to 29.6, showed that diabetes mellitus remitted in 71% of the patients, arterial hypertension remitted in 33%, and sleep apnea in 95% of cases [9]. Table **1** lists the most frequent co morbidities that need to be studied in previously obese patients.

Table 1: Co-Morbidities Following Massive Weight Loss.

Unresolved co-morbidities following massive weight loss
Arterial hypertension
Arteriosclerosis
Varicose veins
Chronic obstructive pulmonary disease
Pulmonary hypertension
Type 2 Diabetes
Hyperlipidemia
Degenerative osteoarthritis
Chronic anemia
Cholelithiasis

3. OBESITY AND CO-EXISTING DISEASES

Anatomy and physiology change in obese individuals. Obesity, especially morbid or super obesity, involves diverse organs and systems, in such a way as to represent a complex pathological entity that challenges anesthetic and surgical procedures. The pulmonary, cardiovascular and metabolic changes are by far the factors that most increase the surgical and anesthetic risks. Other changes involve the airway and the propensity for venous thrombosis and, therefore, the risk of embolism. Table **2** lists some of the pathologies associated with or facilitated by excess weight.

Table 2: Obesity Related Pathologies.

Pathology associated with obesity
Insulin resistance, type 2 diabetes mellitus
Less time off work
Decreased number of general anesthesia inductions
Colon, endometrial, renal, gallbladder cancers
Hyperlipidemia
Infertility, menstrual changes
Coronary artery disease
Complications of pregnancy
Arterial hypertension
Gallstones
Congestive heart failure
Gastroesophageal reflux
Pulmonary hypertension
Fatty liver
Cerebrovascular disease
Osteoarthritis
Obstructive sleep apnea
Urinary incontinence
Hyperuricemia, gout
Depression
Premature death

Endocrine Changes

These are related to peripheral insulin resistance, as well as diminished concentrations of growth hormone, lower testosterone levels in morbid obesity, changes in ovarian cycles and early menopause. Diabetes type 2 and insulin resistance are known to react favorably to weight reduction. Hypothyroidism must be evaluated as a cause of overweight [10, 11].

Cardiovascular Changes

Obesity is a risk factor for cardiovascular diseases. The common co-morbidities of obese patients, such as arterial hypertension, diabetes mellitus and hyperlipidemia, further increase this risk. Where some investigators have concluded that even small (2-3kg) net weight losses might be beneficial for the control of hypertension, Sjöström *et al.* [12] showed that over a long period, not even a maintained 16% weight loss is sufficient to achieve a reduction of the 8-year incidence of hypertension in the severely obese [13].

On echocardiography, morbidly obese patients, after undergoing bariatric surgery, present a reduction in the arterial blood pressure and beneficial structural modifications to the heart, with decrease of the Left Ventricular (LV) hypertrophy translated as the reduction in the LV relative thickness, LV mass and LV mass index (LVM/height$^{2.7}$). They also present improvement in the diastolic function and an apparent beneficial effect on the systolic function.

Obesity-hypoventilation syndrome, which in its most drastic presentation is known as Pickwick's syndrome, is present in 10% of extremely obese patients. It involves chronic daytime hypercapnia and hypoxemia, polycythemia, hypersomnolence, with pulmonary hypertension and right heart failure secondary to increased right ventricular afterload [14].

Fortunately, the majority of obese patients with massive weight loss experience improvement of these cardiopulmonary conditions. However, there is a tendency toward cardiac rhythm changes, probably facilitated by diverse factors such as cardiac hypertrophy, hypoxemia, hypokalemia (due to the use of diuretics), coronary artery disease and a hyperadrenergic state.

Respiratory Changes

Changes in pulmonary function are attributed to an increased intraabdominal pressure, as well as the mechanical effects of excess weight of the thoracic wall, resulting in reduced vital capacity, inspiratory capacity, total lung volume and functional residual capacity. These changes are accompanied by closure of small airways and changes in ventilation/perfusion [15]. These patients have increased metabolic demands due to an excess of metabolically active adipose tissue, and an increase in the consumption of oxygen as well as the production of CO_2. This means they require greater minute ventilation, which may further increase oxygen consumption. A reduced functional residual capacity, in conjunction with an elevated rate of oxygen consumption, results in reduced time to desaturation during apnea produced during anesthetic induction and endotracheal intubation.

In contrast to Obstructive Sleep Apnea Syndrome (OSAS), Obese Hypoventilation Syndrome (OHS) is commonly seen in aged obese patients with high BMI. It is characterized by a daytime derangement of arterial blood gas values, severely restricted lung function and arterial oxygen desaturation during sleep in addition to a high pulmonary artery pressures and pulmonary vascular resistance.

Teichtahl *et al.* showed that 58% of the OHS patients, 36% of the COPD plus OSAS patients, and 9% of the pure OSAS patients had pulmonary hypertension defined as mean pulmonary artery pressure > 20mmHg [16]. The increased respiratory system elastance and flow-resistive loads seen in obesity do not correlate with the degree of daytime hypercapnia, and some patients with OHS can voluntarily hyperventilate and normalize their $PaCO_2$ when requested to do so. These results emphasize the role of respiratory muscle fatigue in patients with OHS [17, 18]. Ventilatory control is abnormal in OHS patients, with blunting of both hypercapnic and hypoxic ventilatory responsiveness. Reduced chest wall compliance seen in OHS patients can lead to increased energy cost of breathing and reduction in inspiratory muscle strength, maximum voluntary ventilation, and maximal inspiratory pressures [19]. Body weight per se does not correlate with chronic daytime hypercapnia, though weight loss in OHS patients can reverse daytime hypercapnia [20, 21]. Also, weight loss in OHS patients is associated with increased maximum voluntary ventilation, FVC, and reduced PaCO2, with little change in respiratory system compliance [22, 23, 20]. Pankow *et al.* [24] showed that noninvasive positive-pressure ventilation unloads the inspiratory muscles in patients with OHS.

Gastro-Intestinal Changes

Patients with gastric bypass have a reduced gastric capacity, with diminished intestinal absorption, and suffer from iron, folate, vitamin K and B_{12} deficiencies. A study of 30 Mexican patients taking multivitamin supplements found that 54.5% still had iron deficiency, 27.3% had cobalamin deficiency and 63.6% had anemia 3 years after Roux-Y gastric bypass. This study did not reveal iron deficiency. Other deficiencies found in bypass patients include vitamin A (11%), vitamin C (34.6%), 25-OH vitamin D (7%) and vitamin B_6 (17.6%) [25, 26]. Patients who had bariatric surgery appear to be at higher risk of pulmonary aspiration on induction of anesthesia than obese [27]. This frequent incidence (6%) may be real, but it may also reflect increased concern by clinicians about aspiration in obese patients.

The risks of pulmonary aspiration may be related to the physiological modifications induced by bariatric surgery. Several studies have suggested that esophageal-gastric peristalsis is altered after gastric banding [27-29]. Moreover, lower esophageal sphincter relaxation impairment is also reported [28, 30]. After vertical banded gastroplasty, a decrease in basal lower esophageal sphincter pressure and an increase in acid reflux were also observed [31]. These changes may increase the risk of esophageal regurgitation and bronchial aspiration during general anesthesia, increasing the possibility of aspiration pneumonia and long-term pulmonary complications [32, 33].

Most patients should receive histamine H2 antagonists and prokinetics, thus decreasing the gastric pH and consequently the risk of the pulmonary aspiration [34-36].

Jean *et al.* [27] compared 66 post bariatric patients (gastric banding and vertical gastroplasty) with 132 controls and found that, among the former group, pulmonary aspiration was significantly increased (4 patients, 6%, $p<0.006$). All four cases of aspiration belonged to the gastric banding group. Despite dramatic weight loss, these patients should be considered at risk for pulmonary aspiration and managed accordingly. When this factor is combined with difficult intubation, induction of anesthesia is associated with very high risk.

Airway Changes

Neither absolute obesity nor increasing BMI was associated with problematic intubation in morbidly obese patients. Therefore, the degree of obesity or neck size that justifies interventions such as elective awake fiberoptic bronchoscopy for intubation remains unknown [37]. Problematic intubation is predicted when a thyromental distance of less than 4 cm is associated to a Mallampati score of ≥ 3.

Renal Changes

After the drastic weight loss 24 months after bariatric surgery, parameters of renal function and blood pressure considerably improved, although a small percentage of patients still had glomerular hyperfiltration, proteinuria, and/or microalbuminuria, given that at 2 yr of follow-up the patients changed to type 1 obesity. The decrease in albuminuria that took place during the first year after bariatric surgery could be attributed mainly to the drastic weight loss that took place during this time, whereas during the second year, other metabolic factors may have played a more important role. Only weight loss decreases GFR and stops the cascade of events that are caused by glomerular hyperfiltration, which slows the evolution toward irreversible renal damage [38].

Immunologic Changes

Post-surgical infections are more common in patients with massive weight loss undergoing body sculpting surgery. One retrospective study of 222 patients undergoing abdominoplasty or panniculectomy showed an incidence of surgical incision infection of 12%, hematoma of 6% and seroma of 14%. Patients with weight loss had incision complications with an incidence of 41% vs. 22% in the control group [39].

Musculoskeletal Changes

A large percentage of the morbidly obese patients complained of various musculoskeletal issues, most commonly chronic back pain. Massive weight loss can result in ptosis of the breasts and excessive laxity of the skin around the arms, back, flanks, abdomen, and proximal legs, which may be causative factors in these complaints [40]. Hooper *et al.* [41] concluded that patients' musculoskeletal complaints significantly decrease after bariatric surgery when compared with their status before surgery. Most likely that these patients develop chronic somatic dysfunctions while morbidly obese, and that although they experience dramatic improvement in their symptoms after bariatric surgery, there continue to complain of musculoskeletal issues secondary to the severe ptosis and weight of the excess skin. Monitor bone changes after gastric bypass surgery. Along with rapid and substantial weight loss, there is an increase in bone turnover markers and a decline in femoral Bone Mineral Density (BMD) [42]. These changes probably reflect a decrease in estrogen, decreased compressive forces on the bones, and a decrease in adipocyte-produced interleukin 6 [43].

4. PREOPERATIVE EVALUATION

The anesthesia process begins with the preoperative work-up. Since contouring body surgery is elective, much attention is paid to getting a thorough medical evaluation and to having all associated medical conditions under optimal control prior to surgery. During the preoperative phase, patients who have had problems with previous anesthesia for bariatric surgery should make an effort to obtain the anesthesia record. Attention should focus on issues unique to the obese patient, particularly cardiorespiratory status and the airway. Patients presenting for post bariatric surgery should be evaluated for systemic hypertension, pulmonary hypertension, signs of right and/or left ventricular failure, and ischemic heart disease. Signs of cardiac failure—such as increased jugular venous pressure, added heart sounds, pulmonary crackles, hepatomegaly, and peripheral edema.

The most common symptoms of pulmonary hypertension include exertional dyspnea, fatigue, and syncope, which reflect an inability to increase cardiac output during activity [44]. Identification of tricuspid regurgitation with echocardiography is the most useful confirmation of pulmonary hypertension [45]. An electrocardiogram may demonstrate signs of right ventricular hypertrophy, such as tall precordial R waves, right axis deviation, and right ventricular strain. The electrocardiogram is more sensitive when the Pulmonary Artery (PA) pressure is higher. Chest radiograph may show evidence of underlying lung disease and evidence of prominent pulmonary arteries [46]. Mild to moderate pulmonary hypertension warrants avoidance of hypoxemia, nitrous oxide, and other drugs that may further worsen pulmonary vasoconstriction. Inhaled anesthetics may be beneficial because they cause bronchodilation and decrease hypoxic pulmonary vasoconstriction [47]. With severe pulmonary hypertension, PA catheterization and monitoring may be necessary.

Peripheral and central venous access and arterial cannulation sites should be evaluated during the preoperative examination and the possibility of invasive monitoring should be discussed with the patient. Baseline arterial blood gas measurements will help evaluate carbon dioxide retention and provide guidelines for perioperative oxygen administration and possible institution of and weaning from postoperative ventilation. Patients scheduled for contouring body surgery may confront the anesthesiologist days, months, or years after the initial bariatric surgery, so the anesthesiologist should be familiar with possible metabolic changes in these patients. Common long-term nutritional abnormalities include vitamin B_{12}, iron, calcium, and folate deficiencies. Vitamin deficiency is uncommon in patients compliant with daily vitamin supplements, especially in patients followed up with regular postoperative visits. With rapid weight loss, patients may also be protein depleted. Electrolyte and coagulation indices should be checked before surgery, particularly if patient compliance has been poor or if the patient is acutely ill. Chronic vitamin K deficiency can lead to an abnormal prothrombin time with a normal partial thromboplastin time because of deficiency of clotting factors II, VII, IX, and X [48]. For this elective surgery, the administration of a vitamin K analog, such as phytonadione, can be used to correct the coagulopathy within 6-24 hours (Table **3**).

Previous Medications

All medications previously taken by the patient should be investigated: physician-prescribed, self-prescribed, homeopathic and herbal products. Some herbal medications and teas contain active ingredients that may interact with drugs administered perioperatively, or may facilitate bleeding or arrhythmias. It is important to determine which medications should be continued until the day of the surgery and which should be discontinued with sufficient anticipation, such as MAOIs, aspirin, warfarin and anorexics. Fen-phen, an anorexic comprised of fenfluramine and phentermine, was popular until 1997, when it was removed from the market because studies showed it produced multiple cardiac valve disease and long term or irreversible pulmonary hypertension [49]. It is therefore imperative to determine if there is a history of using this product.

Table 3: Parameters to Include in the Pre-Anesthetic Evaluation of Patients with Massive Weight Loss Scheduled for Plastic or Reconstructive Surgery.

Parameter	Observations
History and physical	A general clinical review with physical exam done by the anesthesiologist prepares them to anticipate problems such as difficult airway, vertebral abnormalities, mental disorders, disorders of the family environment and the possibility of litigation
Consultation with another specialist	It is prudent to know the opinions of the bariatrician, pneumologist, cardiologist, endocrinologist, surgeon, and psychologist in the search for polypharmacy, drug interactions, *etc.*
Electrocardiogram	Arrhythmias, ischemia, enlargement or dilation
Chest X-ray	Useful in smokers, suspicion of TB, neoplasm, emphysema, kyphosis, pulmonary hypertension
Echocardiogram	Mandatory study in patients with severe arterial hypertension, ischemia, dilated myocardiopathy
Spirometry	Utility has not been demonstrated; however it is recommended in chronic lung disease and smokers
CBC	Diagnosis of subclinical anemia
Coagulation tests	TP, aTPT, INR and bleeding time are mandatory in all patients, especially those using anticoagulants, those with hepatocellular damage and those with malnutrition
Special blood tests	Evaluation of kidney, liver and metabolic function, electrolytes and vitamin deficiencies
Urinalysis	Hematuria, proteinuria and changes in urine density
HIV, hepatitis, drugs, pregnancy	Order based on data in the history and previous experience. It is prudent to request an HIV test to protect medical and paramedical personnel.

Airway

Evaluation of the airway should follow the usual recommendations for any patient, since it has been demonstrated that obesity per se does not significantly increase the difficulty of intubation, except in the case of the morbidly obese. Patients with significant weight loss have not been studied with respect to changes in intubation difficulty. Those patients have already undergone general anesthesia and therefore the situation of unpredicted difficult intubation is very rare. When a difficult airway is documented or suspected it is advisable to proceed to a conscious fiberoptic intubation.

Cardiovascular System

Arterial hypertension is common. Ask the name and dose of all antihypertensive medications, and keep the patient on these medications until the day of surgery. A resting EKG is useful and should be performed on all patients, independent of their age, based on the previously described cardiopulmonary abnormalities. Through the ECG, various parameters can be evaluated, such as right and left ventricular hypertrophy, atrial enlargement, arrhythmias, ischemia and previous infarcts. A stress test, though advisable, is not always possible given the physical conditioning of these patients and osteoarticular limitations which present obstacles to the test. Echocardiography is recommended to evaluate cardiac function, especially ventricular ejection fraction. It is imperative to evaluate the lower extremities for edema and venous insufficiency, since the latter may increase the risk of thrombosis and venous embolism.

Respiratory System

Undiagnosed pulmonary hypertension may cause serious complications during anesthesia, especially during induction. It should be suspected if the patient presents dyspnea, fatigue or syncope during exercise. This clinical picture is secondary to the right ventricle being unable to compensate for the increased demand presented by exercise. An ECG and chest x-ray are useful, though they do not conclusively support the diagnosis. Emphysema, asthma, obstructed sleep apnea and obese hypoventilation syndrome should be investigated, as well as a history of primary or second hand smoking. Keeping in mind that treating OHS patients with weight reduction support, improves daytime hypercapnia and hypoxia without changing the abnormal ventilatory responses in some patients.

Considerations for Anesthetic Management

The surgical procedures that people with massive weight loss need are many and generally are spaced over several surgical interventions, with prolonged surgical times and in unusual surgical positions which may compromise blood supply and/or innervation to certain regions. In addition to the changes described in previous paragraphs, these factors alone are reason enough to consider these patients as having increased risk when compared to healthy patients undergoing aesthetic procedures. It is prudent to follow the recommendations for anesthesia in the obese patient since, as previously discussed, patients with massive weight loss are very similar to overweight patients.

It is advisable to establish a surgical plan according to the physical conditions of each patient, their aesthetic goals and the achievable surgical possibilities, as well as the physical constraints of the surgical center and the abilities of the surgical team. Some patients should be handled in the hospital rather than an outpatient surgical center or short term surgical clinic. This point is frequently undervalued by the surgical team as well as the patients.

These surgical procedures are elective and therefore offer ample opportunity to realize a complete pre-anesthetic evaluation, in which the anesthesiologist should thoroughly review the patient's chart and the patients themselves, actively seeking the previously enumerated co-morbidities. These pathologies should be optimally controlled prior to surgery. It is recommended to deflate the expandable chamber of the gastric band when it is the case before the induction of anesthesia to decrease the risk of pulmonary aspiration.

Any anesthetic technique may be applied to these patients, in accordance with the preoperative evaluation, perioperative position and the surgical plan. Brachioplasty with surgical reduction and mastopexy is a frequent surgical combination that can be executed either with general anesthesia or a cervicothoracic epidural block. Mastopexy alone could be realized under general anesthesia, epidural, paravertebral, intercostal or breast block, depending on the extent of surgery and whether prosthesis are to be put beneath the pectoralis major muscle. Abdominoplasty, lower body lift or thigh lifts are intervention with an intermediate surgical time that can be managed through neuroaxial block reaching a T4-T6 level or general anesthesia. With the face lift in the post bariatric surgery patients, it is recommended to intubate these patients because of the increased risk of bronchopulmonary aspiration and the difficulty to access the airway in this group. However, this procedure could be realized under pure local anesthesia if the patient is cooperative. Liposuction is usually a complementary procedure during other plastic surgeries and rarely is it performed as the only procedure in previously obese patients. When it is the case, tumescent anesthesia can be an option. These cases should be considered as high risk since the patients usually have multiple co-morbid conditions (especially lower limb venous insufficiency), surgical times are prolonged and there tends to be considerable surgical bleeding. The procedure also requires placing the patient in the prone position for a long period of time and special care should be given to the positioning and to securing the airway.

Concurrent and Preoperative Medications

It is recommended that the patient's usual medications, except insulin and oral hypoglycemics, be continued until the time of surgery. Antibiotic prophylaxis is important because of increased risk of postoperative wound infection. The increased incidence of wound infection is due to longer incisions, generally longer operative times because of obesity, tissue trauma from excessive traction, difficulty in dead space obliteration, and inability of adipose tissue to resist infection [50].

Anxiolysis, and prophylaxis against both aspiration pneumonitis and DVT should be addressed during premedication. While the use of benzodiazepines in the obese patient is controversial, we recommend giving these patients oral clonidine (0.1mg) as it facilitates sedation, dry mouth and provide hemodynamic stability during endotracheal intubation and reduces the need for local and general anesthetics. Most patients should receive histamine H2 antagonists and prokinetics, thus decreasing the acidity and risk of the aspirate [32, 33, 34]. Pharmacologic intervention with H2-receptor antagonists (*e.g.,* cimetidine, ranitidine, famotidine) and nonparticulate antacids (e.g., sodium bicitrate) and proton pump inhibitors (*e.g.,* omeprazole, lansoprazole, rabeprazole) will reduce gastric volume, acidity, or both, thereby reducing the risk and complications of aspiration. It is advisable to administer antiemetics and H1 receptor blockers, and avoid the use of opiates.

Low molecular weight heparins (LMWH) have gained popularity in thromboembolism prophylaxis because of their bioavailability when injected subcutaneously [51]. In a survey of members of the American Society for Bariatric Surgery regarding their current practices for thromboprophylaxis [52], small-dose heparin, 5000 U every 8–12 h, was the most preferred method (50% of members), followed by pneumatic compression stockings (33%), LMWH (13%), and other methods (4%). In combination with subcutaneous heparin, we favor placement of pneumatic compression devices on the feet because knee or thigh-length devices tend to slip and fall off.

5. INTRAOPERATIVE CONSIDERATIONS

A. General Anesthesia

Induction, Intubation, and Maintenance of Anesthesia

When general anesthesia is to be used, it is important to pre-oxygenate and de-nitrogenate the patient. Preparation should be made for the possibility of a difficult intubation. A towel or folded blankets under the shoulders and head can compensate for an exaggerated flexed position from posterior cervical fat [53]. The object of this maneuver, known as "stacking," is to position the patient so that the tip of the chin is at a higher level than the chest, to facilitate laryngoscopy and intubation. Brodsky *et al.* [54] used a logistic regression model to quantify the relationship between the ease of intubation and patient characteristics. They predicted that odds of a problematic intubation in a particular patient with a neck circumference 1cm larger than that of another patient are 1.13 times the odds of the patient with a 1cm-smaller neck circumference. Therefore, the probability of a problematic intubation was approximately 5% with a 40cm neck circumference, compared with a 35% probability at 60cm neck circumference. This model identified neck circumference associated to a Mallampati ≥ 3 as the best predictor of problematic intubation.

Venous Access

Venous access may be difficult. When it is not possible to place an 18-gauge catheter, it is advisable to begin with a smaller catheter and change it as early as possible. Another alternative is the placement of a central venous line, which at times can be difficult. Dissection of a vein is the last alternative.

Pharmacology/Weight-Based Dosing

Many physiologic factors influence drug absorption, such as gastric emptying time and the integrity and surface area of the epithelium. The gut has an impressive capacity to compensate for loss of function, so absorption after surgery may eventually normalize. By that time, however, marked weight loss can complicate the clinical picture.

Compared with non-obese persons, obese persons have an increased proportion of adipose tissue, as well as increased total body water, lean body mass, visceral organ mass, and higher glomerular filtration rate. Postoperative bariatric patients often lose more than 100pounds of adipose tissue. This type of weight loss mostly affects lipid-soluble drugs with a large volume of distribution (V_d) that readily cross cell membranes, such as fluoxetine. Drugs with a large V_d, such as fluoxetine, reach all major compartments of distribution, which in a normal weight individual include: plasma (5%), interstitial fluid (16%), intracellular fluid (35%), transcellular fluid (2%), and fat (20%). Because the amount of fat in an overweight patient is initially very high but rapidly decreases after bariatric surgery, drugs with a large V_d can shift into other compartments. For drugs with a small V_d, such as lithium, a lower maintenance dosage may be required because of decreased glomerular filtration following marked weight loss.

Induction is facilitated with hypnotic medications, propofol being the most commonly used. It is advisable to use depolarizing muscle relaxants with a rapid onset and short duration of activity to facilitate endotracheal intubation. Maintenance with desflurane or sevoflurane facilitates hemodynamic stability and rapid anesthetic recovery. General anesthesia is complemented with fentanyl and sufentanil in bolus or remifentanyl drip.

Calculation of the dose of the different medications used in anesthesiology is based on body weight in the majority of cases. This is not always appropriate, especially in the case of lipophylic drugs. Obesity represents a challenge when attempting to establish an optimum dose, and especially difficult challenge when the medication in question has a narrow therapeutic index, as is the case with the majority of the drugs used in anesthesiology. The factors which intervene in the distribution of drugs into various tissues include body composition, regional blood flow and the affinity of the drug for plasma proteins and tissue components. Highly lipophilic substances (Table **4**) [55-66], such as barbiturates and benzodiazepines, show significant increases in volume of distribution (V_d) for obese individuals relative to normal-weight individuals [55, 67, 68]. Less-lipophilic compounds have little or no change in V_d with obesity. Certain exceptions to this rule include digoxin [69], procainamide [70], and remifentanil [65], which are highly lipophilic drugs but which have no systematic relationship between their degree of lipophilicity and their distribution in obese individuals. Consequently, their absolute V_d remains relatively consistent between obese and normal-weight individuals, and their doses should be calculated on the basis of ideal body weight [65, 69, 70].

Drugs with weak or moderate lipophilicity can be dosed on the basis of ideal body weight (IBW) or, more accurately, Lean Body Mass (LBM). These values are not identical, because 20%-40% of an obese patient's increase in total body weight can be attributed to an increase in LBM. Adding 20% to the estimated IBW dose of hydrophilic medications is sufficient to include the extra lean mass. Non-depolarizing muscle relaxants can be dosed in this manner. The majority of anesthetic drugs are strongly lipophilic. Increased V_d is expected for lipophilic substances, but this is not consistently demonstrated in pharmacological studies because of factors such as end-organ clearance or protein binding. Medications with a narrow therapeutic index should be used prudently and the dose should be adjusted to achieve plasma concentrations within the recommended therapeutic window (Tables **4** and **5**) [71].

Desflurane has been suggested as the inhaled anesthetic of choice in this patient population because of its more rapid and consistent recovery profile [72]. Two different studies [73, 74] compared sevoflurane with isoflurane for use during bariatric surgery and favored sevoflurane because of its more rapid recovery, good hemodynamic control, and infrequent incidence of nausea and vomiting, prompt regaining of psychological and physical functioning, early discharge from the hospital, and small cost. Rapid elimination and analgesic properties make nitrous oxide a good inhaled choice during bariatric surgery, but high oxygen demand in the obese limits its use. Obesity increases oxygen consumption and carbon dioxide production [75]. This is due to excess metabolically active tissue and an increased workload on muscles and other supportive tissue. De Divitiis *et al.* [76] performed left and right heart catheterization in 10 morbidly obese but otherwise healthy individuals and noted that the mean oxygen consumption was increased by up to 25% and increased linearly with increasing body weight. The arteriovenous oxygen difference was normal, however, suggesting that cardiac output increases primarily to serve the metabolic requirements of excess fat [77].

Table 4: Weight-Based Dosing of Common IV Anesthetics [54-65]:

Drug	Dosing	Comments
Propofol	IBW Maintenance: TBW	Systemic clearance and V_d at steady-state correlates well with TBW [55]. High affinity for excess fat and other well perfused organs. High hepatic extraction and conjugation relates to TBW.
Thiopental	TBW	Increased V_d. Increased blood volume, cardiac output, and muscle mass [54]. Increased absolute dose. Prolonged duration of action [56].
Midazolam	TBW	Central V_d increases in line with body weight. Increased absolute dose. Prolonged sedation because larger initial doses are needed to achieve adequate serum concentrations [56, 57].
Succinylcholine	TBW	Plasma cholinesterase activity increases in proportion to body weight. Increased absolute dose [56].
Vecuronium	IBW	Recovery may be delayed if given according to TBW because of increased V_d and impaired hepatic clearance [56, 58].
Rocuronium	IBW	Faster onset and longer duration of action. Pharmacokinetics and pharmacodynamics are not altered in obese subjects [59, 60].
Cisatracurium Atracurium	TBW	Absolute clearance, V_d, and elimination half-life do not change. Unchanged dose per unit body weight without prolongation of recovery because of organ-independent elimination [61, 62].
Fentanyl Sufentanil	TBW TBW Maintenance: IBW	Increased V_d and elimination half-time, which correlates positively with the degree of obesity [63]. Distributes as extensively in excess body mass as in lean tissues. Dose should account for total body mass.
Remifentanil	IBW	Systemic clearance and V_d corrected per kilogram of TBW—significantly smaller in the obese. Pharmacokinetics is similar in obese and non-obese patients [64]. Age and lean body mass should be considered for dosing [65].
IBW: Ideal body weight; TBW: Total body weight; V_d: Volume of distribution.		

Table 5: Calculations and Formulas.

Calculation of Ideal Body Weight and Body Mass Index
Estimated Ideal Body Weight (wt) in kilograms (Kg) = Height (ht) in cm - ([100 for men] or [105 for women]). (84)
Body Mass Index (BMI) = Wt in Kg/([Ht in Meters (M)]squared)--(normal is around 24). (85)

Complete muscular relaxation is crucial during the tightening of abdominal musculature. Combined epidural and general balanced anesthesia has been advocated to allow better titration of anesthetic drugs, use of a larger oxygen concentration, and optimal muscle relaxation in the obese [78, 79]. With the introduction of easily titratable drugs such as remifentanil, propofol, and desflurane, decreased intraoperative dosing is not generally required for prompt emergence [80, 81].

Positioning

Particular care should be paid to protecting pressure areas, because pressure sores and neural injuries is frequent in the obese and the diabetic patients. Stretch injuries may be caused by extreme abduction of the arms, thereby stretching the lower roots of the brachial plexus. The upper roots are most likely stretched by excessive rotation of the head to the opposite side [82]. Sciatic nerve palsy may be caused by prolonged ischemic pressure from tilting the table sideways. Lateral femoral cutaneous nerve injury may occur if the lower limb falls and hangs freely. The extent and degree to which a nerve is injured should be well documented so that recovery and prognosis can be discussed with the patient. Electromyography and nerve conduction studies provide valuable clinical information in this respect [82]. Despite careful positioning and appropriate padding, nerve injury may still occur in this at-risk population. Fortunately, most resolve with time.

Monitoring

Clinical and electronic vigilance is the base of good anesthesiology technique. Invasive arterial monitoring should be used with severe cardiopulmonary disease and for those with poor fit of the noninvasive blood pressure cuff because of severe conical shape of the upper arms or unavailability of appropriately sized cuffs. Reliability of NIBP is poor in gynecoid type obese patients, whereas obese patients of the android type tend to give reliable readings. Blood pressure measurements can be falsely increased if a cuff too small for the arm is used [83]. Cuffs with bladders that encircle a minimum of 75% of the upper arm circumference or, preferably, the entire arm, should be used [84]. Comparable and accurate blood pressure readings can be obtained from the wrist [85] or ankle [86] with appropriately sized blood pressure cuffs in situations in which difficulty occurs with upper-arm noninvasive blood pressure measurement. The central venous catheters are used in cases in which peripheral IV access cannot be obtained, whereas PA catheters are reserved for serious cardiopulmonary disease. Oxymetry and capnography are mandatory, as is continuous ECG monitoring, with all precordial leads whenever possible. It is important to monitor airway pressure and calculate pulmonary compliance. Airway pressures should not exceed 40cm H_2O.

Mechanical Ventilation

Tidal volume should be calculated based on ideal body weight. A tidal volume of 10-12 mL/kg with a respiratory rate of 8 to 14 will avoid hyper and hypocapnia. Tidal volumes ≥13mL/kg IBW offer no added advantage during ventilation of morbidly obese patients during anesthesia. Also, in light of evidence that the lung can be injured by excessive expansion (volutrauma) from large tidal volumes leading to pulmonary edema and that Positive End-Expiratory Pressure (PEEP) actually reduced lung water content in this type of edema [87], it seems prudent to use moderate levels of PEEP (enough to preserve hemodynamic stability) with prolong inspiration times to improve oxygenation by recruiting and maintaining open more alveoli rather than large tidal volumes in an attempt to improve oxygenation.

Intraoperative Fluid

Requirements are usually larger if postoperative acute tubular necrosis is to be prevented. Patients usually require up to 4-5L of crystalloid for an average 2-hours operation. This adds up to twice the calculated maintenance fluid requirement plus the calculated deficit based on a 12-hours fasting period for an average 70-kg patient for the first hour by using the 4-2-1 formula (4 mL x kg^{-1} x h^{-1} for the first 10 kg; 2 mL x kg^{-1} x h^{-1} for the next 10 kg; then 1mL x kg^{-1} x h^{-1} for every kilogram thereafter). The next hour usually requires the same amount of crystalloid, after which the amounts are reduced to approximately twice the calculated maintenance requirement, based on LBM, for the next 12 h (200 mL/hr overnight).

B. Regional Anesthesia

Regional anesthesia offers distinct advantages over general anesthesia for these patients. A regional anesthetic allows minimal airway manipulation, avoidance of anesthetic drugs with cardiopulmonary depression, and reduced Postoperative Nausea and Vomiting (PONV), as well as greater postoperative pain control. Regional anesthesia may also reduce perioperative and postoperative opioid requirements, which is of critical importance in a patient population prone to postoperative pulmonary complications. It reduces the likelihood of atelectasis, venous thrombosis and possibly pulmonary embolism. Bleeding is also reduced in comparison with general anesthesia.

Establishing neuraxial blockade in the previously morbid obese patient can also be challenging. In those patients, there may be difficulty in palpating bony landmarks or even identifying the midline because of skin laxity and residual excess adipose tissue. The presence of fat pockets may result in false-positive loss of resistance during needle placement [88].

Drug distribution may also be altered [89, 90]. Hood and Dewan [90] described an initial success rate of only 42% for placing epidural catheters in obese patients compared with an initial 94% success rate in non-obese controls. Overall, the success rate was similar for both groups, but obese patients required more placement attempts to achieve success.

Obese patients require less local anesthetic in their epidural and subarachnoid spaces in order to achieve the same level of block when compared with non-obese controls. After 3ml of 0.5% bupivicaine was injected into the subarachnoid space at the L3–4 interspace, Taivainen *et al.* [89] demonstrated a higher cephalad spread in obese vs. non-obese individuals. Similarly, Hodgkinson and Husain [91] demonstrated a higher cephalad spread of 20ml 0.75% bupivicaine injected into the L_{3-4} epidural in obese vs. non-obese individuals. The addition of coadjuvant medication to local anesthetics optimizes the quality and duration of anesthetic block and neuroaxial analgesia.

Although the apparent lower spinal anesthetic dose requirement may be explained by the fact that obese patients have smaller cerebrospinal fluid volumes than do non-obese individuals [92]. The reason for the lower epidural anesthetic dose requirement is less clear, the epidural fat and dilated epidural veins could be a cause.

Ultrasound-guided techniques have been promoted to aid epidural catheter placement [93-96]. Even with the use of ultrasound, there have been reports of accidental dural puncture during attempted epidural placement in morbidly obese patients [97]. However, increased difficulty of performing neuraxial blocks in obese patients must be taken into consideration. Longer spinal and epidural needles may be necessary, and landmarks may be concealed by excess body tissue. When an epidural block is planned, it is prudent to insert the catheter a few extra centimeters into the epidural cavity since lax skin leads to increased mobility. Epidural anesthesia in obese patients undergoing thoracic and upper abdominal surgery decreases opioid requirements and reduces postoperative pulmonary complications [98, 99]. When combined with a general anesthetic, epidural anesthesia may result in earlier time to tracheal extubation than with a balanced anesthetic alone [100].

A malabsorptive procedure such as RYGB is associated with postoperative nutritional deficiencies that can present unique problems to the anesthesiologist performing regional anesthesia. Vitamin K deficiency occurs in 50–68% patients following RYGB, even in those taking daily vitamins [101]. There have been case reports describing the adverse effects of vitamin K deficiency on coagulation after gastric surgery [102]. The anesthesiologist considering a neuraxial block in a patient who has had RYGB surgery in the past should be concerned about potential vitamin K deficiency. In addition, water-soluble vitamin deficiencies may also be present, most notably vitamins B_{12} and folate. B_{12} deficiency is present in an estimated 6–70% of postbariatric surgery patients [103]. Manifestations of B_{12} deficiency include peripheral neuropathy and subacute combined degeneration with white matter lesions in the posterior column and pyramidal tract that can manifest as demyelination and can progress to axonal degeneration and neuronal death [104]. This may lead to weakness, loss of motor function, and proprioception. The risk/benefit of a neuraxial block should be considered in the context of whether the patient has a peripheral neuropathy.

In addition to nutritional deficiencies and malabsorption, rapid weight loss may also lead to peripheral neuropathy in postbariatric surgical patients. A recent study [105] found that significant weight loss is correlated to a higher risk of peroneal nerve injury after bariatric surgery. Mechanical injury to peripheral nerves may occur during positioning as a decreased fat pad after weight loss may leave nerves more susceptible to compression [103, 106].

Therefore, careful consideration should be taken before performing regional anesthesia in a patient following bariatric surgery. A thorough history and physical examination directed at any potential nutritional deficiencies or neurologic dysfunction are mandatory before a block is performed. A high index of suspicion for coagulopathy must be present, and a coagulation profile should be obtained.

C. Tumescent Anesthesia

Large volumes of dilute local anesthetic and adrenaline are now used for subcutaneous infiltration, enabling up to 10% of the body surface to be anesthetized locally and perioperative bleeding to be significantly reduced. Infiltration with a large volume of local anesthetic furthermore reduces the need for perioperative analgesics. Patients undergoing liposuction of large volumes receive high doses of epinephrine, lidocaine and IV solutions that produce important cardiovascular and thermoregulatory changes. Kendel *et al.* [107] found a change in the cardiac index (57%), heart rate (47%) and mean pulmonary artery pressure (44%). Central venous pressure remained unchanged. The unchanged central venous pressure levels indicate that young healthy patients with compliant right ventricles can accommodate the fluid loads of large-volume liposuction. Maximum epinephrine levels were observed 5 to 6 hours after induction of anesthesia, and a significant correlation was found between the epinephrine levels and transoperative cardiac index (r=0.75). All patients suffered hypothermia, with a mean temperature of 35.5°C. Overall hemodynamic parameters remained within safe limits. Within these surgical parameters, patients should be clinically screened for cardiovascular and blood pressure disorders before liposuction is undertaken, and preventative measures should be taken to limit intraoperative hypothermia.

6. POSTOPERATIVE CONSIDERATIONS

Undesirable anesthetic incidents are related to the physical condition of each individual, the changes previously described, the prolonged time of anesthesia, positions which impede circulation, ventilation, pain and hypothermia are the most relevant complications.

The majority of anesthetic accidents occur during the postoperative period; hypoxia, respiratory failure, recurarization, hypothermia and acute myocardial infarction are some of the reported incidences. All patients should be monitored (ECG, oximetry, NIBP and clinical vigilance) in the post-anesthesia care unit, especially those who have undergone extensive surgery. The patient should receive supplementary oxygen using nasal prongs, adequate analgesia, and the head should be elevated to 30 degrees to facilitate ventilation. Special postoperative positioning should be respected in certain operations for example the sitting position with knee flexed on a pillow after an abdominoplasty. Although incentive spirometry has been recommended by some authors to prevent postoperative atelectasis [108, 109], adequate evidence is lacking in this respect. Patients with a history of severe sleep apnea may require overnight observation in the intensive care unit because prolonged obstructive apnea is a real possibility, especially when parenteral narcotics are used.

Postoperative Analgesia

The pain from a surgical procedure with aponeurosis manipulation can be quite significant. Epidural local anesthetics and/or narcotics *via* the thoracic route are a safe and effective form of postoperative analgesia in these patients. Intrathecal narcotics are also a viable option. Potential advantages of thoracic epidural analgesia in the setting of contouring body surgery include prevention of DVT, improved analgesia, and earlier recovery. Investigators have been unable to document a difference in the incidence of thrombophlebitis and PE with continuous epidural analgesia [110, 111]. Less oxygen consumption and decreased left ventricular stroke work have, however, been documented benefits of local anesthetic epidural analgesia [112]. After an abdominoplasty, patients may avoid taking deep breaths because of pain. Adequate analgesia and a properly fitted elastic binder for abdominal support may encourage patients to cooperate with early ambulation and incentive spirometry. Most of these patients do well with local anesthetic infiltration beneath the anterior rectus sheath preoperatively with basic parenteral narcotics, such as PCA postoperatively. Avoid prescribing nonsteroidal anti-inflammatory drugs, even selective cyclooxygenase 2 (COX-2) inhibitors, in these patients because of the risk of gastrointestinal bleeding, which, if it occurs distal to the stapling or banding, requires major surgery and has a high morbidity rate. Patients can use acetaminophen, tramadol, or opiates.

Pulmonary Embolism

Precautions against and observation for signs of venous embolism should be continued. Thrombotic events are a very real phenomenon in this group of patients, whether or not they involve embolism. For example, abdominoplasty has a 1.1% risk of deep vein thrombosis, a complication attributed to the plication of the rectus abdominis, which in turn increases the intra-abdominal pressure, which reduces venous return, causing venous distention and thrombosis. Huang *et al.* [113] found other factors that increased intra-abdominal pressure, such as degree of bed flexing, abdominal bandages and general anesthesia. A prospective study by Gravante *et al.* [114] involving 103 patients who underwent abdominoplasty and flank liposuction found an incidence of pulmonary embolism of 2.9%. All patients had received thromboembolism prophylaxis. These authors found a significant relationship between fat extraction of more than 1500g and time of anesthesia exceeding 140 minutes. Another group of investigators [115] studied 138 post-bariatric patients undergoing body contour surgery (abdomen, back, arms and thorax) and found three patients with deep vein thrombosis requiring anticoagulation and one death due to pulmonary embolism. This implies a 2.9% risk of thromboembolism, the same as reported by Gravante [114]. The BMI of patients with thrombosis was 48.5 vs. 31.8 for patients who did not present this complication. It is vitally important to establish a plan of venous thromboembolism prophylaxis that includes elastic stockings, intermittent compression pneumatic pump, heparin and early ambulation. The program should be applied commensurate with the degree of risk factors [115].

Hypothermia

Hypothermia is a common event in plastic surgery patients due to the long period of time of the surgery and where the application of a warming matrix is difficult because of the large exposed area to the cold environment of the operating room. Shivering, increased oxygen consumption, myocardial ischemia and changes in coagulation are some of the complications reported due to a critical perioperative drop in body temperature. One study of plastic surgery patients showed that maintaining transoperative normothermia helps patients maintain normal coagulation. In unprotected patients the temperature dropped 2°C and both the thromboplastin time as well as bleeding time increased significantly [116].

7. CONCLUSION

Body contouring after massive weight loss is a significant undertaking on the part of both the surgeon and the anesthesiologist. As stated earlier, many of these patients are chronically malnourished which predisposes them to increased surgical morbidity. The surgery itself is laborious and sometimes involves considerable blood loss. For these reasons, most surgeons stage the procedures. Although severe complications such as pulmonary embolism or even death are relatively rare (0.02% or less), the post-bariatric surgery is associated with a greater frequency of complications than "traditional" body contouring. Physiologic, anatomic and pharmacologic changes paint a complex clinical scenario with infinite anesthesiologic considerations.

The popularity of bariatric surgery has created this new group of patients with massive weight loss who still maintain many of the co-morbidities of obesity. Anesthetic management of these patients should take into consideration the specific problems associated with obesity and optimize them before surgery: physical changes, co-morbid medical conditions and changes related to bariatric surgery as well. Much of what has been learned from studies in obese patients is that the pharmacokinetic alterations of medications are variable. Broad application of dosing guidelines even among medications within the same therapeutic class is likely not appropriate. Teamwork is the cornerstone of success in the integral treatment of these patients.

The anesthesiologist surgeon-patient interaction should be scrupulously planned, and alternatives should always be considered, for the safety of the patient.

Procedures should be performed in accredited facilities with appropriately trained staff, and special intraoperative consideration should be given to patient positioning and avoidance of hypothermia. Postoperative management is optimized by the development of clinical pathways involving the surgeon, anesthesiologist, patient, and support staff.

Although obvious physical and psychological improvement may be attained by such surgical procedures, the prospective patient must be fully informed of the multiple risks and complications inherent to such procedures.

REFERENCES

[1]   Haslam DW, James WPT. Obesity. Lancet 2005; 366: 1197-1209.
[2]   Buchwald H, Williams SE. Bariatric surgery worldwide 2003. Obes Surg 2004; 14: 1157-64.
[3]   Adams JP, Murphy PG. Obesity in anaesthesia and intensive care. Br J Anaesth 2000; 85: 91-108.
[4]   Presutti RJ, Gorman RS, Swain JM. Primary care perspective on bariatric surgery. Mayo Clin Proc 2004; 79: 1158-1166.
[5]   Gusenoff JA, Messing S, O'Malley W, et al. Temporal and demographic factors influencing the desire for plastic surgery after gastric bypass surgery. Plast Reconstr Surg 2008; 121: 2120-26.
[6]   Gusenoff JA, Messing S, O'Malley W, et al. Patterns of plastic surgical use after gastric bypass: who can afford it and who will return for more. Plast Reconstr Surg 2008; 122: 951-58.
[7]   Cintra W, Modolin ML, Gemperli R, et al. Quality of life after abdominoplasty in women after bariatric surgery. Obes Surg 2008; 18: 728-32.
[8]   Mittermair RP, Weiss H, Nehoda H, et al. Laparoscopic Swedish adjustable gastric banding: 6-year follow-up and comparison to other laparoscopic bariatric procedures. Obes Surg 2003; 13: 412-17.
[9]   Champault A, Duwat O, Polliand C, et al. Quality of life after laparoscopic gastric banding: Prospective study (152 cases) with a follow-up of 2 years. Surg Laparosc Endosc Percutan Tech 2006; 16: 131-36.
[10]   Ashley FW, Kannel WB. Relation of weight change to changes in atherogenic traits: the Framingham study. J Chronic Dis 1974; 27: 103-14.
[11]   Sjöström CD, Lissner L, Sjöström L. Relationships between changes in body composition and changes in cardiovascular risk factors: the SOS Intervention Study: Swedish Obese Subjects. Obes Res 1997; 5: 519-30.
[12]   Davis BR, Blaufox MD, Oberman A, et al. Reduction in long-term antihypertensive medication requirements: effects of weight reduction by dietary intervention in overweight persons with mild hypertension. Arch Intern Med 1993; 153: 1773-82.
[13]   Sjöström CD, Markku P, Hans W, et al. Differentiated long-term effects of intentional weight loss on diabetes and hypertension. Hypertension 2000; 36: 20-25.
[14]   Achincloss JH, Cook E, Renzetti AD. Clinical and physiological aspects of a case of polycythemia and alveolar hypoventilation. J Clin Invest 1955; 34: 1537-45.
[15]   Pelosi P, Croci M, Ravagnan I. et al. Respiratory system mechanics in sedated, paralyzed, morbidly obese patients. J Applied Physiol 1997; 82: 811-18.
[16]   Teichtahl H. The Obesity-Hypoventilation Syndrome Revisited. Chest 2001; 120: 336-39.
[17]   Sharp JT, Henry JP, Sweany SK, et al. The total work of breathing in normal and obese men. J Clin Invest 1964; 43: 728-739.
[18]   Leech J, Onal E, Aronson R, et al. Voluntary hyperventilation in obesity hypoventilation. Chest 1991; 100: 1334-38.
[19]   Zwillich CW, Sutton FD, Pierson DJ, et al. Decreased hypoxic ventilatory drive in the obesity-hypoventilation syndrome. Am J Med 1975; 59: 343-48.
[20]   Rochester DF, Enson Y. Current concepts of the obesity hypoventilation syndrome. Am J Med 1974; 57: 402-20.
[21]   Lopata M, Onal E. Mass loading, sleep apnea, and the pathogenesis of obesity hypoventilation. Am Rev Respir Med 1982; 126: 640-45.
[22]   Garay SM, Rapaport DM, Sorkin B, et al. Regulation of ventilation in the obstructive sleep apnea syndrome. Am Rev Respir Dis 1981; 124: 451-57.
[23]   Sugerman HJ, Fairman PR, Sood RK, et al. Long-term effects of gastric surgery for treating respiratory insufficiency of obesity. Am J Clin Nutr 1992; 55: 597S-601S.
[24]   Pankow W, Hijjeh N, Schuttler F, et al. Influence of noninvasive positive pressure ventilation in obese subjects. Eur Respir J 1997; 10: 2847-52.
[25]   Vargas-Ruiz AG, Hernández-Rivera G, Herrera MF. Prevalence of iron, folate, and vitamin B12 deficiency anemia after laparoscopic Roux-en-Y gastric bypass. Obes Surg 2008; 18: 288-93.

[26]   Clements RH, Katasani VG, Palepu R, *et al.* Incidence of vitamin deficiency after laparoscopic Roux-en-Y gastric bypass in a university hospital setting. Am Surg 2006; 72: 1196-02.

[27]   Jean J, Compe`re V, Fourdrinier V. The risk of pulmonary aspiration in patients after weight loss due to bariatric surgery. Anesth Analg 2008; 107: 1257-59.

[28]   Weiss HG, Nehoda H, Labeck B, *et al.* Treatment of morbid obesity with laparoscopic adjustable gastric banding affects esophageal motility. Am J Surg 2000; 180: 479-82.

[29]   Klaus A, Gruber I, Wetscher G, Nehoda H, Aigner F, Peer R, Margreiter R, Weiss H. Prevalent esophageal body motility disorders underlie aggravation of GERD symptoms in morbidly obese patients following adjustable gastric banding. Arch Surg 2006 Mar;141(3):247-51.

[30]   Dixon JB, O'Brien PE. Gastroesophageal reflux in obesity: the effect of lap-band placement. Rapid and major improvement in symptoms of GERD occurs after Lap-Band placement. Obes Surg 1999; 9: 527-31.

[31]   Weiss HG, Nehoda H, Labeck B, *et al.* Adjustable gastric and esophagogastric banding: a randomized clinical trial. Obes Surg 2002; 12: 573-78.

[32]   Kocian R, Spahn DR. Bronchial aspiration in patients after weight loss due to gastric banding. Anesth Analg 2005; 100: 1856-57.

[33]   Alamoudi OS. Long-term pulmonary complications after laparoscopic adjustable gastric banding. Obes Surg 2006; 16: 1685-88.

[34]   Di Francesco V, Baggio E, Mastromauro M, *et al.* Obesity and gastro-oesophageal acid reflux: physiopathological mechanisms and role of gastric bariatric surgery. Obes Surg 2004; 14: 1095-02.

[35]   Schwartz DJ, Wynne JW, Gibbs CP, *et al.* The pulmonary consequences of aspiration of gastric contents at pH values greater than 2.5. Am Rev Respir Dis 1980; 121: 119-26.

[36]   Narchi P, Edouard D, Bourget P, *et al.* Gastric fluid pH and volume in gynaecologic outpatients. Influences of cimetidine and cimetidine-sodium citrate combination. Eur J Anaesthesiol 1993; 10: 357-61.

[37]   Bond A. Obesity and difficult intubation. Anaesth Intensive Care 1993; 21: 828-30.

[38]   Navarro-Diaz M, Serra A, Romero R, *et al.* Effect of drastic weight loss after bariatric surgery on renal parameters in extremely obese patients: long-term follow-up. J Am Soc Nephrol 2006; 17: S213-S217.

[39]   Greco JA 3rd, Castaldo ET, Nanney LB, *et al.* The effect of weight loss surgery and body mass index on wound complications after abdominal contouring operations. Ann Plast Surg 2008; 61: 235-42.

[40]   Hurwitz DJ. Total body lift: reshaping the breasts, chest, arms, thighs, hips, back, waist, abdomen and knees after weight loss, aging and pregnancies. New York, NY: MDPublish.com; 2005.

[41]   Hooper MM, Stellato TA, Hallowell PT, *et al.* Musculoskeletal findings in obese subjects before and after weight loss following bariatric surgery. Int J Obes 2007; 31: 114-20.

[42]   von Mach MA, Stoeckli R, Bilz S, *et al.* Changes in bone mineral content after surgical treatment of morbid obesity. Metabolism 2004; 53: 918-21.

[43]   Manolagas SC, Jilka RL. Bone marrow, cytokines and bone remodeling: emerging insights into the pathophysiology of osteoporosis. N Engl J Med 1995; 332: 305-11.

[44]   Nauser TD, Stites SW. Diagnosis and treatment of pulmonary hypertension. Am Fam Physician 2001; 63: 1789-98.

[45]   Schiller NB. Pulmonary artery pressure estimation by Doppler and two-dimensional echocardiography. Cardiol Clin 1990; 8: 277-87.

[46]   Widimsky J. Noninvasive diagnosis of pulmonary hypertension in chronic lung diseases. Prog Respir Res 1985; 20: 69-75.

[47]   Konduri GG, Garcia DC, Kazzi NJ, *et al.* Adenosine infusion improves oxygenation in term infants with respiratory failure. Pediatrics 1996; 97: 295-300.

[48]   Stoelting RK, Dierdorf SF. Anesthesia and co-existing disease. 4th ed. Philadelphia: Churchill Livingstone, 2002.

[49]   Fleming RM, Boyd LB. The longitudinal effects of fenfluraminephentermine use. Angiology 2007; 58: 353-59.

[50]   Fisher A, Waterhouse TD, Adams AP. Obesity: its relation to anaesthesia. Anaesthesia 1975; 30: 633-47.

[51]   Pineo GF, Hull RD. Unfractionated and low-molecular weight heparin: comparison and current recommendations. Med Clin North Am 1998; 82: 587-99.

[52]   Wu EC, Barba CA. Current practices in the prophylaxis of venous thromboembolism in bariatric surgery. Obes Surg 2000; 10: 7-13.

[53]   McCarroll SM, Saunders PR, Brass PJ. Anesthetic considerations in obese patients. Prog Anesthesiol 1989; 3: 1-12.

[54]   Brodsky JB, Lemmens HJM, Brock-Utne JG, *et al.* Morbid obesity and tracheal intubation. Anesth Analg 2002; 94: 732-36.

[55]   Jung D, Mayersohn M, Perrier D, *et al.* Thiopental disposition in lean and obese patients undergoing surgery. Anesthesiology 1982; 56: 269-74.

[56]   Servin F, Farinotti R, Haberer JP, *et al.* Propofol infusion for maintenance of anesthesia in morbidly obese patients receiving nitrous oxide: a clinical and pharmacokinetic study. Anesthesiology 1993; 78: 657-65.

[57]   Adams JP, Murphy PG. Obesity in anaesthesia and intensive care. Br J Anaesth 2000; 85: 91-108.

[58]   Greenblatt DJ, Abernethy DR, Locniskar A, *et al.* Effect of age, gender, and obesity on midazolam kinetics. Anesthesiology 1984; 62: 27-35.

[59]   Schwartz AE, Matteo RS, Ornstein E, *et al.* Pharmacokinetics and pharmacodynamics of vecuronium in the obese surgical patient. Anesth Analg 1992; 74: 515-18.

[60]   Puhringer FK, Khuenl-Brady KS, Mitterschiffthaler G. Rocuronium bromide: time-course of action in underweight, normal weight, overweight, and obese patients. Eur J Anaesthesiol 1995; 11(Suppl): 107-10.

[61]   Puhringer FK, Keller C, Kleinsasser A, *et al.* Pharmacokinetics of rocuronium bromide in obese female patients. Eur J Anaesthesiol 1999; 16: 507-10.

[62]   Weinstein JA, Matteo RS, Ornstein E, *et al.* Pharmacodynamics of vecuronium and atracurium in the obese surgical patient. Anesth Analg 1988; 67: 1149-53.

[63]   Varin F, Ducharme J, Theoret Y, *et al.* Influence of extreme obesity on the body disposition and neuromuscular blocking effect of atracurium. Clin Pharmacol Ther 1990; 48: 18-25.

[64]   Schwartz AE, Matteo RS, Ornstein E, *et al.* Pharmacokinetics of sufentanil in obese patients. Anesth Analg 1991; 73: 790-93.

[65]   Egan TD, Huizinga B, Gupta SK, *et al.* Remifentanil pharmacokinetics in obese versus lean patients. Anesthesiology 1998; 89: 562-73.

[66]   Minto CF, Schnider TW, Shafer SL. Pharmacokinetics and pharmacodynamics of remifentanil. II. Model application. Anesthesiology 1997; 86: 24-33.

[67]   Blouin RA, Warren GW. Pharmacokinetic considerations in obesity. J Pharm Sci 1999; 88: 1-7.

[68]   Abernethy DR, Greenblatt DJ, Divoll M, *et al.* Prolonged accumulation of diazepam in obesity. J Clin Pharmacol 1983; 23: 369-76.

[69]   Abernethy DR, Greenblatt DJ, Smith TW. Digoxin disposition in obesity: clinical pharmacokinetic investigation. Am Heart J 1981; 102: 740-44.

[70]   Christoff PB, Conti DR, Naylor C, *et al.* Procainamide disposition in obesity. Drug Intell Clin Pharm 1983; 17: 516-22.

[71]   Cheymol G. Effects of obesity on pharmacokinetics implications for drug therapy. Clin Pharmacokinet 2000; 39: 215-31.

[72]   Juvin P, Vadam C, Malek L, *et al.* Postoperative recovery after desflurane, propofol, or isoflurane anesthesia among morbidly obese patients: a prospective randomized study. Anesth Analg 2000; 91: 714-19.

[73]   Sollazzi L, Perilli V, Modesti C, *et al.* Volatile anesthesia in bariatric surgery. Obes Surg 2001; 11: 623-26.

[74]   Torri G, Casati A, Albertin A, *et al.* Randomized comparison of isoflurane and sevoflurane for laparoscopic gastric banding in morbidly obese patients. J Clin Anesth 2001; 13: 565-70.

[75]   Luce MJ. Respiratory complications of obesity. Chest 1980; 78: 626-31.

[76]   De Divitiis O, Fazio S, Pettito M, *et al.* Obesity and cardiac function. Circulation 1981; 64: 477-82.

[77]   Rexrode KM, Manson JE, Hennekens CH. Obesity and cardiovascular disease. Curr Opin Cardiol 1996; 11: 490-95.

[78]   Buckley FP, Robinson NB, Simonowitz DA. Anaesthesia in the morbidly obese: a comparison of anaesthetic and analgesic regimens for upper abdominal surgery. Anaesthesia 1983; 38: 840-51.

[79]   Michaloudis D, Fraidakis O, Petrou A, *et al.* Continuous spinal anesthesia/analgesia for perioperative management of morbidly obese patients undergoing laparotomy for gastroplastic surgery. Obes Surg 2000; 10: 220-29.

[80]   Shenkman YS, Shir Y, Brodsky JB. Perioperative Management of the Obese Patient. Br J Anaesth 1993; 70: 349-59.

[81]   Buckley, FP. *Anesthesia and Obesity and Gastrointestinal Disorders.* In: *Clinical Anesthesia,* 2nd edition. eds: PG Barash, BF Cullen, RK Stoelting. JB Lippincott Co., Philadelphia. pp.1169-1175, 1992.

[82]   Sawyer RJ, Richmond MN, Hickey JD, et al. Peripheral nerve injuries associated with anaesthesia. Anaesthesia 2000; 55: 980-91.

[83]   Maxwell MH, Waks AU, Schroth PC, *et al.* Error in blood pressure measurement due to incorrect cuff size in obese patients. Lancet 1982; 2: 33-36.

[84]   Mann GV. The influence of obesity on health. N Engl J Med 1974; 291: 178-85.

[85]   Emerick DR. An evaluation of non-invasive blood pressure (NIBP) monitoring on the wrist: comparison with upper arm NIBP measurement. Anaesth Intensive Care 2002; 30: 43-47.

[86]   Block FE, Schulte GT. Ankle blood pressure measurement, an acceptable alternative to arm measurements. Int J Clin Monit Comput 1996; 13: 167-71.

[87]   Dreyfuss D, Soler P, Basset G, *et al.* High inflation pressure pulmonary edema: respective effects of high airway pressure, high tidal volume and positive end-expiratory pressure. Am Rev Respir Dis 1988; 137: 1159-64.

[88]   Ingrande J, Brodsky JB, Lemmens HJM. Regional anesthesia and obesity. Current Opinion in Anaesthesiology 2009; 22: 683-86.

[89]   Taivainen T, Tuominen M, Rosenberg PH. Influence of obesity on the spread of spinal analgesia after injection of plain 0.5% bupivacaine at the L3-4 or L4-5 interspace. Br J Anaesth 1990; 64: 542-46.

[90]   Hood DD, Dewan DM. Anesthetic and obstetric outcome in morbidly obese parturients. Anesthesiology 1993; 79: 1210-18.

[91]   Hodgkinson R, Husain FJ. Obesity and the cephalad spread of analgesia following epidural administration of bupivacaine for Cesarean section. Anesth Analg 1980; 59: 89-92.

[92]   Hogan QH, Prost R, Kulier A, *et al.* Magnetic resonance imaging of cerebrospinal fluid volume and the influence of body habitus and abdominal pressure. Anesthesiology 1996; 84: 1341-49.

[93]   Grau T, Bartusseck E, Conradi R, *et al.* Ultrasound imaging improves learning curves in obstetric epidural anesthesia: a preliminary study. Can J Anaesth 2003; 50: 1047-50.

[94]   Grau T, Leipold RW, Conradi R, *et al.* Efficacy of ultrasound imaging in obstetric epidural anesthesia. J Clin Anesth 2002; 14: 169-75.

[95]   Grau T, Leipold RW, Conradi R, *et al.* Ultrasound control for presumed difficult epidural puncture. Acta Anaesthesiol Scand 2001; 45:766-71.

[96]   Wallace DH, Currie JM, Gilstrap LC, *et al.* Indirect sonographic guidance for epidural anesthesia in obese pregnant patients. Reg Anesth 1992; 17: 233-36.

[97]   Whitty RJ, Maxwell CV, Carvalho JC. Complications of neuraxial anesthesia in an extreme morbidly obese patient for Cesarean section. Int J Obstet Anesth 2007; 16: 139-44.

[98]   Buckley FP, Robinson NB, Simonowitz DA, *et al.* Anaesthesia in the morbidly obese: a comparison of anaesthetic and analgesic regimens for upper abdominal surgery. Anaesthesia 1983; 38: 840-51.

[99]   Fox GS, Whalley DG, Bevan DR. Anaesthesia for the morbidly obese: experience with 110 patients. Br J Anaesth 1981; 53: 811-16.

[100]   Gelman S, Laws HL, Potzick J, *et al.* Thoracic epidural vs balanced anesthesia in morbid obesity: an intraoperative and postoperative hemodynamic study. Anesth Analg 1980; 59: 902-08.

[101]   McMahon MM, Sarr MG, Clark MM, *et al.* Clinical management after bariatric surgery: value of a multidisciplinary approach. Mayo Clin Proc 2006; 81: S34-S45.

[102]   Van Mieghem T, Van Schoubroeck D, Depiere M, *et al.* Fetal cerebral hemorrhage caused by vitamin K deficiency after complicated bariatric surgery. Obstet Gynecol 2008; 112: 434-36.

[103]   Thaisetthawatkul P. Neuromuscular complications of bariatric surgery. Phys Med Rehabil Clin N Am 2008; 19: 111-124.

[104]   Kumar N. Nutritional neuropathies. Neurol Clin 2007; 25: 209-55.

[105]   Weyns FJ, Beckers F, Vanormelingen L, *et al.* Foot drop as a complication of weight loss after bariatric surgery: is it preventable? Obes Surg 2007; 17: 1209-12.

[106]   Elias WJ, Pouratian N, Oskouian RJ, *et al.* Peroneal neuropathy following successful bariatric surgery: case report and review of the literature. J Neurosurg 2006; 105: 631-35.

[107]   Kenkel JM, Lipschitz AH, Luby M, *et al.* Hemodynamic physiology and thermoregulation in liposuction. Plast Reconstr Surg 2004; 114: 503-13.

[108]   Yao F-SF, Savarese JJ. *Morbid obesity.* In: *Anesthesiology: problem oriented patient management.*Ed. Yao F-SF. Lippincott-Raven, Philadelphia. pp. 1001-1018, 1998.

[109]   Byrne TK. Complications of surgery for obesity. Surg Clin North Am 2001; 81: 1181-93.

[110]   Fox GS, Whalley DG, Bevan OR. Anaesthesia for the morbidly obese: experience with 110 patients. Br J Anaesth 1981; 53: 811-16.

[111]   Rawal N, Sjo"strand U, Christofferson E, *et al.* Comparison of intramuscular and epidural morphine for postoperative analgesia in the grossly obese: influence on postoperative ambulation and pulmonary function. Anesth Analg 1984; 63: 583-92.

[112]   Gelman S, Laws HL, Potzick J, *et al.* Thoracic epidural vs. balanced anesthesia in morbid obesity: an intraoperative and postoperative hemodynamic study. Anesth Analg 1980; 59: 902-08.

[113]   Huang GJ, Bajaj AK, Gupta S, *et al.* Increased intraabdominal pressure in abdominoplasty: delineation of risk factors. Plast Reconstr Surg 2007; 119: 1319-25.

[114]   Gravante G, Araco A, Sorge R, *et al.* Pulmonary embolism after combined abdominoplasty and flank liposuction: a correlation with the amount of fat removed. Ann Plast Surg 2008; 60: 604-08.

[115]   Shermak MA, Chang DC, Heller J. Factors impacting thromboembolism after bariatric body contouring surgery. Plast Reconstr Surg 2007; 119: 1590-96.

[116]   Whizar-Lugo VM, Cisneros-Corral R, Reyes-Aveleyra MA, *et al.* Anesthesia for Plastic Surgery Procedures in Previously Morbidly Obese Patients. Anestesia en México 2009; 21: 186-93.

Index

A

Abdominoplasty 48, 63, 65, 70-73, 129
Appetite control 23-24
Appetite signaling peptides 23-24

B

Bariatric surgery procedures 15-20
Body image 29, 33-34
Body mass index 3-5, 51, 54-55, 60
Belt lipectomy 48, 89
Brachioplasty 48, 119, 132-138

C

Cholecystokinin 23-24
Colle's fascia 80-81
Compression garment 56

D

Diabetis 9, 24-25, 51, 149
Duodenal switch 18-19

F

Fleur-de-lys abdominoplaty 66-68, 73-75

G

Gastric banding 19-20
Gherlin 23-24
Gluteal augmentation 91-100

H

Hernia 64
Hypothermia 56, 162

I

Insulin resistance 9

L

Lateral thigh lift 84-85
Lockwood's medial thighplasty 81-83
Lower body lift 48, 119

N

Nutritional deficiency 24-26, 52, 58, 152

O

Obesity associated diseases 8-10
Obesity epidemic 5-8

Control Engineering

Control System Power and Grounding Better Practice

Control Engineering
Control System Power and Grounding Better Practice

David Brown, David Harrold, and Roger Hope

ELSEVIER

CONTROL
ENGINEERING.

Newnes

AMSTERDAM • BOSTON • HEIDELBERG • LONDON • NEW YORK • OXFORD
PARIS • SAN DIEGO • SAN FRANCISCO • SINGAPORE • SYDNEY • TOKYO
Newnes is an imprint of Elsevier

Newnes is an imprint of Elsevier
200 Wheeler Road, Burlington, MA 01803, USA
Linacre House, Jordan Hill, Oxford OX2 8DP, UK

Library of Congress Cataloging-in-Publication Data
An application has been submitted.

British Library Cataloguing-in-Publication Data
A catalogue record for this book is available from the British Library.

ISBN: 0-7506-7826-7

For information on all Newnes publications
visit our Web site at www.newnespress.com

07 08 09   10 9 8 7 6 5 4 3 2 1

Transferred to digital printing 2009

Contents

Figures

Tables

Introduction

Because of inadequate installation practices, control system reliability is often jeopardized before power is applied. Frequently, causes of poor control system reliability are unexamined with the result of the control system receiving a "black eye" in the judgement of the end–user. When causes of poor control system reliability are examined, poor power and grounding practices are often the root cause.

When control systems were proprietary, hardware designs remained unique for each control system. The unique, proprietary designs required manufacturers to develop complete and detailed documentation for every aspect of the system, including installation.

Open, hybrid systems make use of commercial off-the-shelf devices, such as power supplies, carriers, mounting methods, etc., almost to the point where it's difficult to identify a control system manufacturer without first looking at the logos and markings.

Most control system manufacturers still develop and distribute documentation, however the detail and clarity is often missing. *Control Engineering*, working in conjunction with several control system manufacturers and recognized power and grounding experts, has developed this single control system power and grounding resource.

When this resource is used with control system manufacturer installation documentation, users can expect robust, reliable control system installation; one that remains free of "phantom" problems caused by power and grounding glitches.

CE Mark Statement

C E

If you intend to have your control system certified for compliance to appropriate European Union directives, you must carefully consider the installation procedures described in the control system manufacturer's installation manual. Normally, products must be installed in accordance with the procedures to obtain certification.

Warnings, Cautions, and Notes

Warnings, Cautions, and Notes attract attention to essential or critical information in this manual. The types of information included in each are explained in the following:

Warning... All warnings have this form and symbol. Do not disregard warnings. They are installation, operation, or maintenance procedures, practices, conditions, statements, and so forth, which if not strictly observed, may result in personal injury or loss of life.

Caution... All cautions have this form and symbol. Do not disregard cautions. They are installation, operation, or maintenance procedures, practices, conditions, statements, and so forth, which if not strictly observed, may result in damage to, or destruction of, equipment or may cause long term health hazards.

Note ... *Notes have this form and symbol. Notes contain installation, operation, or maintenance procedures, practices, conditions, statements, and so forth, that alert you to important information which may make your task easier or increase your understanding.*

Liability

Every effort has been made to ensure the information here provides appropriate guidance. However, applications, standards, and equipment change regularly. Therefore, Reed Business Information does not assume and hereby disclaims any liability to any person for any loss or damage caused by errors or omissions in the material contained herein, regardless of whether such errors results from negligence, accident, or other cause whatsoever.

Chapter 1 Site Preparation Overview

Before you install your control system, your site must be adequately prepared to accept the system. Such preparation is of prime importance for personnel safety and system availability. Preparation includes proper power and grounding and satisfactory signal shielding to maximize reliable and efficient operation of your control system.

This guide provides information you need to help adequately prepare your site. It is believed that the recommendations and guidelines in this manual meet or exceed applicable local electrical codes and regulations. If differences occur between this guide and local codes and regulations, then the local codes and regulations should be given preference.

While the recommendations and guidelines in this guide cover most situations, there will, no doubt, be peculiarities at some installations that require alternate approaches. In these cases, it is recommended that you contact your control system manufacturer for assistance.

System Availability

Basic requirements exist to assure system availability at levels you require, and include single-point grounding, clean AC and DC power, and other considerations as described below. Beyond these, you may be able to make tradeoffs between system availability required by your process and economic concerns.

If a process is critical to plant operation and process interruptions must be minimized, secondary power sources, signal redundancy, and other forms of backup controls should be considered. If less stringent requirements are acceptable, you can make certain cost control decisions. However, basic requirements cannot be compromised. This guide provides basic requirements and ideas on backup controls.

Impact of New Technologies on System Availability

Control system availability means that a system is available to perform its intended function. System availability can be defined mathematically as the percent of time that the system is available (operating as intended) compared to the total time that the system is in use.

The percent of time of a system's availability is expressed as Mean Time Between Failures (MTBF) divided by the sum of the MTBF and the Mean Time to Repair (MTTR) times 100. In equation form:

Percent available time = MTBF/(MTBF+MTTR)*100

System "Boundary" and "Operational" conditions make the availability definition specific to a system. Typically, boundary conditions are determined by operator consoles, system communications, AC and DC power equipment, controllers, and I/O assemblies.

Operational conditions traditionally limit the unavailability of display or control to one loop. Operational condition are based on historical analysis, and assumes that a final control element has the highest component failure rate. By definition, control strategies that can survive a final control element failure are available if no more than one final control element fails.

However, the use of new technologies may impact these definitions. Fieldbus technologies, which can connect multiple final control elements on a single bus, may extend the system boundary conditions from only an I/O assembly to include multi-drop fieldbus segments. Examples of conditions which can change boundary conditions are: loss of power in a bus-powered fieldbus (which may cause the loss of multiple instruments) and loss of a fieldbus intrinsic safety barrier (which may cause loss of multiple elements).

Importance of Mean Time to Repair

Mean time to repair (MTTR) is an important consideration for system availability; the shorter the MTTR, the more the system is available. MTTR includes identifying a fault after its first indication, location of the fault, and the time required to correct the fault. For information about methods to minimize MTTR, refer to ANSI/ISA-84.01-1996, *Application of Safety Instrumented Systems for the Process Industries.*

Electrical Noise Influences

The reliability of a control system can be reduced when it is subjected to unusually high amounts of electrical noise, but proper power and grounding can minimize the effects.

Various components of a system can be affected to different degrees. The following observations have been made during system troubleshooting:

- Properly installed, standard, non-communicating (called traditional or classic) I/O products that use 4–20 mA analog I/O signals only are typically not affected by electrical noise, except high frequency noise.

- High frequency electromagnetic noise generated by AC drives, insulated gate bipolar transistor (IGBT) power switching devices, and other sources can affect all control products without bias to a particular vendor.

- Digital data signals used in bus communication systems are susceptible to electrical noise.

Minimizing Influences

These effects can be minimized by:

- Using isolated AC power sources

- Grounding at single points

- Minimizing undue influence on signal wiring from stray magnetic fields

- Selecting appropriate cables and pathways, including adequate cable separation.

AC Power Considerations

Good AC power installation practices can improve the operation of control systems.

Voltage Regulation

Typically, primary AC power is supplied to a control system through an isolation transformer or other filtering means from the commercial AC source to obtain adequate voltage regulation. The source and its power distribution network should be able to handle initial inrush currents (lasting at least ten cycles) and still maintain voltage regulation within the operating specifications of the control system.

Electrical Noise Isolation

For isolation from source electrical noise, it is recommended that the control system be supplied from an isolation transformer or power source that is separate from sources used for room lighting and other power loads. Ideally, this separate power source should be supplied from the highest primary source voltage available.

Backup Power

Often, the process being controlled requires backup power to achieve defined levels of reliability and availability. Backup power should be considered if the following conditions exist:

- The primary power source has a history of failure or of fluctuations beyond the specified tolerance of equipment input power.

- The process requires electrical power for control, and there are no non-electrical means available for control.

- Loss of power requires manual setting of relays or solenoids.

- Loss of power can cause the process to enter an uncontrollable cycle.

- Loss of power causes a process reaction that creates a hazardous condition or an extensive loss of product.

You may want to use power backup of the source only, or you may prefer power backup of every power supply in the control system, such as backup power for every controller. The extent of backup affects system availability and, therefore, is a consideration when determining the level of system availability required.

System Ground

The ground network for a control system is a vital consideration. Proper ground networks can eliminate the affects of potential deficiencies in the power source and distribution network that affect personnel safety and control system operation.

Poor or faulty grounds are among the most common causes of control system faults. For example, improper grounding of AC neutrals at two or more points can cause stray currents that, in turn, can induce noise and operational errors into the system.

Many times, glitches in systems are traced to electrical ground faults only after hours are spent checking the system itself. The extra time and effort spent in laying out a good ground network is rewarded by easier startup and more reliable operation. It is suggested that you review IEEE Standard 1100-1992, *Recommended Practice for Power and Grounding Sensitive Electronic Equipment*, for industry-accepted methods to obtain good power and grounding.

Earth Ground

Proper earth grounding is also important to user safety and efficient operation of a control system. A good earth grounding system safely conducts stray electrical currents into the ground. It can also considerably reduce electrical noise caused by static discharge, lightning charges, and electrical components of electro-mechanical devices.

Building steel must be part of a good earth ground system. For protection from both lightning and electrical noise caused by plant equipment, the steel and earth must form a conductive network and equal voltage potential ground system. For example, electric motor frame grounds are normally connected to building steel as supplemental grounding. Therefore, the steel must contain a good, low impedance, electrical path to earth to provide adequate electrical noise protection for control systems.

Signal Wiring for Traditional I/O Systems

To achieve optimum I/O subsystem operation, the system should use industry-prescribed cables (as described below) that are properly installed to negate electromagnetic interference (EMI) and other electrical noise that can adversely affect the instruments controlling the process.

Selecting Analog Current or Voltage Signals

Among analog signals, current signals generally have higher noise immunity than voltage signals. It is recommended that two-wire current transmission be used whenever possible.

Separating Signal Lines From Power Lines

Signal cabling should be kept away from AC power lines, transformers, rotating electrical equipment, or other high power machinery to reduce the possibility of electromagnetic interference being induced on analog and discrete signals. Normally, power lines and signal cables should be carried in separate cable trays. If necessary, power lines and signal cables can be carried in the same tray if separated by a tray divider. Some national electric codes may also require separation. Table 5-1 lists recommended distances between signal cables and power lines.

Multi-Conductor Cables

Signal circuits rarely contain enough power to generate noise in adjacent circuits. Typically, if your plant design practices allow, traditional I/O signals can be run through multi-pair cables without generating interfering noise. Using multi-pair cables instead of individual wires or pairs usually reduces installation costs and provides adequate noise immunity. Multi-pair cables should use color-coding, numbering, and so forth to identify pairs of twisted wires for each signal.

For added protection from electrical noise, standard, 4–20 mA, 1–5 VDC, and 24 VDC discrete signals may use multi-pair cable in which each twisted pair contains a metallic shield and drain wire plus an overall metallic shield and drain wire for the cable. Normally specified cable is shielded, twisted pair per National Electric Code (NEC) Type tray cable (TC), instrumentation tray cable (ITC), power limited tray cable (PLTC), or metal clad (MC). Multi-core cable with a single, overall shield is generally accepted for wiring 120 VAC discrete signals and is often used for 240 VDC discrete signals.

Protection from electrical noise must be given to low amplitude signals, such as those derived from millivolt, thermocouple, and resistance-temperature devices. Commercially available, individually shielded, single-pair or multi-pair cables should be used for such devices. However, even with their use, the cables should be kept away from AC circuits, especially those containing motor or generator control solenoids, similar types of circuits with relatively high inrush currents, and solid-state switching circuits.

Various industry standards, such as IEEE 518-1982, *IEEE Guide for the Installation of Electrical Equipment to Minimize Noise Inputs to Controllers from External Sources*, describe noise identification and classification as well as recommended wiring practices. It is recommended that you obtain appropriate standards.

Signal Wiring for I/O Bus Systems

To achieve optimum system operation, digital I/O bus systems, such as AS-Interface, DeviceNet, FOUNDATION Fieldbus, and Profibus, must be properly installed. Otherwise, electromagnetic interference (EMI) and other electrical noise can adversely affect the capabilities of the instruments controlling the process.

When using digital I/O systems, cable types, lengths, and sizes; grounding requirements; power; and other factors must be considered. Information about these factors is usually available from the control system manufacturer. Once a digital I/O system is installed, its operation should be verified before it is commissioned.

For most digital I/O systems, important design information is available on the manufacturer's web sites. Sites for commonly used digital I/O systems are listed in Chapter 5 of this manual.

Intrinsically Safe I/O Systems

Most control system manufacturers offer a selection of intrinsically safe I/O systems, including I/O cards, isolating power supplies, and local bus isolators. When using the control system manufacturers intrinsically safe devices, consult the manufacturers installation documentation.

Most control systems can be used with intrinsically safe I/O systems from third-party manufacturers. When using intrinsically safe devices from multiple manufacturers, it is suggested that you review intrinsic safety information in ANSI/ISA-RP12.6-1995, *Wiring Practices for Hazardous (Classified) Locations —Instrumentation Part 1: Intrinsic Safety.*

Lightning Protection

In an area where damage from electrical storms may occur, a lightning protection system should be installed to protect both personnel and equipment. The protection system should include buildings, the power distribution system, copper wire LAN systems, and any cables that run outdoors to other locations.

For detailed information, refer to:

■ *National Fire Protection Association Inc. (NFPA) Lightning Protection Codes NFPA-78*

■ *IEEE Recommended Practices for Grounding of Industrial and Commercial Power Systems IEEE Standard 142-1991*

Chapter 2   AC Power Distribution

This chapter provides recommendations for powering control systems from AC sources, including information on single-phase power, three-phase power, transformer isolation, distribution, power supply connections, and AC connections to workstations.

AC Power Considerations

Commercial AC power utilities normally provide power that meets the voltage and frequency requirements of the control system. However, plant distribution networks may drop 5 percent or more of the input AC power between the plant's service entrance point and the final power connection to various portions of the control system.

Additionally, power disturbances can be introduced on power lines by starting transients induced from large motors and other loads connected to the distribution system. These disturbances can cause momentary line-voltage reductions as well as possible waveshape distortions. For example, starting large motors may momentarily drop line voltage as much as 15%. Therefore, assessing AC power requirements for a control system and then designing a plant AC distribution system that meets the requirements is critical to reliable, efficient control system operation.

Maintaining Power Quality

To maintain good AC power quality, items such as power loss, intermittent noise, low voltage, transients, and surges on power lines must be designed around or otherwise controlled. Process application and process availability are critical factors for determining the amount of noise protection desired.

It is recommended that AC power supplied to a control system be taken from an AC power distribution system isolated from the AC power supplied to all other functions in the plant area. As a minimum, an isolation transformer should be installed between the commercial power source and the main AC power distribution panel for the control system. The power source should be as close to the system as possible.

Note ... *Isolation transformers are recommended because they inherently provide good line regulation and transient filtering. In addition, each building or site containing control system devices should have a separate power source or backup power source, or both.*

A separate distribution system is recommended for each building containing control system devices. In addition, if system devices are located in the same building, but in different rooms that are more than 200 cable feet from the AC power source, an isolation transformer for the devices in each room is generally beneficial.

Loads, such as AC drives and solid-state switching electronics, on AC power systems other than the control system cannot be ignored. These loads can induce electrical noise on nearby power systems. Protection for electrical noise generated by these loads may be required.

To suppress electrical noise, a dedicated feeder between the main distribution panel and the control system branch panel is recommended. Other devices that may be used to provide protection, either alone or in combination, include:

- Noise filters

- Line conditioners

- Voltage regulating power sources

- Motor-generator sets

- Uninterruptible power supply (UPS)

If loss of power from a commercial power source is a probability, a backup power source, such as an uninterruptible power supply for critical portions of the control system, is recommended.

Primary Power Requirements

Most control systems use single-phase power in a voltage range of 85 to 264 Volts and a frequency range of 47 to 63 Hz. These ranges allow for AC input power from normal 120 VAC and 240 VAC sources. Frequently, a plant uses three-phase power. If it is, the load between phases should be balanced at each power panel, minimizing any voltage differentials between the AC neutral and the ground conductors. Typically, one phase powers control system devices, another powers associated field devices, and the other powers associated support equipment.

The power source and its power distribution network should be sized to handle initial inrush currents that can last up to 10 cycles and still be able to regulate the voltage to the control system.

Note ... *The primary power source must be free of non-repeating power interruptions greater than 20 milliseconds. A power interruption greater than 20 milliseconds can cause loss of control, system configuration data, and process data in a control system.*

If your existing plant power does not meet the 20-millisecond requirement, uninterruptible power supplies (UPS) are available from a variety of sources. Small supplies are available to provide backup to selected plant areas. Large supplies can be used for a single common unit to back up an entire system. If a common unit is used, each plant area should be isolated in a manner duplicating the primary source isolation method.

When using isolation transformers, the input power to the transformers should be supplied from the highest line voltage available from the commercial source. Using the highest power makes distribution easier. In addition, an isolation transformer should be dedicated to the control system. Other systems, such as emergency shutdown (ESD) systems and programmable logic controllers (PLCs), should have their own source or sources and also should be isolated.

Primary Power Calculation

Manufacturer product data sheets usually list the power input requirements. Review these sheets and add the individual power requirements to obtain the total power required for control systems.

Recommended Wire Sizes

The wiring from a power source to equipment should be large enough to maintain the specified voltage range at equipment input terminals when all equipment is energized. Wiring must conform to NFPA and NEC as well as applicable local, regional (state, province), and national codes to ensure that it can conduct the current load safely and without overheating.

Recommended wire sizes for various load currents and run lengths are usually determined from formulas that take into account ampacity, temperature, power factor (for switching loads), and allowable voltage drop limits. Formulas and further wiring recommendations are found in appropriate NEC, CSA, and European standards. Refer to these standards to determine proper wiring lengths and sizes for your plant.

Power Cable Run Alternatives

Power cables may be run underground or above ground. Each method offers advantages. Underground runs are generally better physically protected than above ground runs and are safer from fire. However, underground runs can be subject to faults caused by moisture and are difficult to repair or modify unless underground cable vaults are installed. In addition, if the control system is being installed into an existing plant, adding underground cable may be impossible. If you choose above-ground runs, cables should be physically supported in conduits or on cable trays.

AC Distribution System

Input AC power is supplied through an isolation transformer or UPS, and the AC ground point for the control system is established at or near the transformer or UPS. The AC circuit conductors are routed through the main distribution panel (containing the main disconnect switch) into the circuit breaker panel or panels.

This design meets or exceeds the requirements for grounding of the NEC. For large control systems, multiple circuit breaker panels should be used. Separate panels are dedicated to system enclosures and to control system workstations.

Figure 2-1 provides details for single-phase wiring between an isolation transformer and the main distribution panel.

Figure 2-2 shows an uninterruptible power supply (UPS) used in a single-phase system with an isolation transformer.

Figure 2-3 provides details for three-phase wiring between an isolation transformer and the main distribution panel.

Figure 2-4 shows an uninterruptible power supply (UPS) used in a three-phase system with an isolation transformer.

Figure 2-5 provides an overview of an adequate grounding network for an AC distribution system.

Figure 2-6 shows a method of wiring grounded type AC receptacles through a circuit breaker panel to obtain a good ground system.

In Figure 2-6, the neutral and ground conductors are shown bonded to separate bus bars inside the circuit breaker panel. The bus bars are electrically isolated from the panel and from each other. Throughout the distribution system, all AC circuit conductors (line, neutral, and ground) are electrically isolated from their conduits and circuit breaker panels.

This conductor isolation is maintained from the isolation transformer or UPS to the point of final connection at the instrumentation. The only connection between neutral, isolated ground, and earth ground is at the main bonding jumper. The insulated ground conductor should be the same size or larger than the phase and neutral conductors. The connection should never be made to building steel, but building steel can be connected to the control system ground (CSG), as shown in Figure 2-1.

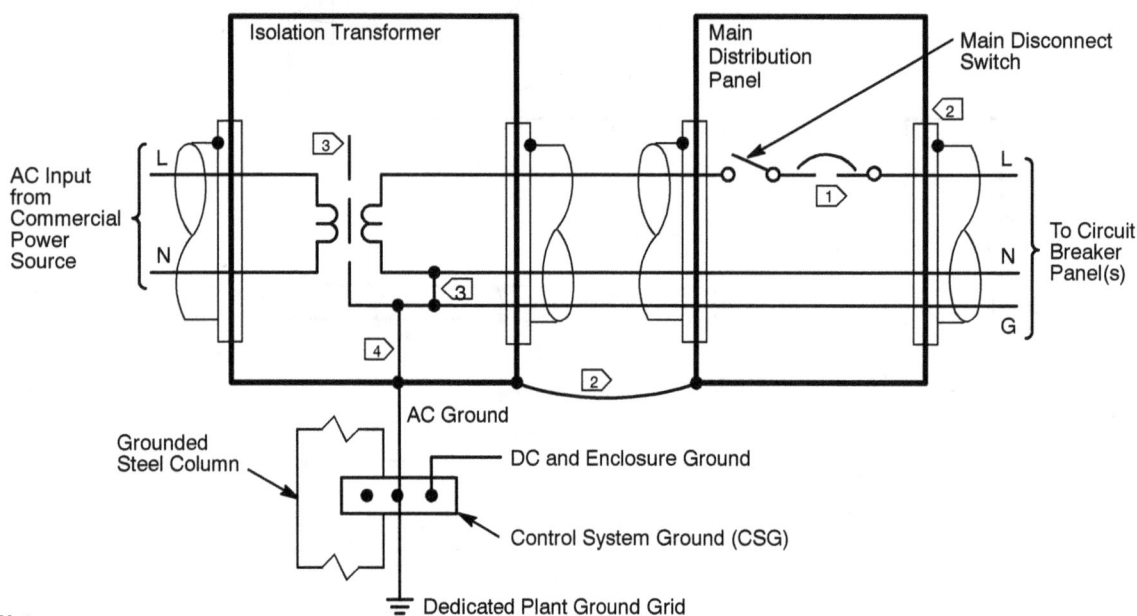

Notes:

1 > Circuit breaker, as required by national codes and regulations

2 > Conduit provides a safety ground connection for individual equipment housings. If no conduit or non-metallic conduit is used, bond housings together with cable. See NEC table 250-122.

3 > All isolation transformers should be shielded.

4 > The conductor between the neutral and ground leads and the dedicated AC ground should be as short as physically possible.

Legend:
L Line
N Neutral
G Ground

Figure 2-1   Single Phase, Separately Derived, AC Power Input System

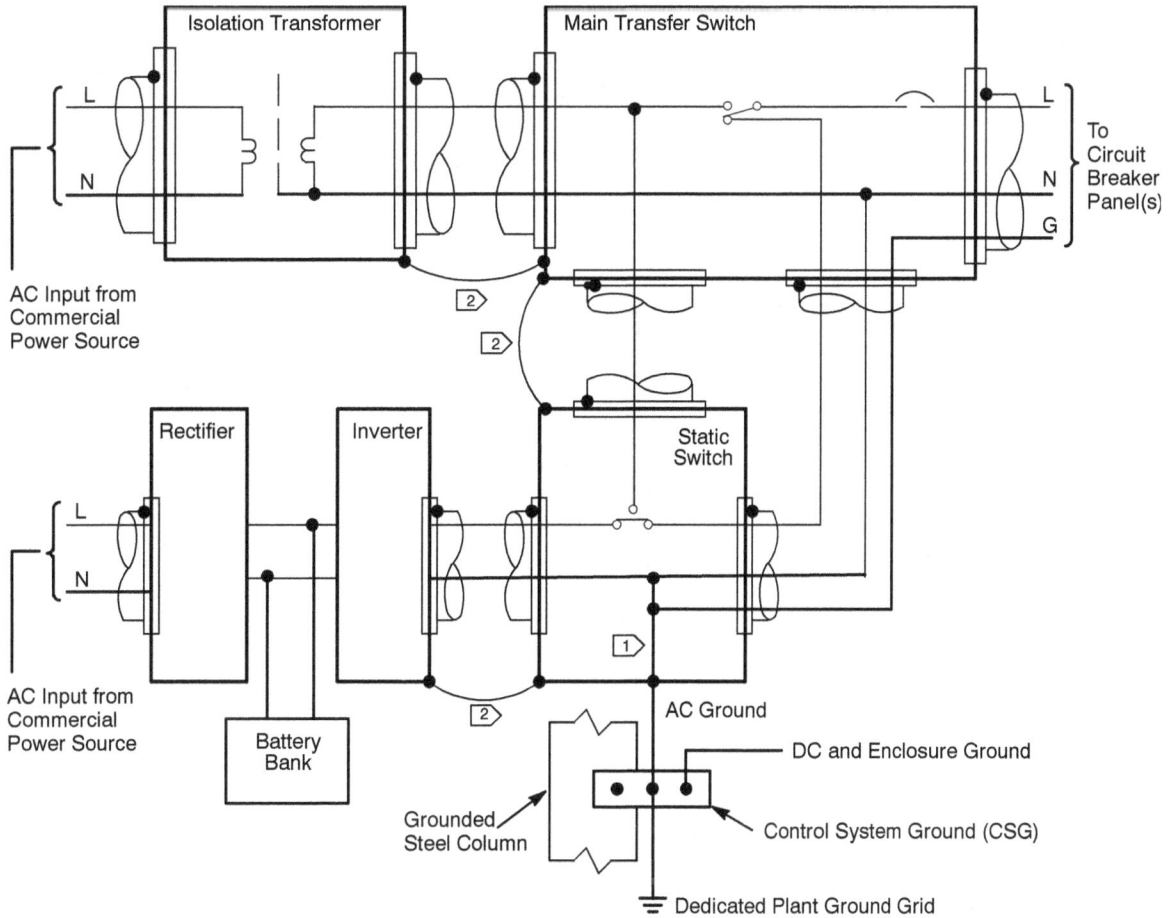

Notes:

1 The conductor between the neutral and ground leads and the dedicated AC ground should be as short as physically possible.

2 Conduit provides a safety ground connection for individual equipment housings. If no conduit or non-metallic conduit is used, bond housings together with cable. See NEC table 250-122.

Legend:
L Line
N Neutral
G Ground

Figure 2-2   Single-Phase, Separately Derived, Reverse Transfer Uninterruptible Power Supply (UPS) with a Manual Transfer Switch

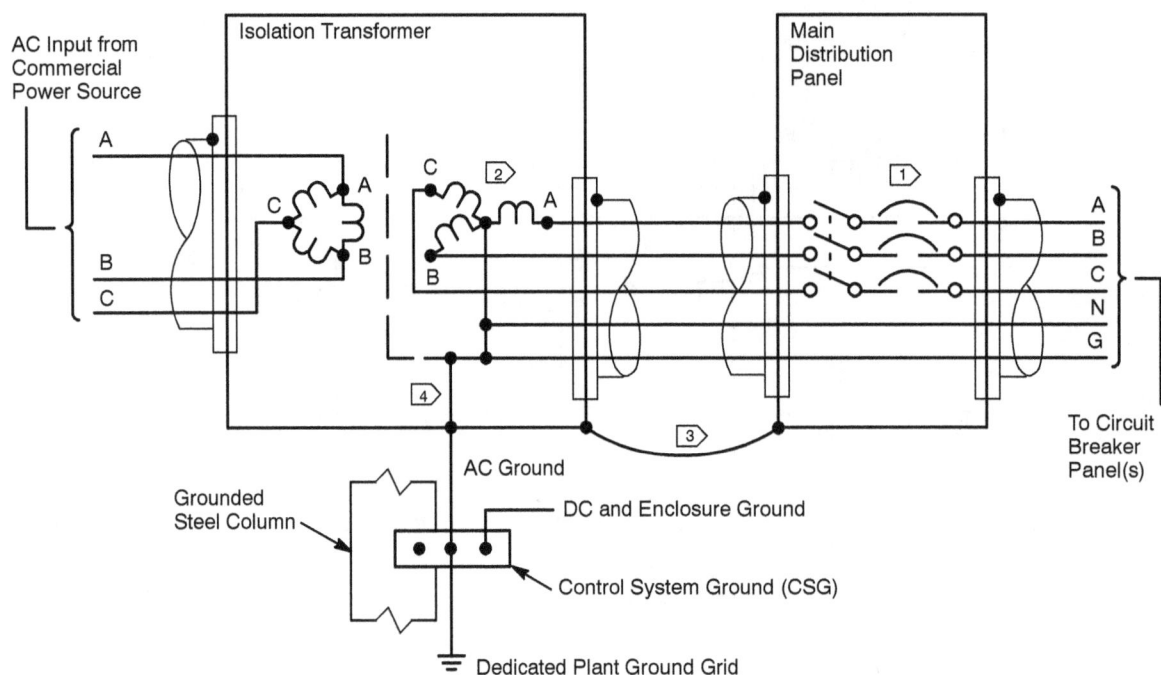

Notes:

1. Circuit breaker, as required by local codes and regulations

2. The isolation transformer secondary can be a 208Y/120 Volt, 120 Volt, 120/240 Volt, or European 230/240 Volt output.

3. Conduit provides a safety ground connection for individual equipment housings. If no conduit or non-metallic conduit is used, bond housings together with cable. See NEC table 250-122.

4. The conductor between the neutral and ground leads and the dedicated AC ground should be as short as physically possible.

Legend:
L   Line
N   Neutral
G   Ground

Figure 2-3   Three Phase, Separately Derived, AC Power Input System

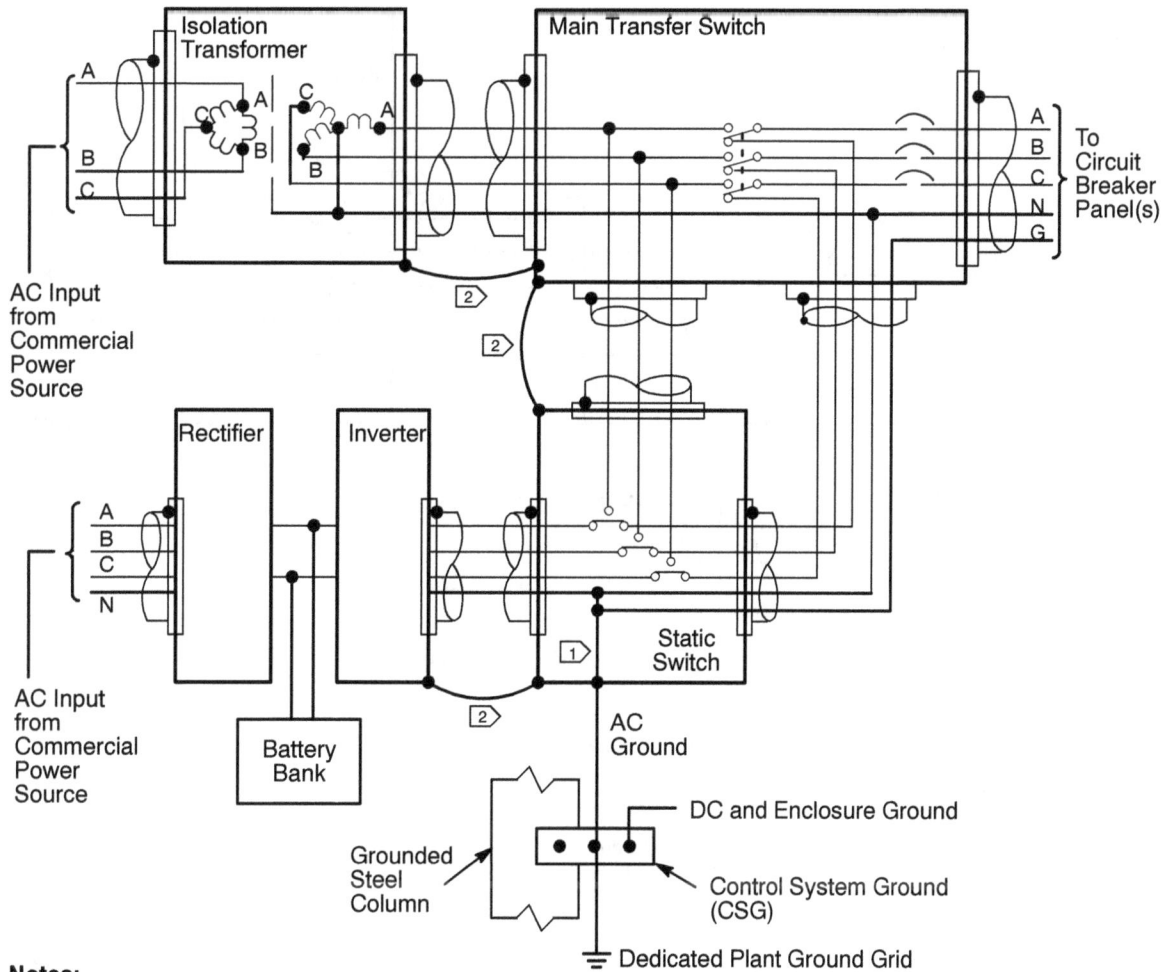

Notes:

1. The conductor between the neutral and ground leads and the dedicated AC ground should be as short as physically possible. Table C-2 provides length and size information.

2. Conduit provides a safety ground connection for individual equipment housings. If no conduit or non-metallic conduit is used, bond housings together with cable. See NEC table 250-122.

Legend:
L Line
N Neutral
G Ground

Figure 2-4   Three-Phase, Separately Derived, Reverse Transfer Uninterruptible Power Supply (UPS) with a Manual Transfer Switch

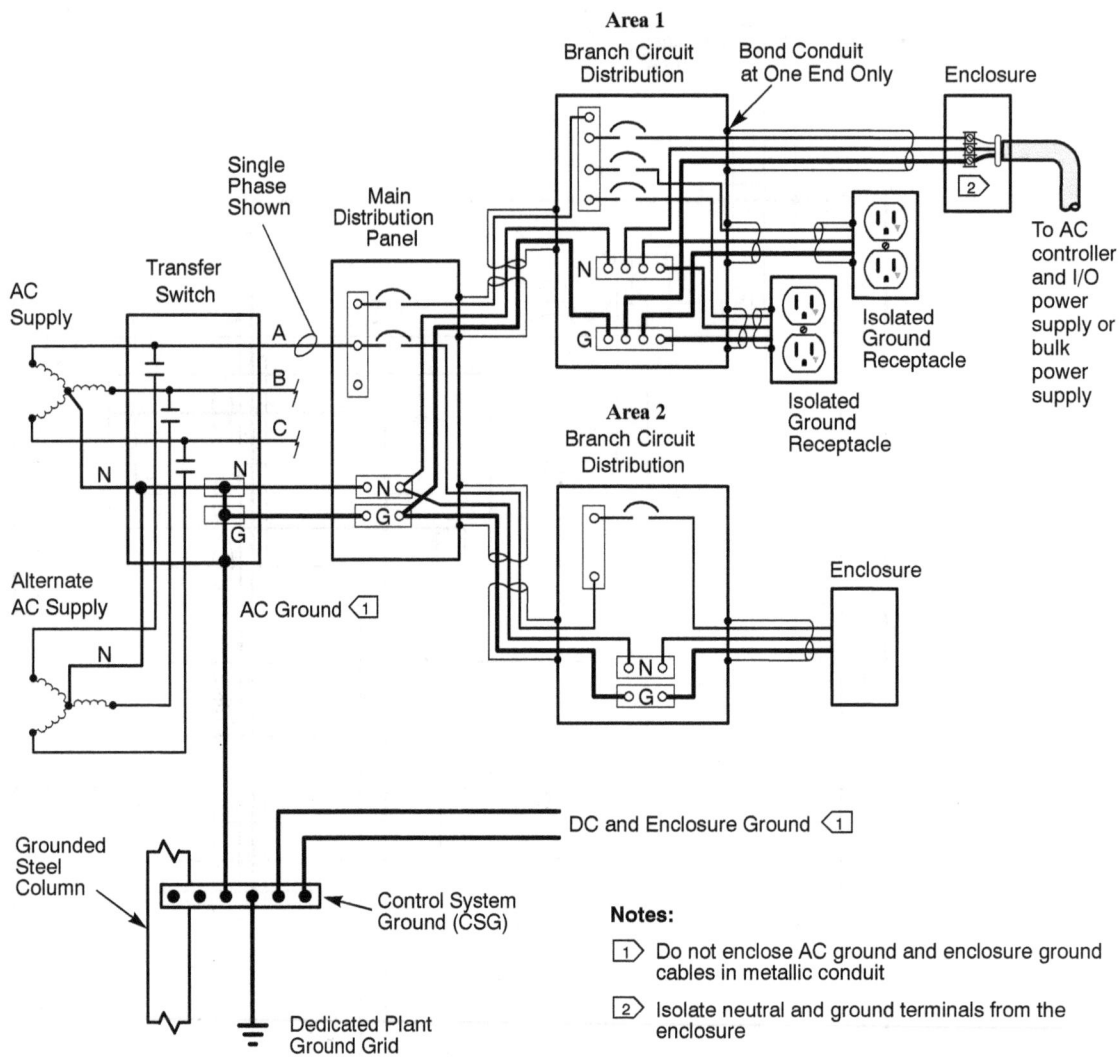

Figure 2-5   AC Distribution System Grounding

Area 1
Branch Circuit Distribution
Bond Conduit at One End Only
Enclosure

Single Phase Shown
Main Distribution Panel
Transfer Switch
AC Supply
A
B
C
N

To AC controller and I/O power supply or bulk power supply

N
G

Isolated Ground Receptacle

Area 2
Branch Circuit Distribution

Isolated Ground Receptacle

N
G

AC Ground ⟨1⟩

Alternate AC Supply
N

Enclosure

N
G

DC and Enclosure Ground ⟨1⟩

Grounded Steel Column

Control System Ground (CSG)

Dedicated Plant Ground Grid

Notes:

⟨1⟩ Do not enclose AC ground and enclosure ground cables in metallic conduit

⟨2⟩ Isolate neutral and ground terminals from the enclosure

Conduit provides a safety ground connection for individual equipment housings. If no conduit or non-metallic conduit is used, bond housings together with cable. See NEC table 250-122.

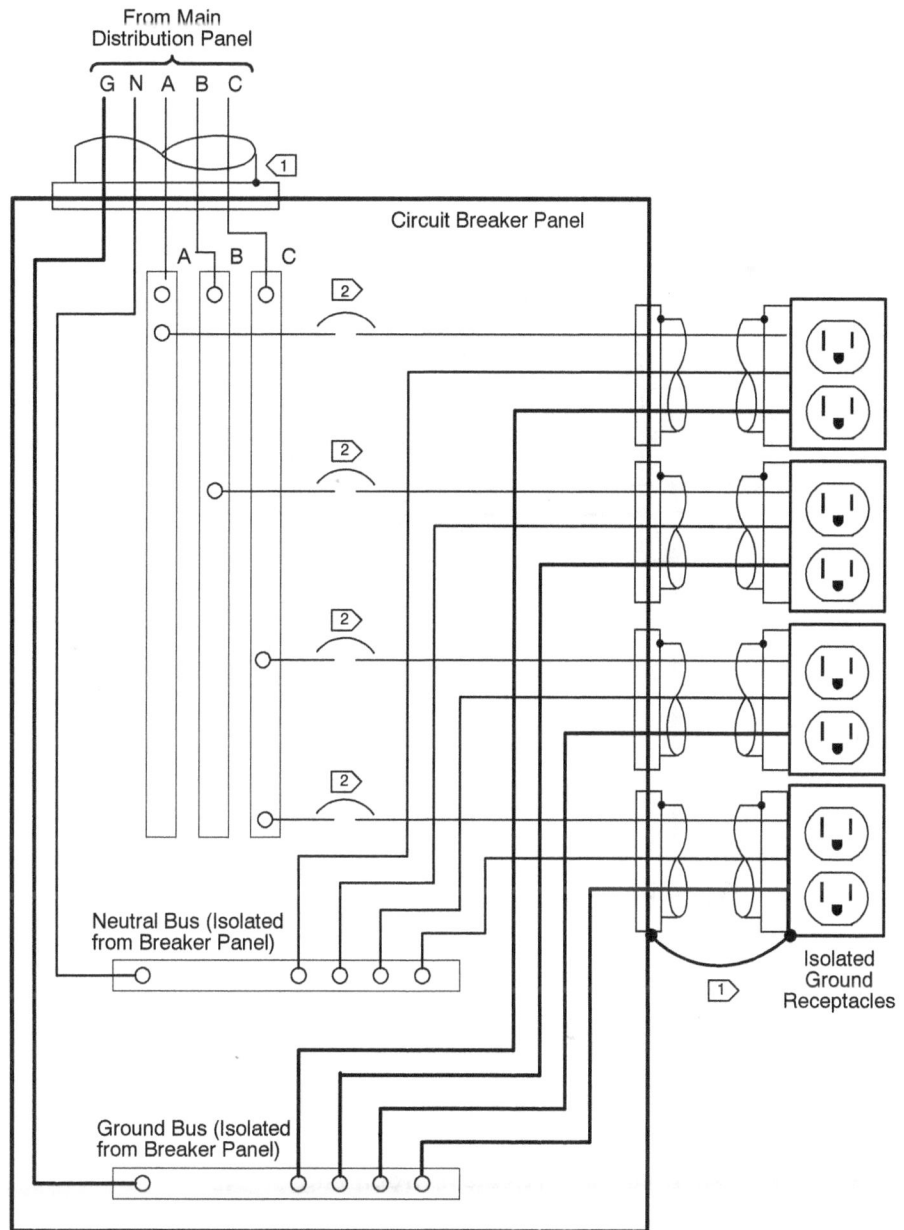

From Main
Distribution Panel

G N A B C

Circuit Breaker Panel

A B C

Neutral Bus (Isolated
from Breaker Panel)

Isolated
Ground
Receptacles

Ground Bus (Isolated
from Breaker Panel)

Notes:

1. Conduit provides a safety ground connection for individual panels. If no conduit or non-metallic conduit is used, bond panels together with cable. See NEC table 250-122.

2. Circuit breaker as required by national codes and regulations.

Figure 2-6   Single Circuit Breaker Panel Wiring for Isolated Ground Recept

AC Power Supply Connections

It is recommended that AC power from a breaker be connected to a terminal block inside the enclosure, as illustrated in Figure 2-5. AC power for control system power supplies is connected from the enclosure terminal blocks to terminals on the supplies.

Figure 2-7 shows the line, neutral, and ground connections for a control system AC to DC controller and I/O power supply. Figure 2-8 shows line, neutral, and ground connections for a single bulk power supplies. Figure 2-9 shows line, neutral, and ground connections for multiple, grouped, bulk power supplies.

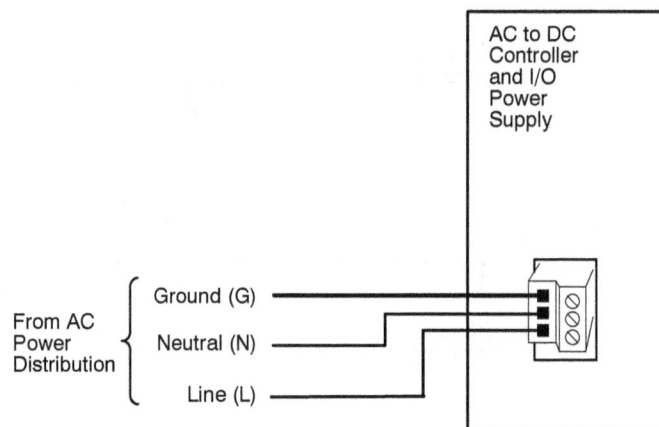

Figure 2-7   AC to DC Controller and I/O Power Supply Input Connections

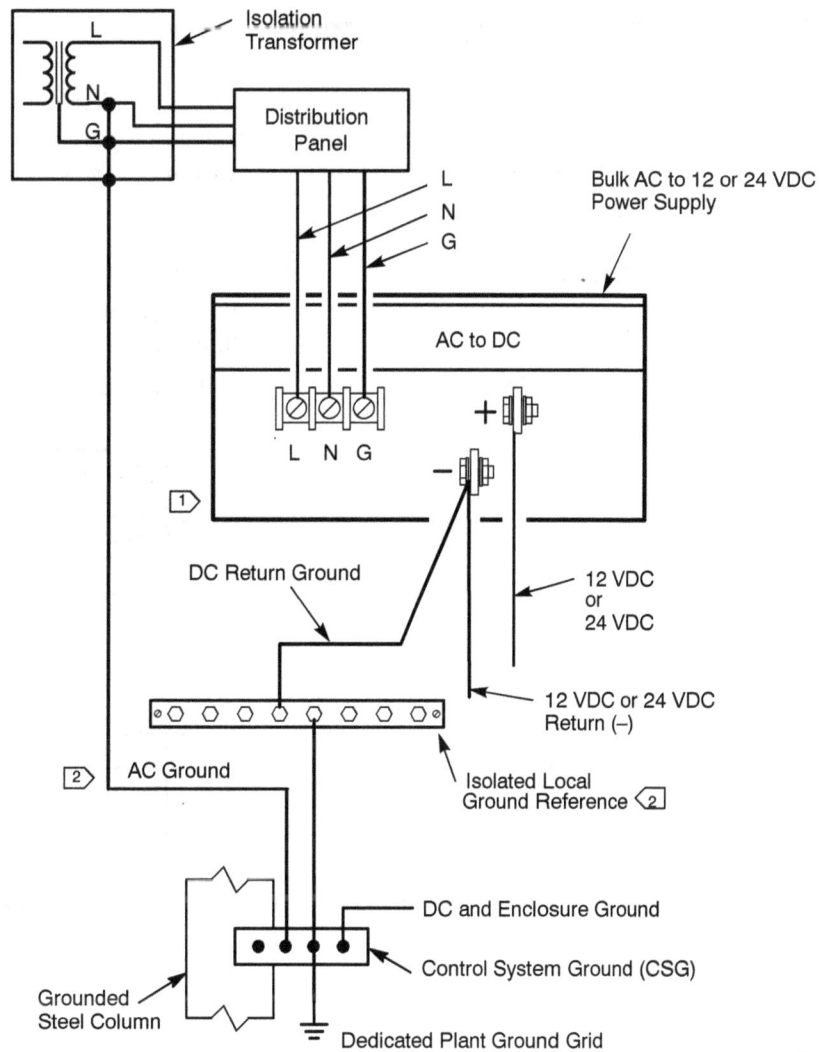

Figure 2-8   Connections for Single Bulk Power Supply, Including Ground

In the figure:

- Isolation Transformer (L, N, G)
- Distribution Panel
- L, N, G
- Bulk AC to 12 or 24 VDC Power Supply
- AC to DC
- L N G
- ⟨1⟩
- DC Return Ground
- 12 VDC or 24 VDC
- 12 VDC or 24 VDC Return (–)
- ⟨2⟩ AC Ground
- Isolated Local Ground Reference ⟨2⟩
- DC and Enclosure Ground
- Control System Ground (CSG)
- Grounded Steel Column
- Dedicated Plant Ground Grid

Notes:

⟨1⟩ Power supply chassis is grounded through the mounting enclosure.

⟨2⟩ See Figure 4-1 and Figure 4-2 for ground connection details.

Legend:
L Line
N Neutral
G Ground

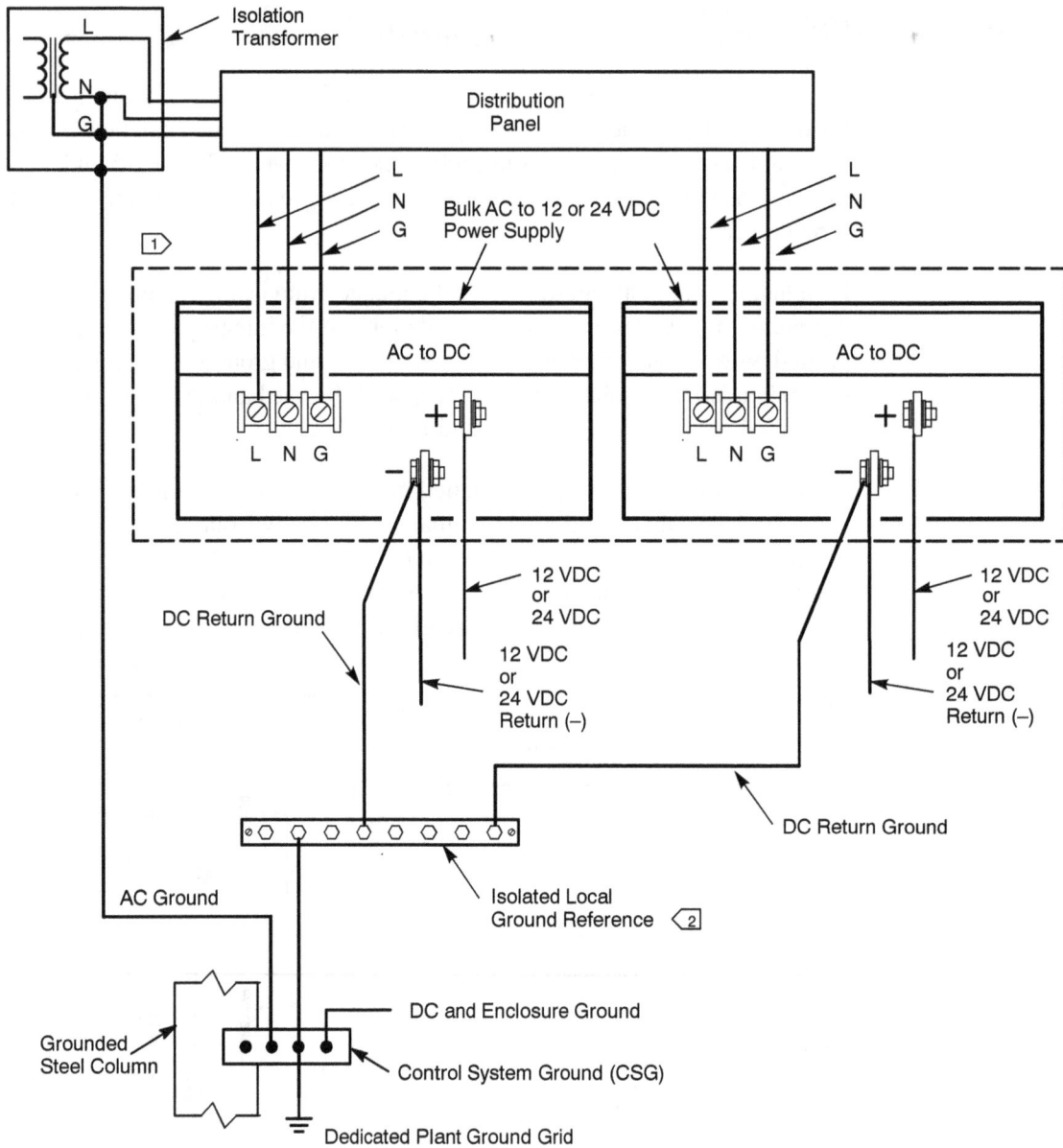

Figure 2-9   Connections for Multiple, Grouped Bulk Power Supplies, Including Ground

Notes:

1 Dashed line indicates that bulk power supplies are grouped in an enclosure or plant area.

2 See Figure 4-1 and Figure 4-2 for ground connection details.

Legend:
L Line
N Neutral
G Ground

Workstations and Peripheral Equipment

All AC power for control system workstations should be routed from a circuit breaker panel to dedicated isolated AC receptacles. The panel can be the same one that is used for equipment in enclosures, as shown in Figure 2-5 and Figure 2-6. These figures illustrate a good safety ground setup, using isolated ground receptacles.

Peripheral equipment can sometimes be plugged into nearby receptacles, depending on the equipment power isolation and the integrity of the safety ground wire. To assure proper equipment power and ground connections, it is recommended that isolated ground receptacles be used and that they are connected as shown in Figure 2-5 and Figure 2-6.

Isolated ground receptacles are detailed in Figure 2-10. They must be constructed and installed so that the ground terminal is electrically isolated from the conduit and the box in which the receptacle is mounted. If non-isolated communications or signal wiring is used, connected peripheral units and electronic units must receive AC power from the same circuit breaker panel.

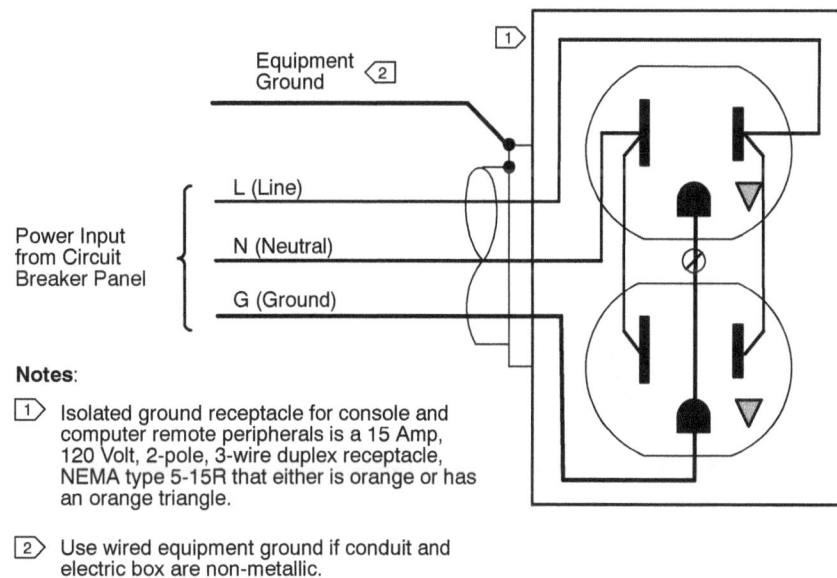

Notes:

1 Isolated ground receptacle for console and computer remote peripherals is a 15 Amp, 120 Volt, 2-pole, 3-wire duplex receptacle, NEMA type 5-15R that either is orange or has an orange triangle.

2 Use wired equipment ground if conduit and electric box are non-metallic.

Figure 2-10 Isolated Ground Receptacle Details

Auxiliary Equipment

You may want to use auxiliary equipment, such as fans, modems, and other small AC loads in your enclosures. AC power for such equipment is typically routed through a 1-pole, 15-ampere circuit breaker to AC receptacles. You may use circuit breakers capable of handling current up to the required load.

However, do not use breakers greater than 30 Amperes. Most NEC codes limit these circuits to 30 Amperes and most wiring used for auxiliary equipment is not rated for current greater than 30 Amperes.

Be sure to use circuit breakers for auxiliary equipment so that faults in the auxiliary equipment do not cause the main breaker to trip. Figure 2-11 shows a possible design, including receptacles for the primary and redundant power supplies. A typical North American receptacle is shown; use receptacles appropriate for each world area.

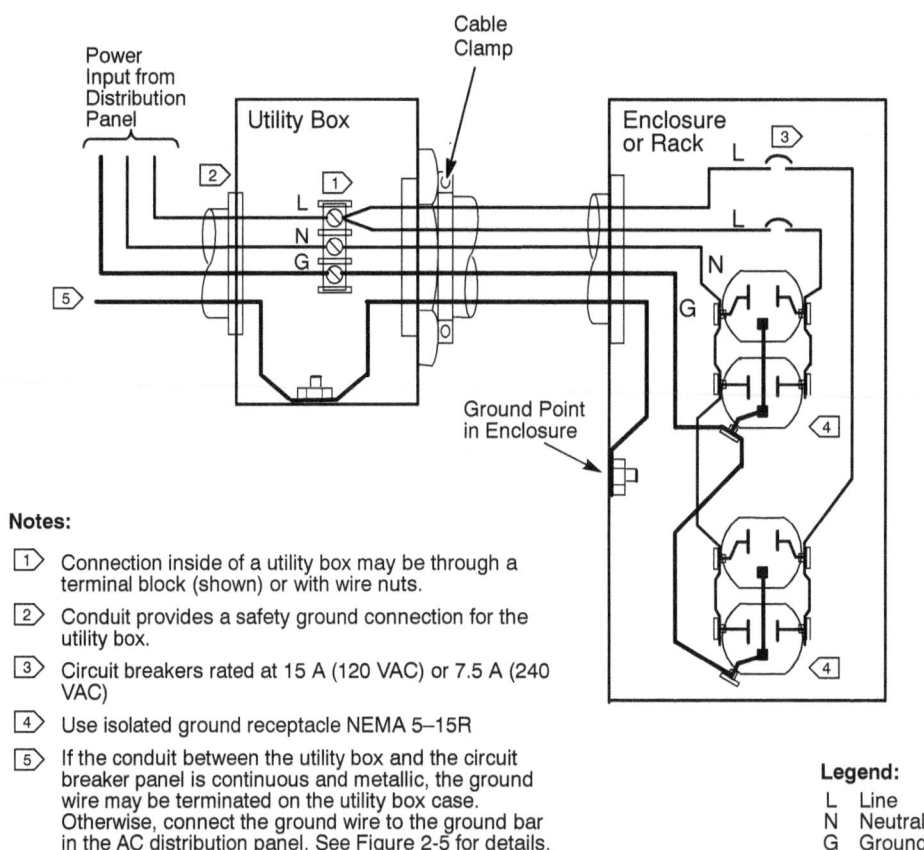

Notes:

[1] Connection inside of a utility box may be through a terminal block (shown) or with wire nuts.

[2] Conduit provides a safety ground connection for the utility box.

[3] Circuit breakers rated at 15 A (120 VAC) or 7.5 A (240 VAC)

[4] Use isolated ground receptacle NEMA 5–15R

[5] If the conduit between the utility box and the circuit breaker panel is continuous and metallic, the ground wire may be terminated on the utility box case. Otherwise, connect the ground wire to the ground bar in the AC distribution panel. See Figure 2-5 for details.

Legend:
L Line
N Neutral
G Ground

Figure 2-11 AC Power to Receptacles in Enclosures

Chapter 3 DC Power Distribution

Control system controllers and I/O systems are usually powered by controller and I/O power supplies, providing a variety of DC voltages such as 12 VDC, 5 VDC, and others for controllers and I/O cards. This section describes distribution requirements for DC supplies.

DC Power Considerations

When designing DC power distribution system, the design should be reviewed to ensure that the DC distribution system can sustain the required reliability. A good review reveals whether or not DC power redundancy is required. Backup of both bulk power supplies and DC to DC controller and I/O supplies can be applied.

If a backup bulk supply is used, the backup supply must be connected to the same local ground point as the associated primary supply. A backup supply can back up a primary supply in a separate enclosure.

DC Power Connections

Many control systems use DC/DC passthrough power supplies for controllers and I/O. The supplies receive DC power from bulk power supplies or from other in-plant DC sources. Often, DC/DC passthrough supplies accept 12 VDC input. Figure 3-1 shows typical connections.

Some systems include dual DC/DC supplies. These supplies can accept both 12 VDC and 24 VDC voltages, depending on input terminal connections. Figure 3-2 shows these connections. Usually, the connections are made between source and supply with stranded wire of a minimum size of 8 AWG (8.4 mm^2). The size and length of wire must be chosen to minimize the voltage drop to 2% or less.

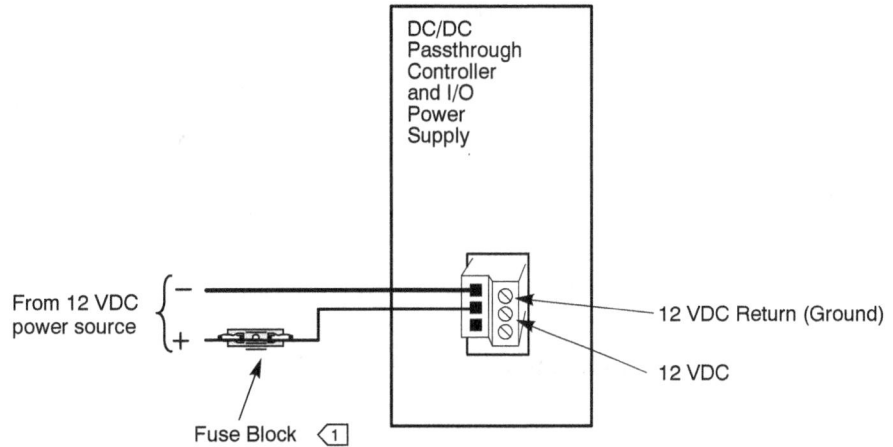

Note:

1▷ A fuse block and fuse are optional. If you are providing bulk DC power to several DC/DC passthrough controller and I/O power supplies, it is recommended that you fuse the line to each supply.

Figure 3-1   DC to DC Controller and I/O Power Supply Connections

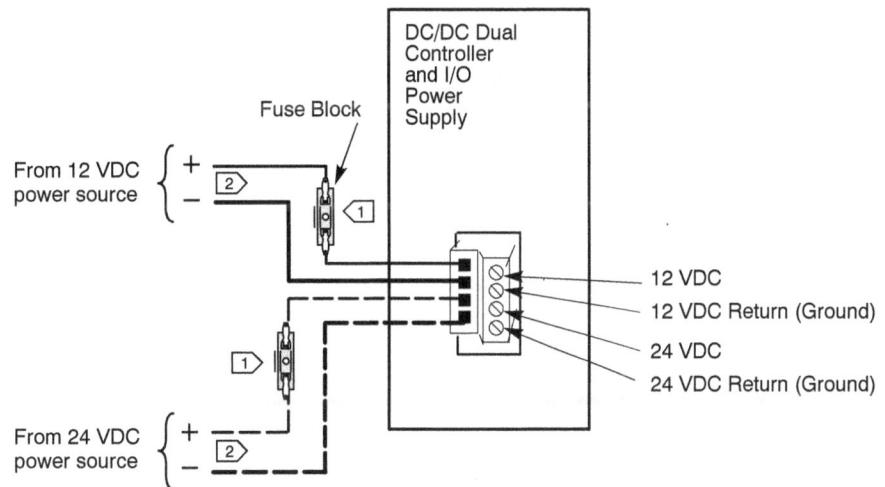

Note:

1▷ A fuse block and fuse are optional. If you are providing bulk DC power to several DC to DC controller and I/O supplies, it is recommended that you fuse the line to each supply.

2▷ Use either 12 VDC or 24 VDC. Only one input voltage is used; not both simultaneously.

Figure 3-2   DC/DC Dual Power Supply Input Connections

Bussed Field Power

Bussed field power is connected directly to the I/O interface carrier, as shown in Figure 3-3. For 24 VDC field devices, either bulk DC power supplies or DC plant power can be used. For 120 VAC or 240 VAC field devices, power from the AC power distribution system can be used.

The AC distribution system used for bussed field power should not be the same one as used for the control system. Using a separate system isolates electrical noise from the control system. The noise can be fed back on field power lines from heavy switching relays and other loads that typically cause spikes.

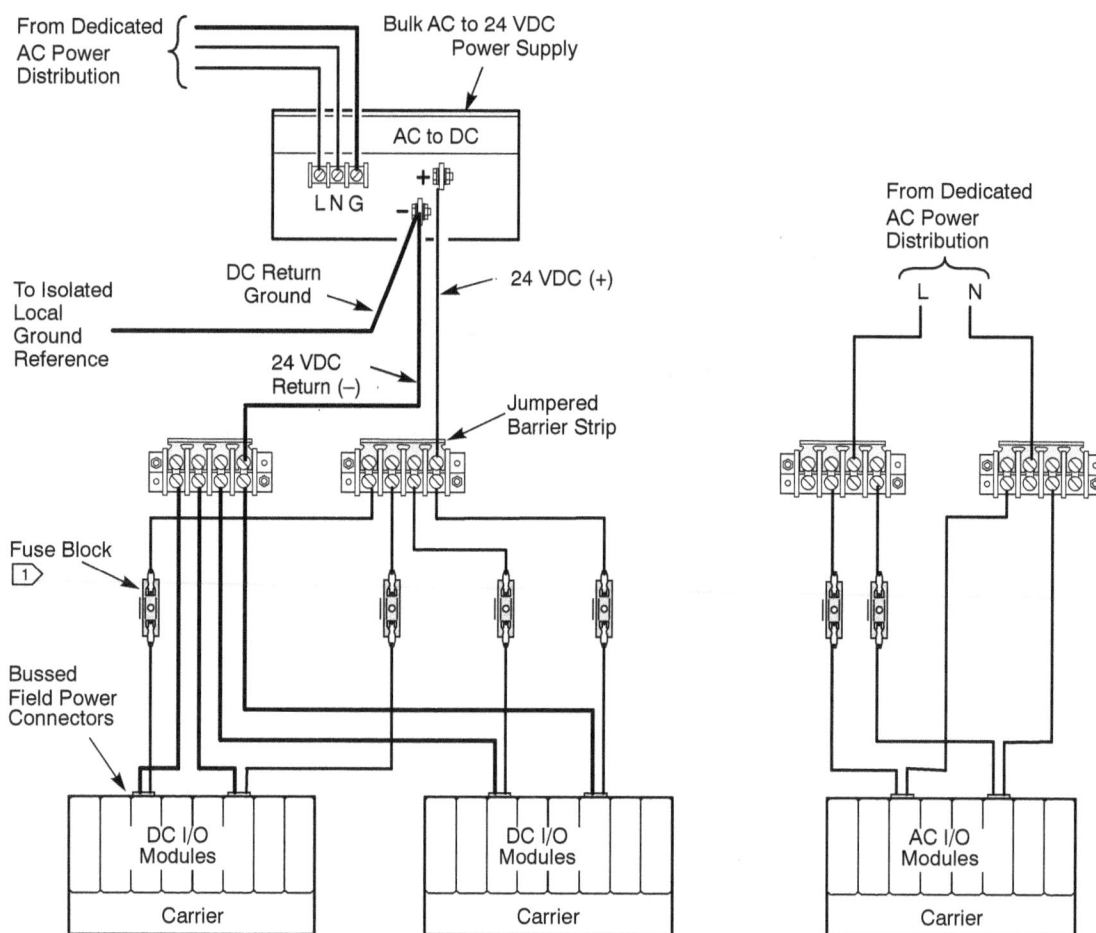

Note:

[1] A fuse block and fuse are optional. If you are providing bussed field power to several carriers from one power supply, it is recommended that you fuse the line to each carrier.

Legend
L Line
N Neutral
G Ground

Figure 3-3   Bussed Field Power Connections

Barrier Strips

I/O interface carriers contain bussed field power connections. Figure 3-3 shows a connection being powered through a barrier strip. When connecting field device power to several I/O interface carriers, barrier strips should be used. Carriers should not be daisy chained together because it is very difficult to remove power from one carrier without affecting all other carriers in a daisy chain. You may daisy chain the connectors on the same carrier if you wish, but separate connections, as shown in Figure 3-3, are recommended.

Carrier Connections

Figure 3-4 shows connector details for power connections on typical bussed field power connectors of I/O carriers.

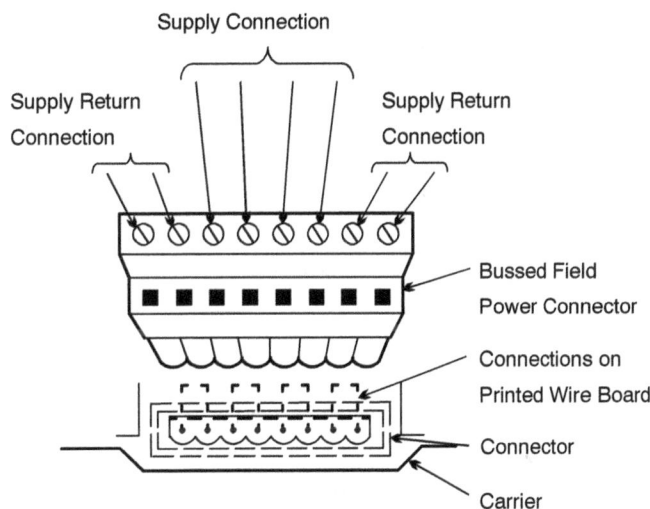

Figure 3-4   I/O Carrier Bussed Field Power Terminal Block

As shown in Figure 3-4, two terminals provide supply connections, and two terminals provide return connections. The connections route power to two adjacent I/O cards. Cards 1 and 2 are paired and must use the same field voltage level. Similarly, cards 3 and 4 must use the same voltage level, although the level can differ from that of cards 1 and 2. However, it is recommended to supply the same field voltage to all cards on the same carrier.

Figure 3-5 illustrates typical connections for adjacent I/O cards. Figure 3-6 illustrates typical connections for extending power from power supplies used for adjacent I/O cards.

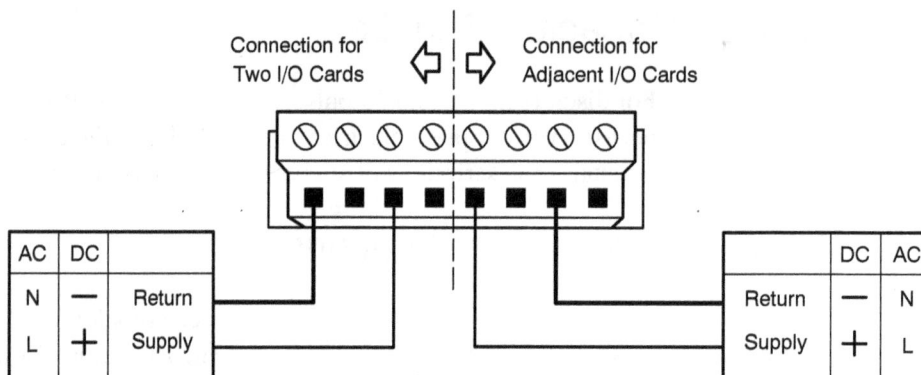

Figure 3-5   Bussed Field Power Wiring (Non-Extended Power)

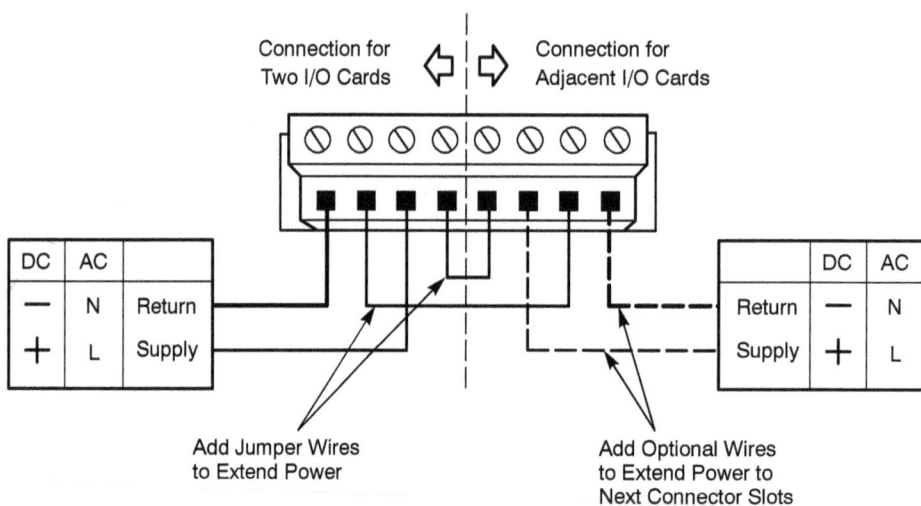

Figure 3-6   Bussed Field Power Wiring (Extended Power)

Note ...   *It is neither recommended, nor is it good engineering practice, to connect both AC and DC power to the same connector.*

If AC and DC power are connected to the same block, the block must be marked with a warning label to indicate that 120 VAC or 240 VAC and 24 VDC are found on the block. In addition, precaution must be taken that both AC and DC power are disconnected when the block is removed.

The bussed field power connector provides power to the field through the I/O cards only. Extending power from this connection to additional field devices is not supported or appropriate.

Resistor-Capacitor Snubbers

For discrete inputs and outputs, resistor-capacitor (R-C) snubbers may be necessary to suppress electrical arcing. Refer to the control system manufacturers manual for descriptions of snubbers you can use for suppression.

Intrinsic Safety Considerations

Most control systems do not require special or additional DC power supplies to power Intrinsic Safety (IS) I/O cards. The DC output for the control system intrinsic safety power supply connects directly to the IS I/O cards through the backplane. The local bus isolator electrically isolates intrinsic safety I/O cards from non-IS I/O cards. See Figure 4-14 for the location of the power supply and isolator on the IS I/O carrier.

Chapter 4   System Grounding

To ensure best performance and reliability of your control system, an adequate electrical ground network must be installed. Poor or faulty grounds are among the most common causes of control system faults. The extra time and effort spent in installing a good ground network is rewarded by easier system startup and more reliable system operation.

Single-Point Ground Techniques

The ground network is an organized system of ground wiring that terminates in a single, dedicated point on the plant ground grid. With single-point ground, you can achieve a clean reference for the control signals in your control system, thereby ensuring increased system reliability over systems with poorly designed ground networks. Figure 4-1 illustrates a typical, single-point ground network for a small control system.

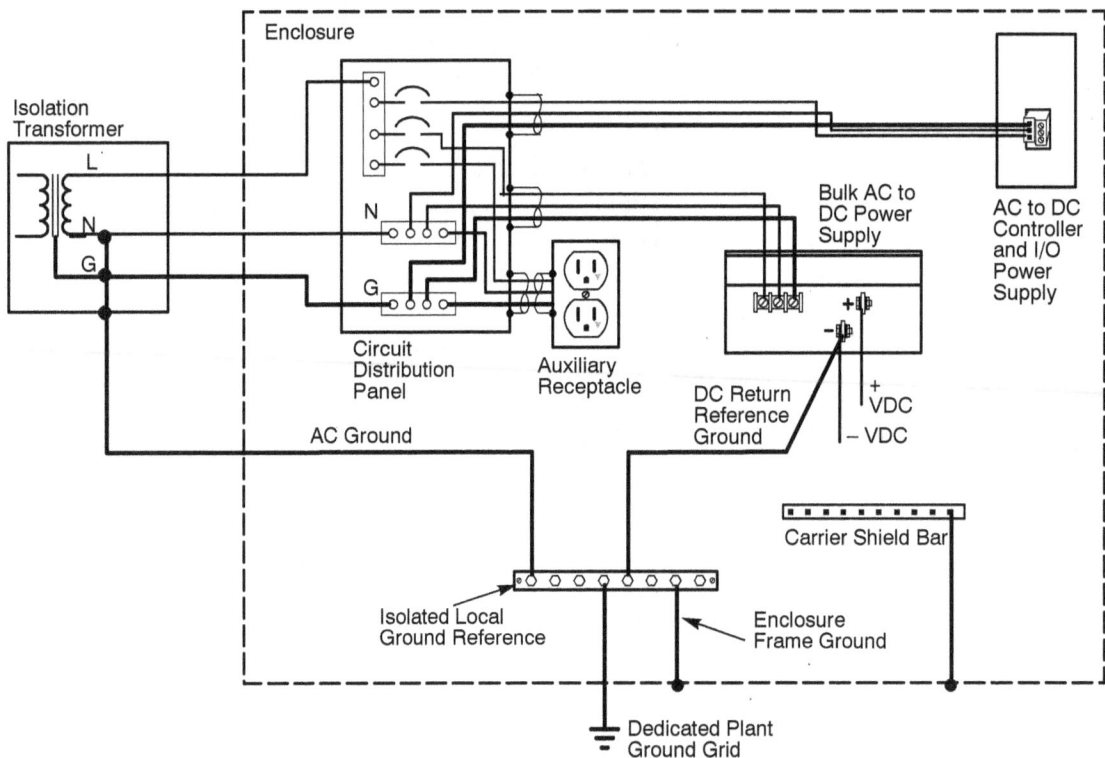

Figure 4-1   **Typical, Single-Point Ground Network for a Control System Housed in a Single-Enclosure**

The system illustrated in Figure 4-1 usually consists of a single enclosure with the isolation transformer included on the enclosure. For heat control purposes, the transformer is usually mounted outside of the enclosure. The transformer only provides power for the equipment in the enclosure. Its AC ground is connected to the control system ground (CSG).

Figure 4-2 illustrates a typical, single-point ground network for either a large control system housed in several adjacent enclosures or for a system spread over a local area. It is important that the term "local area" is well understood in relationship to power and grounding.

Each local area is defined as having one isolation transformer, a ground reference for the area, and cable-feet limitations. By this definition, all instruments powered by the transformer are located in the same local area, whether the instruments are located in one enclosure, in multiple adjacent enclosures, or spread in remote enclosures. All instrumentation and enclosure grounds, and the transformer ground, terminate at one ground reference for the area. Additionally, each power and ground path is typically limited to 200 cable-feet.

By using a separate isolation transformer, an isolated common ground reference for each area, and cable length limitations, each local area is electrically isolated from other areas. Therefore, ground loops between areas are eliminated, electrical noise is minimized, and common reference voltage is maintained.

Using Figure 4-2 as an example, there is one isolation transformer for the area. All control system instruments powered by the transformer are grounded to the same common ground reference through the local ground reference in each enclosure. The isolation transformer is connected to the same control system ground (CSG) as the common ground reference. The AC Ground is 200 cable-feet and the path from any isolated local ground reference to the CSG is 200 cable-feet.

The same isolated local ground reference can be used as follows:

■ Enclosures are located in the same local area.

■ Enclosures are bolted together to form one continuous unit.

■ Separate enclosures share the same single-ended signal. This case occurs if one field transmitter is connected to two points in separate enclosures (Figure 4-3) or an output from a device is used as an input for a device in another enclosure (Figure 4-4).

■ Enclosures are located in separate buildings or at distant from one another, but the AC power to the enclosures is taken from the same power source where transformer isolation is not used at both locations (Figure 4-5).

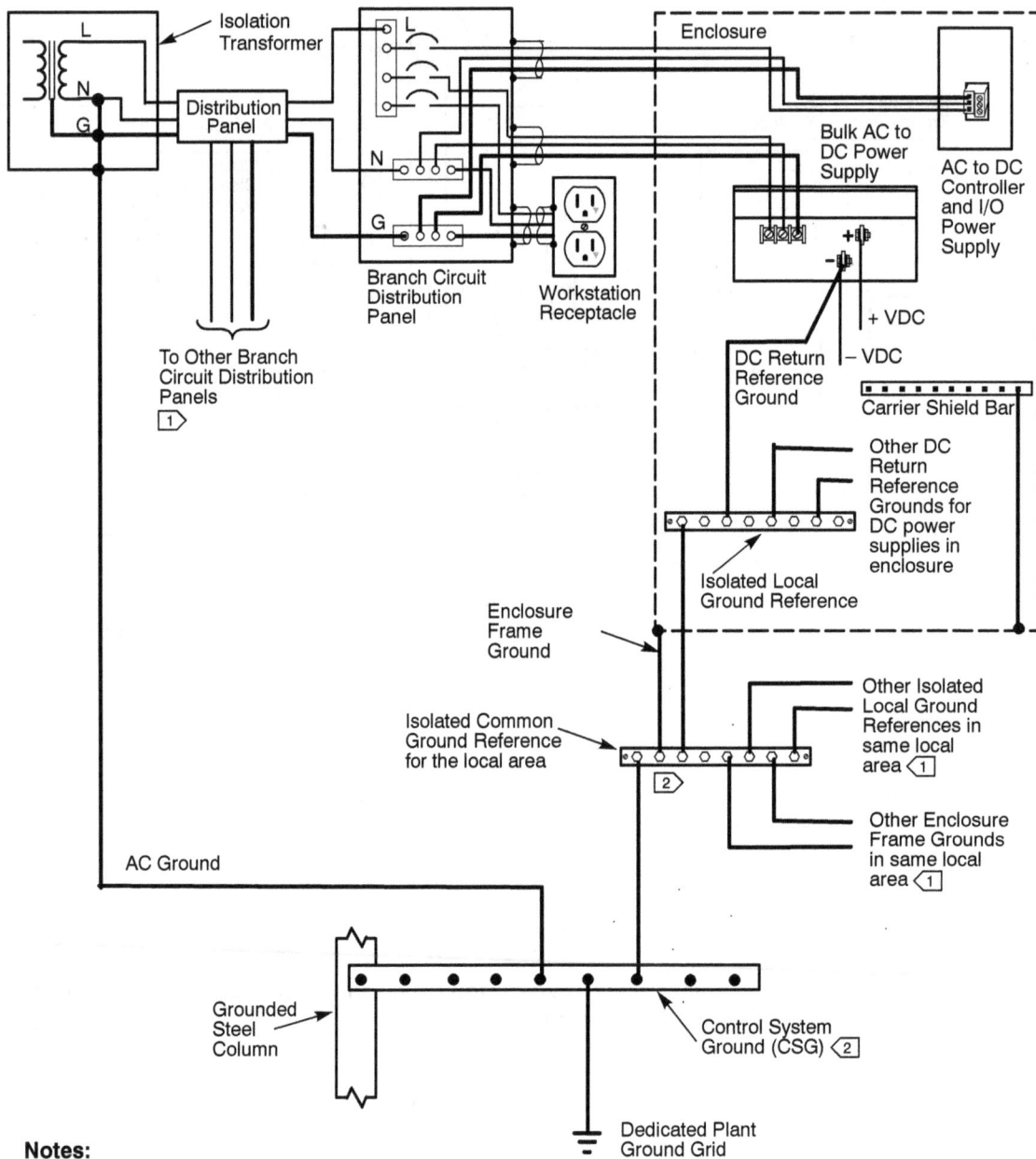

Notes:

1. > Applicable for 200 cable-feet or less, other branch distribution panels located in the same room, and having a common isolation transformer (same AC source)

2. > If space and distance permit, the Isolated Common Ground Reference bar and the Common System Ground may be the same bar.

Figure 4-2   Typical, Single-Point Ground Network for Control System Housed in Multiple Enclosures

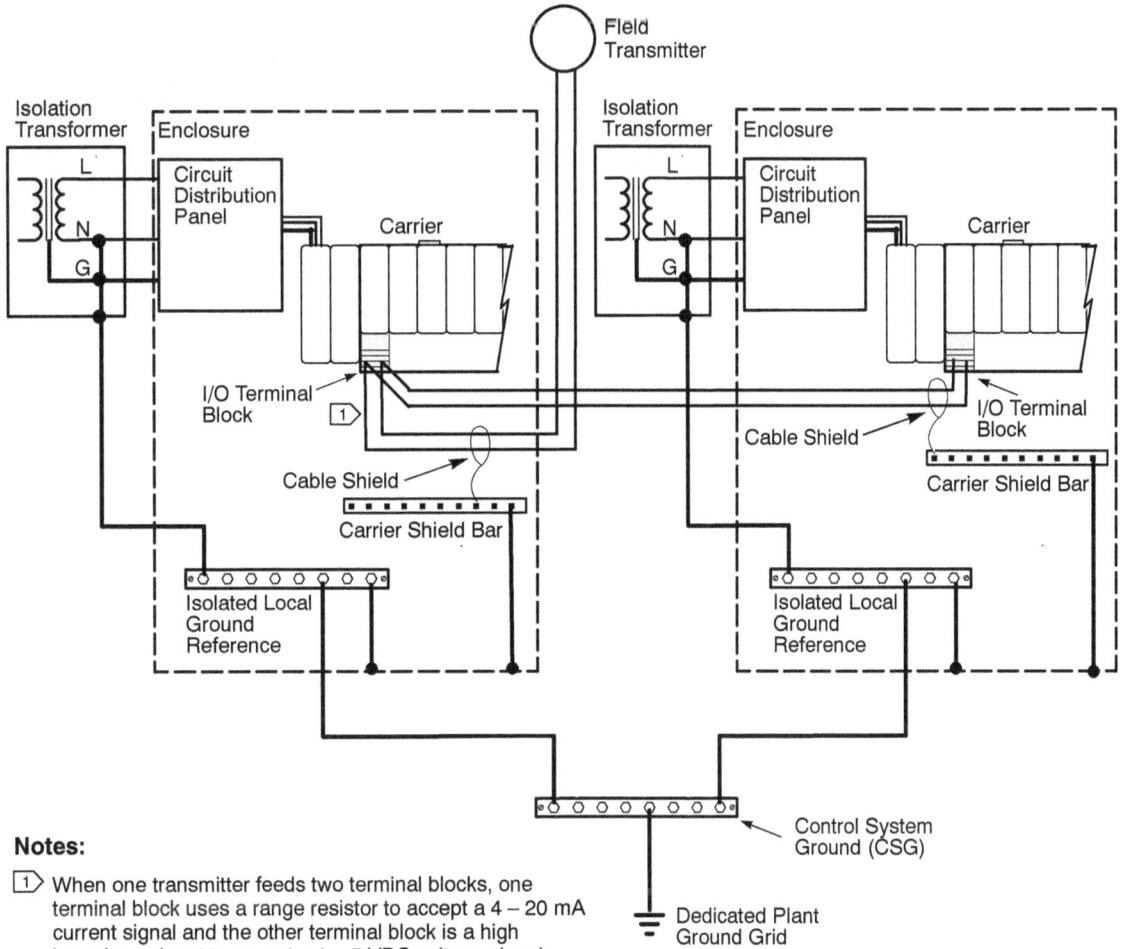

Notes:

1️⃣ When one transmitter feeds two terminal blocks, one terminal block uses a range resistor to accept a 4 – 20 mA current signal and the other terminal block is a high impedance input to accept a 1 – 5 VDC voltage signal.

Power and ground cables are limited to 200 cable-feet maximum from last point to the dedicated plant ground grid.

Figure 4-3   Single-Point Ground When One Field Transmitter Connects to Two Points in Separate Enclosures

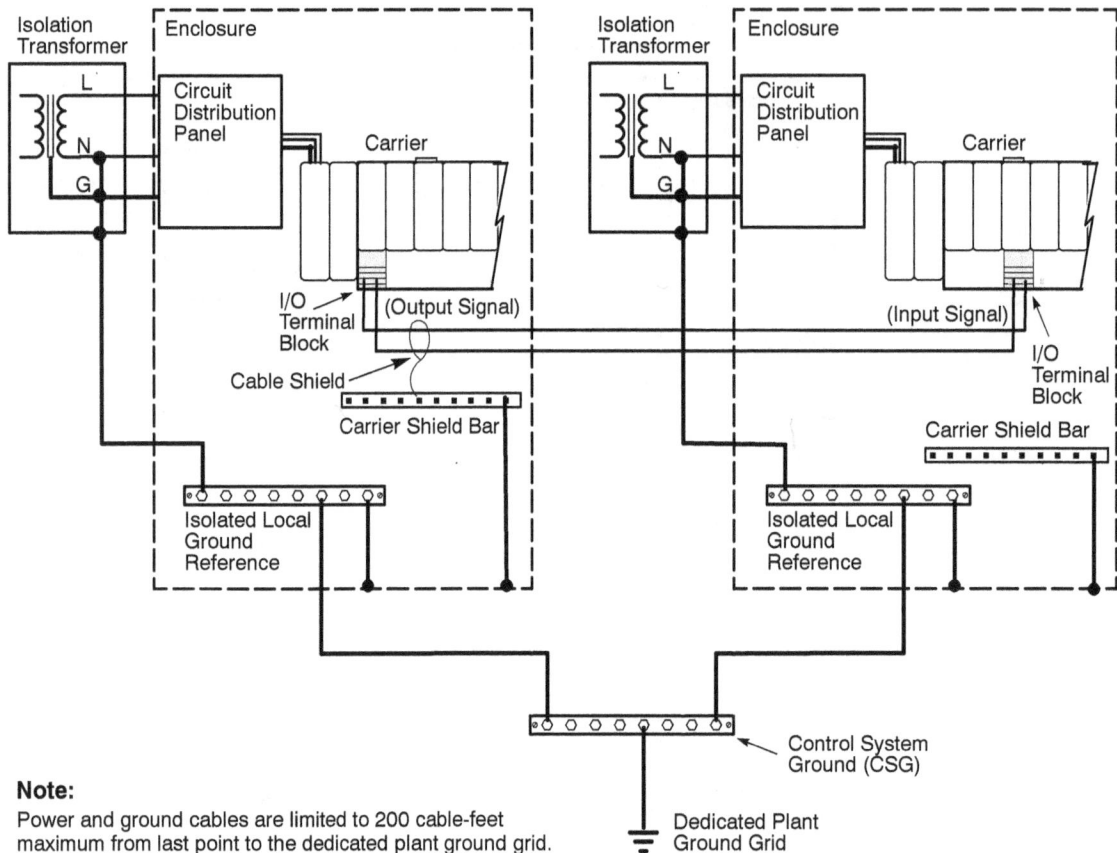

Figure 4-4 Single-Point Ground When Output From a Device in an Enclosure is Used as an Input to a Device in Another Enclosure in a Single-Ended Mode

Devices connected **only** by a control system communication network (e.g., Ethernet) normally do not require connection to the same ground system because the network usually provides isolation between devices. Systems isolated in this manner may have separate power sources.

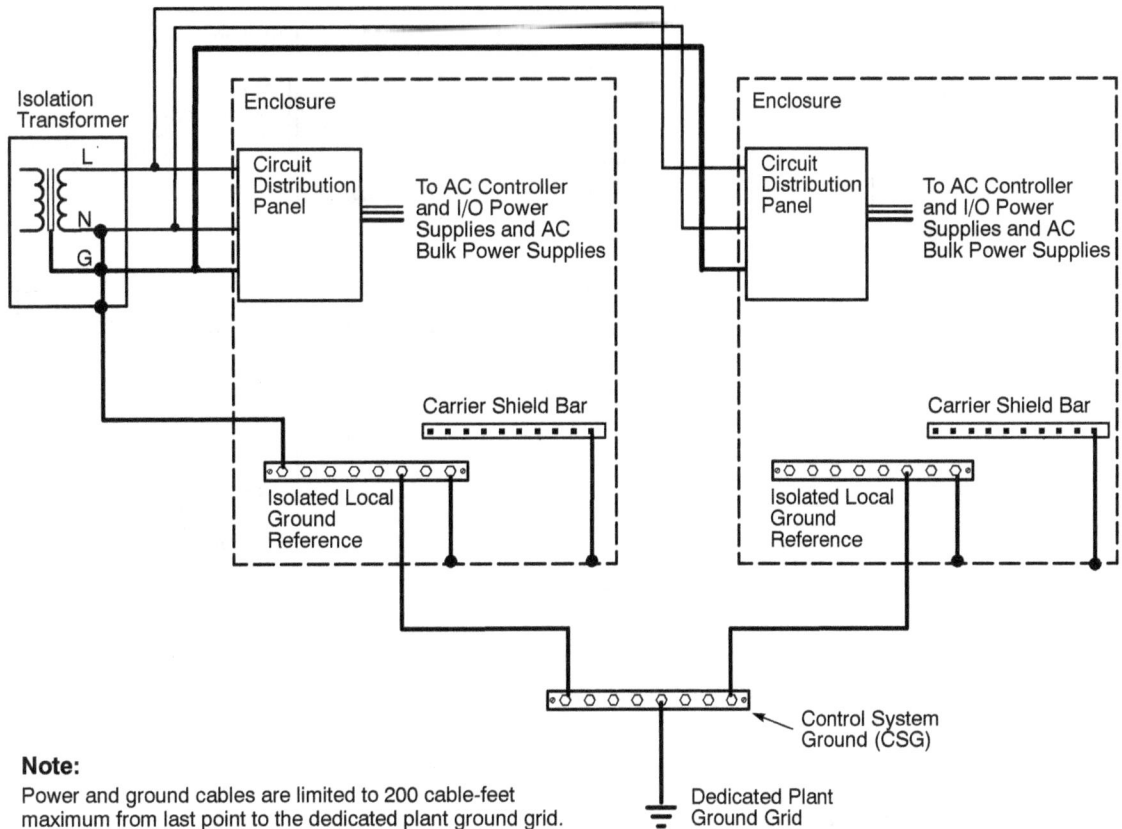

Figure 4-5   Single-Point Ground When Distant Enclosures are Powered from a Single Isolation Transformer

Enclosure Ground Considerations

An enclosure ground provides protection to equipment by minimizing effects from electromagnetic interference (EMI) and ground loops and to personnel by minimizing accidental shock hazards. Enclosures may be isolated from building steel and grounded by cable to the local ground reference point, or they may be grounded to building steel if it provides a low impedance path to true earth ground.

If enclosures are grounded to building steel, an isolation transformer that is closely located to the enclosures (preferably mounted on them) should be used. By using a transformer, possible ground loops between the transformer AC ground and the enclosure ground has a negligible effect on the control system. By keeping power and ground cables to within 200 cable-feet, EMI noise coupling is normally kept within adequate levels for proper system operation.

An enclosure ground must remain separate from all other DC ground connections. Figure 4-6 shows enclosure and DC grounds that retain the proper separation. In a group of enclosures, grounds are routed to the center enclosure.

Effective Grounding

The following guidelines help you provide effective grounding for your control system:

- Provide a ground network dedicated to the control system. Do not share a ground network with other plant systems.

- Design the ground network so that it is accessible for testing.

- Ground enclosures through the single-point ground network.

- Connect all enclosures within a group to the same ground point.

- Provide a single-point ground for all enclosures interconnected by non-isolated signals. Also, provide a single-point ground for all enclosures sharing a backup power supply.

- Ground workstations by the ground conductor that is included in their AC power circuits.

- Provide a low impedance, high integrity, ground path between all instrumentation and control system plant ground connections.

- Keep ground wiring as short as possible.

Explanations of the guidelines are included in following subsections.

Using Existing Ground Systems

With the installation of a new control system, an effective ground network can be installed at the beginning. However, the expansion of an older system often uses existing ground networks. If the existing network does not meet the criteria of a single-point ground scheme, it may be more cost effective to install a new ground network to ensure proper control system operation. In all cases, care must be taken so that ground loops are not created through signal connections or metal chassis connections.

Separating AC and DC Grounds

Separate ground networks are used for AC power and DC power to provide a safety return path to earth to protect from faults occurring in the control system and to isolate noise between AC and DC circuits. Figure 4-6 shows the separate ground networks. Both networks terminate at a single point on the control system ground (CSG).

Figure 4-6 shows a primary and secondary source with isolation transformer. If the desired level of system availability permits, you may use a single source. If you use two sources, both sources must use a separate AC ground connection to the control system ground, as shown.

AC Ground Network

The single-point termination shown in Figure 4-6 provides the AC ground for AC powered devices in the control system. The transformer shield and case as well as the AC neutral and ground are connected to AC ground. The AC ground must conform to all applicable local, state, and federal electrical code requirements.

DC Ground Network

Figure 4-6 also shows the DC ground network for DC power and I/O signals in the control system. The DC ground serves as the final termination point for all signal common and power supply common wiring. The power supply common is the power return for all DC power connections in the system. Note that the I/O cable shield grounds are connected to the enclosure frame, from which the shield grounds are connected with the enclosure case ground to the common ground reference for the plant area.

Special Cases

In special cases, such as on oil and gas platforms, floating 24 VDC power may be used to power the control system. To obtain proper electrical ground for the system, a single-point of ground on the platform is needed for controller power supples, enclosure mounted power supplies, signal commons, shield grounds, case grounds, and so forth. This need means that some type of isolating device, such as an inverter-rectifier, must be used with the isolated 24 VDC power to maintain its floating design.

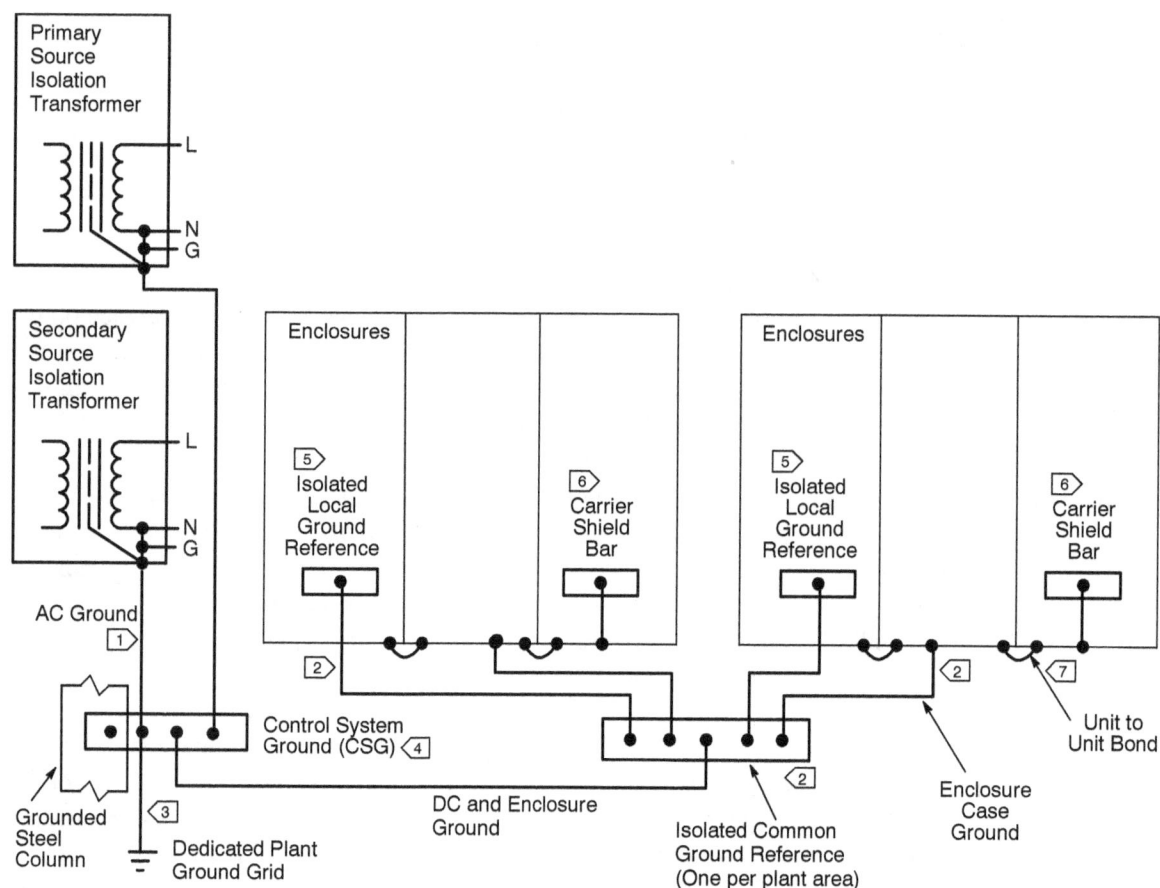

Notes:

1> Stranded, insulated AWG 1/0 to 4/0 cable — Conductor used to connect the grounding electrode to the neutral ground bond at the source of a separately derived instrumentation power system [per NEC 250-30(a)(2)] (CSA C22.1 Section 10)

2> Stranded, insulated AWG 1/0 to 4/0 cable — Conductors used to provide a low impedance ground reference for the DC power system (Logic, Transmitter, Output) and enclosure ground for EMI/RFI noise protection of the enclosures, carriers, and field wiring shields

3> Supplemental conductor used to connect the grounding electrode for the source of a separately derived instrumentation power system directly to the plant ground grid system. This conductor is used to provide low impedance ground reference to EMI/RFI noise [per NEC 250-54-Supplemental Grounding Electrode 250-50(a)(2)].

4> The DIG must be effectively bonded to building steel [NEC 250-50(b)], and building steel must be properly grounded to earth ground. Refer to Chapter 7 for earth ground information.

5> Isolated local ground reference is used when several devices in the enclosure grouping require ground reference. Use stranded, insulated AWG 1/0 to 4/0 cable between isolated local ground reference and isolated common ground reference.

6> Use stranded, insulated AWG 1/0 to 4/0 cable between carrier shield bar and enclosure frame.

7> Be sure to strap (not bolt) all grouped enclosures. Half-inch (12 mm) wide, braided, ground strap can be used up to six inches (15 cm). For longer bonding, use insulated 10 AWG wire. Refer to NEC Table 250-122 and Article 250-119.

Figure 4-6   Details of AC and DC Ground System for Transformer-Isolated Area with Multiple Cabinets with All Grounds Tied to a Single Point

Ground Wiring

Proper connections, wire sizing, and overall ground impedance are important to effective grounding.

Local and Isolated Common Reference Points

An organized ground system includes reference points that are local to an enclosure (or groups of enclosures) and also an isolated common reference point for the plant area. A local reference point provides a central termination point for all power supply common connections within an enclosure. The isolated common reference point ties together the local reference points.

Note ...   *Some figures in this manual show DC Return Reference Grounds connected directly to an isolated common ground reference. This scheme is applicable when there are only a few reference ground returns. Where many reference grounds exist, as in the case of grouped enclosures, use an isolated local ground reference in each group and connect these to the isolated common ground reference for the area.*

Good reference points should be highly conductive, copper, copper-clad, or brass bars. Such bars are available from local industrial electronics suppliers, or you can fabricate your own ground bars. The bars mount on isolated brackets inside an enclosure. Bars normally allow screw connections for wiring lugs. Use lugs that accept wire sizes of AWG 1/0 to 4/0 (53.5 to 107.2 mm^2).

Figure 4-7 shows the details of a typical ground bar. Table C-2 lists recommended wire sizes based on cable length. If you fabricate your own ground bars, ensure that the following conditions are met:

■   Copper/copper clad steel or hard brass (B16)

■   Minimum of 3/8 inch (9.5 mm) thick and 1-3/4 (44.5 mm) inch wide

■   Holes for lugs

■   Double-bolted lugs

■   Bus isolated from mounting bracket with standoffs

Isolated Mounting for Ground Bars

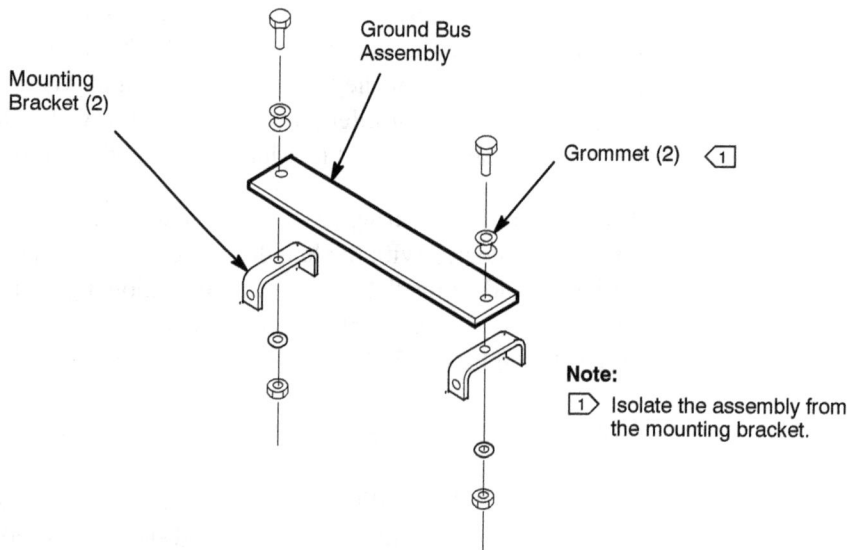

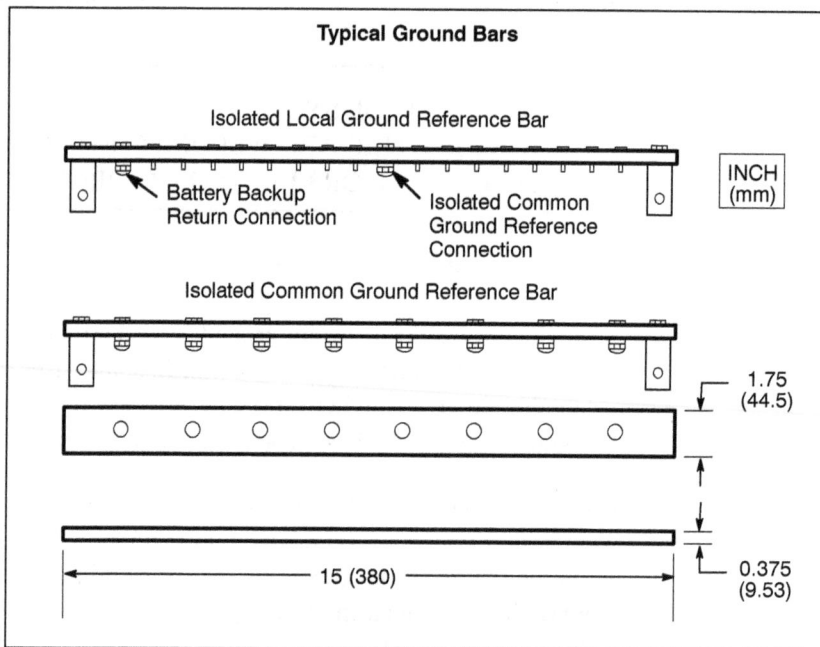

Mounting
Bracket (2)

Ground Bus
Assembly

Grommet (2)

Note:
1 Isolate the assembly from
the mounting bracket.

Typical Ground Bars

Isolated Local Ground Reference Bar

Battery Backup
Return Connection

Isolated Common
Ground Reference
Connection

INCH
(mm)

Isolated Common Ground Reference Bar

1.75
(44.5)

15 (380)

0.375
(9.53)

Figure 4-7   Typical Reference Point Ground Bars

DC Return Ground Wiring

Ensure that DC return grounds are on separate wires from each power supply to the enclosure local ground reference point. The DC return ground connection at the power supply is the negative (–) DC terminal. Where bulk power supplies feed a DC-to-DC controller and I/O power supply, the DC return is placed between the carrier ground terminal and ground reference.

The recommended wiring for these ground points is AWG 8 (8.4 mm^2) copper, stranded, insulated with the lengths being as short as possible. The return ground wiring should be insulated to avoid unintentional ground loops that can occur if bare wires touch the metal enclosure frame, or each other. Proper isolation from metal should be verified by electrical measurement after installation.

Marking Grounds

To aid in ground identification, identifiable insulation colors (green or green with a yellow stripe) or some labeling method should be used that meets U.L.-listed connections. All system ground points should be labeled as follows:

> FOR CONTROL SYSTEM GROUND ONLY. DO NOT USE
> FOR ELECTRIC ARC WELDER CONNECTION OR OTHER
> ELECTRICAL OR ELECTRONIC CIRCUITS.

Ground Impedance

A high quality control system ground should provide a ground point that measures one ohm or less to true earth. In some cases, three ohms may be acceptable. In an area where soil does not provide a good ground, it may be necessary to select the best ground impedance available.

There are several methods that can be used to obtain a high quality earth ground system; these methods vary, depending on the soil type and moisture content at the individual location. Refer to Chapter 7 for information about soils and earth grounds. Testing an earth ground per the procedure described in Chapter 7 is highly recommended.

Shield Grounds

Proper field-wiring shield grounds help ensure proper system operation by reducing electromagnetic interference in signal wiring. A proper shield ground path is to ground the shield at one end only. Grounding at both ends can cause ground loops. The preferred end is at the carrier shield bar. If field transmitters or final control elements are well-grounded four-wire units, shield wires can be grounded to the units instead of the carrier shield bar.

Note ...   *The carrier shield bar is connected to the enclosure frame unless the control system occupies only one enclosure, in which case, the shield bar may be connected directly to the isolated common ground reference.*

The recommended wiring between the shield bar and its ground termination on the enclosure frame, or the isolated common ground reference, is AWG 8 (8.3 mm^2), copper, stranded with length as short as possible. The wiring should be insulated to avoid unintentional ground loops that can occur if bare wires touch the metal enclosure frame, or each other. Proper isolation from metal should be verified by electrical measurement after installation.

To obtain the intended electromagnetic interference (EMI) protection, it is important that the insulation on shield wiring be properly stripped at the ground end and that it is properly wrapped at the ungrounded end. Figure 4-8 illustrates proper insulation stripping and wrapping.

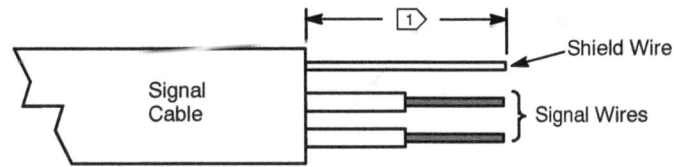

1 Preferred length of non-insulated portion of shield wire is 1 inch (25 mm); 2 inch (50 mm) may be used if needed to reach terminations on the carrier shield bar.

Stripped Insulation Length on Grounded End of Shield Wire

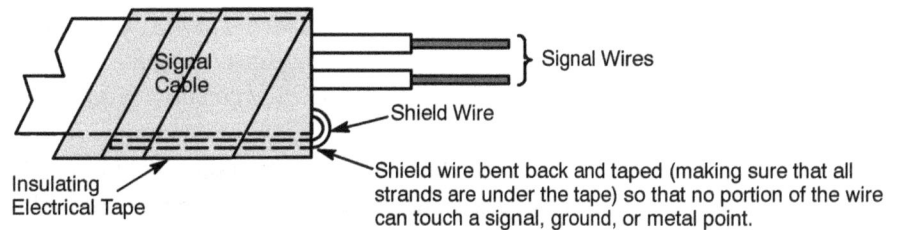

Wrapped Insulation on Non-Grounded End of Shield Wire

Figure 4-8   Proper Length for Stripped Shield Ground Wire and Proper Wrapping

Workstations and Peripheral Devices

Workstations and peripheral devices, such as network hubs, switches, and printers are grounded to the control system AC ground through the neutral wire and the safety ground wire in the AC power cord. The conduit carrying the AC power conductors to the workstations and peripheral devices must be electrically isolated from the workstations, but it must be bonded at the power supply. The best way to isolate the conduit is to install an isolated AC ground receptacle (NEMA Type 5-15R) near the equipment and run the conduit to the receptacle [per NEC 250-146(d)].

The AC ground wire in the AC conductors should be the same size or one size larger than the current-carrying conductors. For example, the line and neutral lines should be AWG 12 (3.3 mm^2), copper, stranded wire, and the AC ground should be AWG 10 (5.3 mm^2), copper, stranded wire. For best noise control, use AC ground wire made of a large number of small strands rather than coarse, stranded wire.

AC/DC Controller and I/O Power Supply

Figure 4-9 shows how ground connections should be made with a control system which uses an AC to DC controller and I/O power supply.

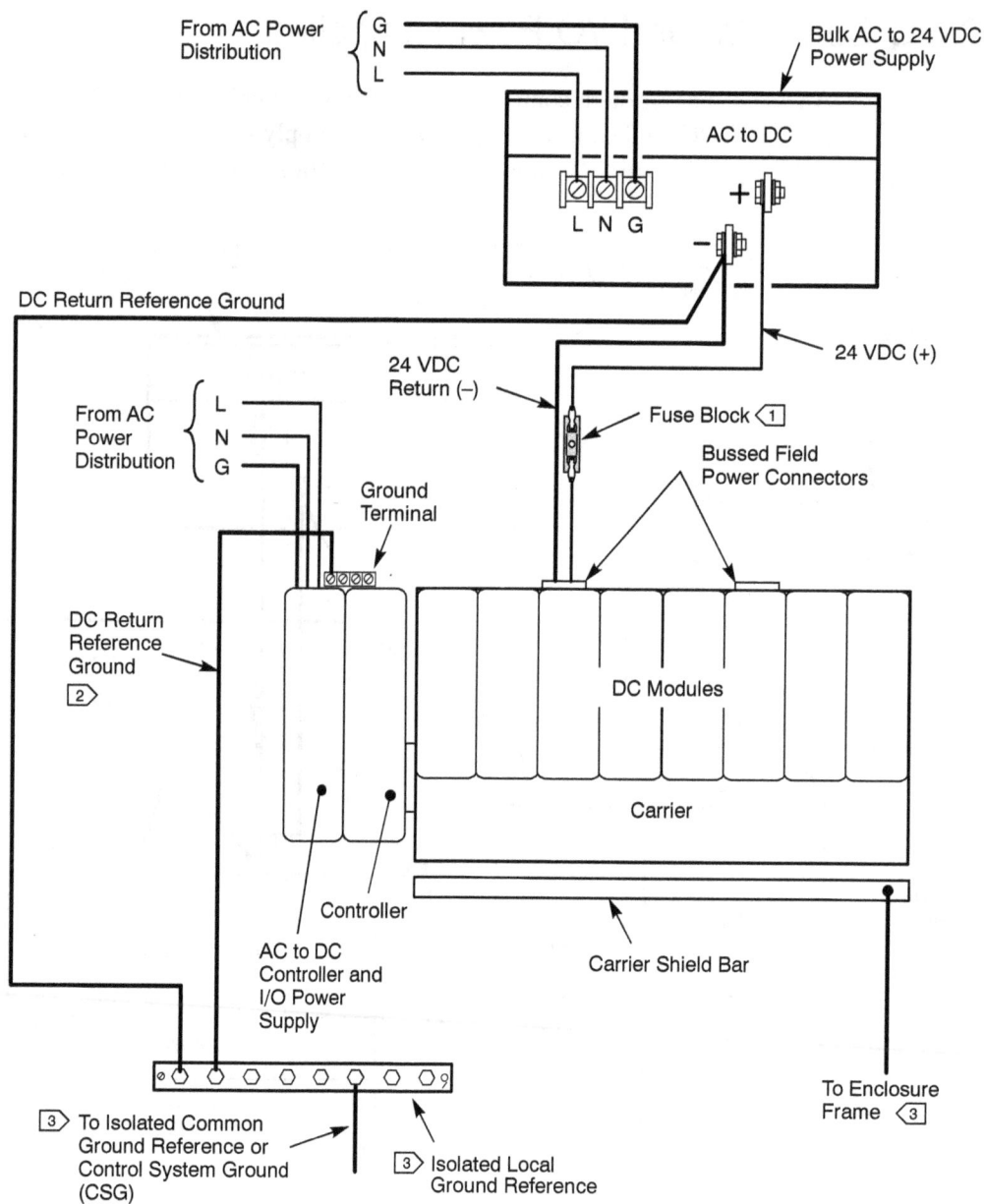

From AC Power Distribution { G N L

Bulk AC to 24 VDC Power Supply

AC to DC

L N G

+

−

DC Return Reference Ground

24 VDC Return (−)

24 VDC (+)

From AC Power Distribution { L N G

Fuse Block ◁1

Bussed Field Power Connectors

Ground Terminal

DC Return Reference Ground 2▷

DC Modules

Carrier

Controller

AC to DC Controller and I/O Power Supply

Carrier Shield Bar

To Enclosure Frame ◁3

3▷ To Isolated Common Ground Reference or Control System Ground (CSG)

3▷ Isolated Local Ground Reference

Notes:

1▷ A fuse block and fuse is optional. If you are providing bussed-field power to several carriers from one power supply, it is recommended that you fuse the line to each carrier.

2▷ When AC-DC controller and I/O power supplies are used, a DC Return Reference Ground should always be connected at the carrier ground terminal.

3▷ Refer to Figure 4-1 and Figure 4-2 for ground connection details.

Legend:
L   Line
N   Neutral
G   Ground

Figure 4-9   Ground Connections for AC to DC Controller and I/O Power Supply

DC/DC Controller and I/O Power Supplies

A DC/DC passthrough controller and I/O power supply accepts 12 VDC input, and a DC/DC dual controller and I/O power supply accepts either 12 VDC or 24 VDC input. Figure 4-10 shows ground connections for either of these supplies.

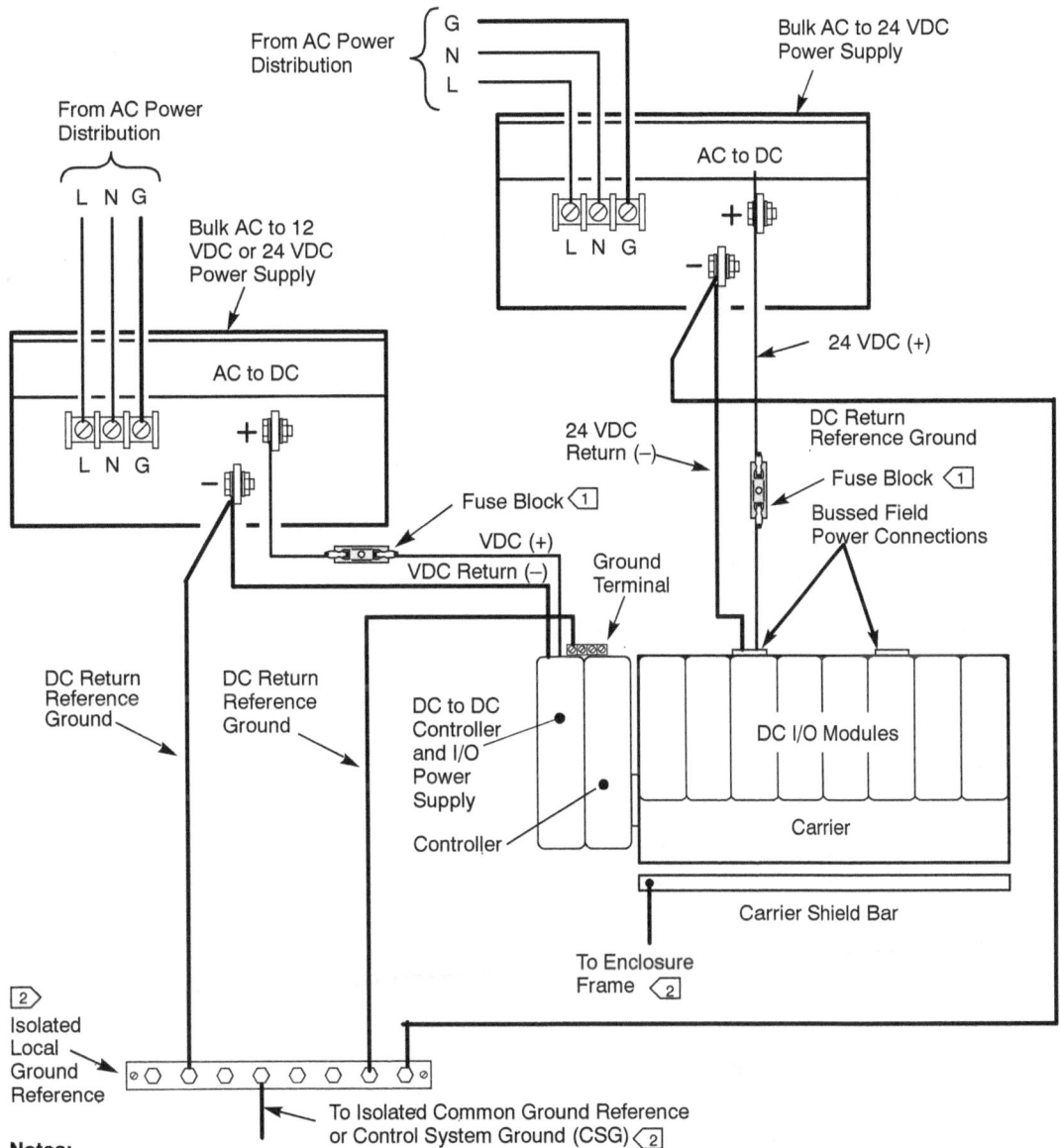

Notes:

1. A fuse block and fuse is optional. If you are providing bussed-field power to several carriers from one power supply, it is recommended that you fuse the line to each carrier.

2. Refer to Figure 4-1 and Figure 4-2 for ground connection details.

Legend:
L Line
N Neutral
G Ground

Figure 4-10 Ground Connections for DC/DC Controller and I/O Power Supplies

Bulk Power Supplies

When using a bulk power supply, it is recommended that an isolated ground be used for DC reference. Figure 4-11 shows the recommended connections for all bulk power supplies. For ground connections when several bulk supplies are grouped in an enclosure, refer to Figure 2-9.

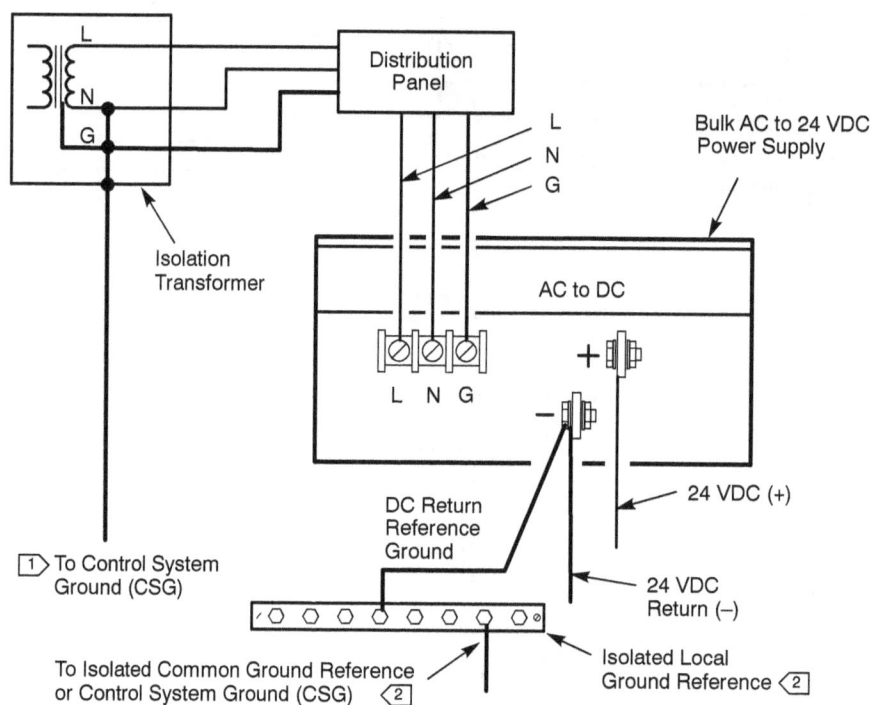

Notes:

1 If the power supply is not located in an enclosure, the DC Return Reference Ground may be connected directly to the Control System Ground.

2 Refer to Figure 4-1 and Figure 4-2 for ground connection details.

Legend:
L Line
N Neutral
G Ground

Figure 4-11   Power and Ground Connections for a Bulk AC to 12 VDC or 24 VDC Power Supply

Redundant Power Connections

Figure 4-12 (load-sharing bulk supplies) and Figure 4-13 (non-load-sharing bulk supplies) show power and ground connections when primary and backup power supplies are used. Barrier strips include shorting bars across the top set of terminals. One barrier strip is used for each source and another barrier strip is used to consolidate ground wiring before making a ground connection to the isolated common ground reference.

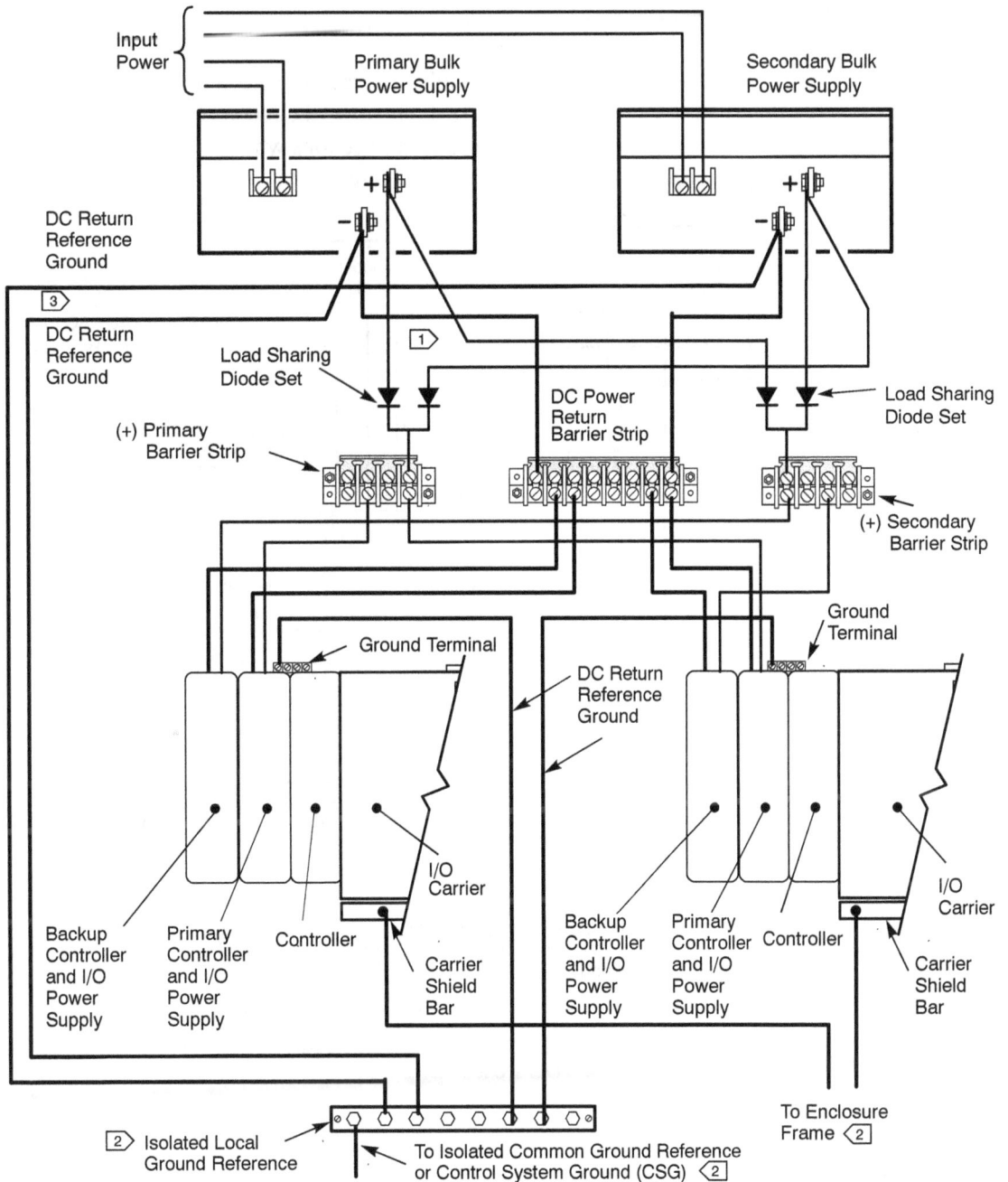

Notes:

1. Size power and ground wiring large enough to carry the total current drawn from all loads.

2. Refer to Figure 4-1 and Figure 4-2 for ground connection details.

3. DC Return Reference Grounds should be a short as possible and as close to each other as possible.

Figure 4-12 Power and Ground Connections for Primary and Backup Controller and I/O Power Supplies with Load-Sharing Bulk Power Supplies

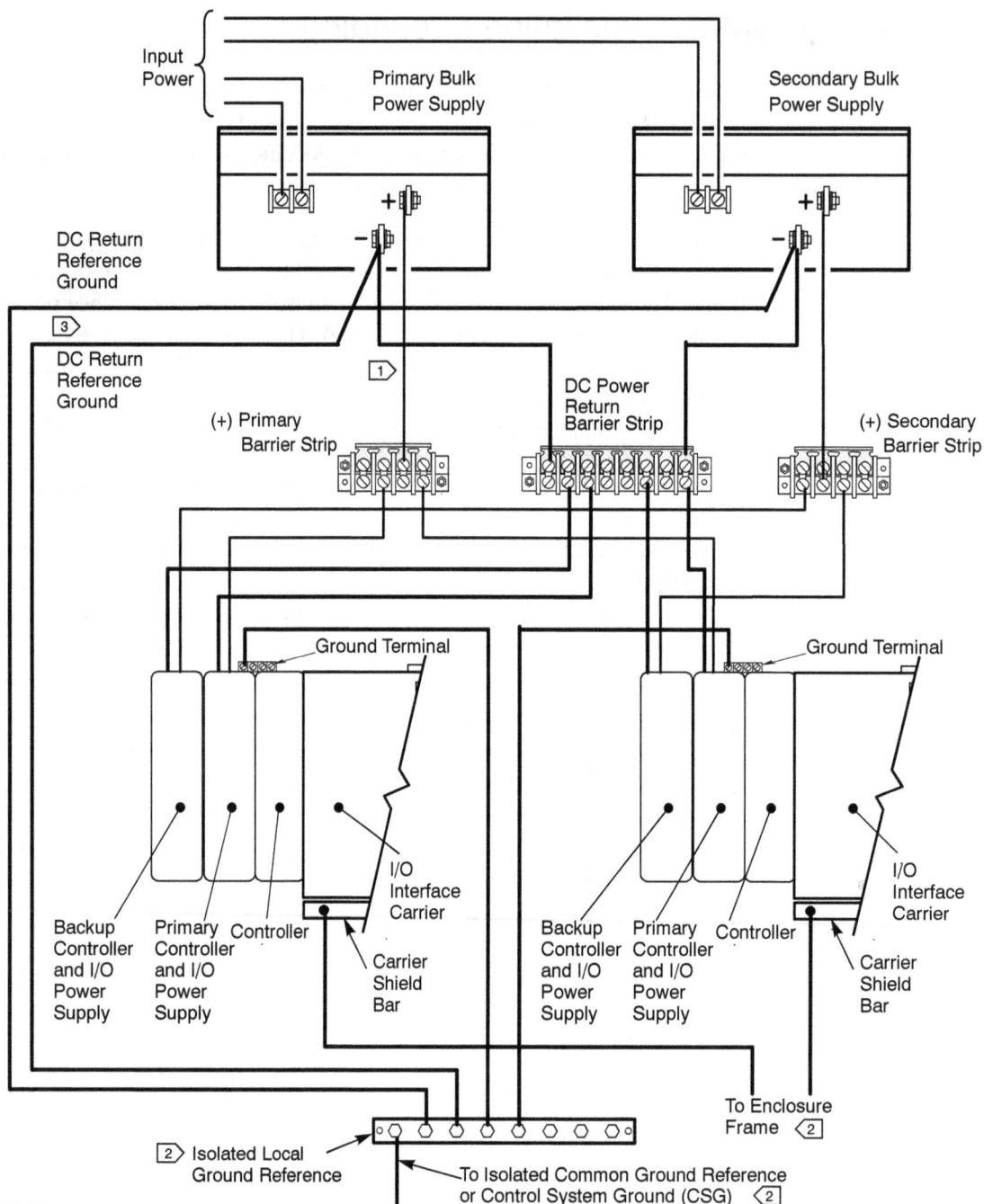

Notes:

1. Size power and ground wiring large enough to carry the total current drawn from all loads.

2. Refer to Figure 4-1 and Figure 4-2 for ground connection details.

3. DC Return Reference Grounds should be a short as possible and as close to each other as possible.

Figure 4-13 Power and Ground Connections for Primary and Backup Controller and I/O Power Supplies with Non-Load-Sharing Bulk Power Supplies

Intrinsic Safety Installation Grounding

In some applications where hazardous gases are present, special wiring practices or other special handling must be used. Conformity with local codes and regulations is essential. Several industry documents are available that describe the requirements for hazardous area instrumentation use or code guidelines.

Figure 4-14 shows typical grounding requirements for a control system using an intrinsically safe (IS) I/O system. The shield bar is connected to the enclosure frame. The carrier is connected to either an isolated local reference point or the isolated common reference point (shown). These grounds provide the electrical ground required for the I/O cards and the IS power supply. Power leads to the supply are connected from a 24 VDC bulk power supply. The negative (–) output terminal is grounded at the bulk power supply output (as shown in Figure 4-11).

For complete I/O system grounding, refer to the intrinsic safety barrier manufacturer's manuals. For additional information, refer to the intrinsic safety information in ANSI/ISA-RP12.6-1995, *Wiring Practices for Hazardous (Classified) Locations —Instrumentation Part 1: Intrinsic Safety,* and in NEC Section 504 for grounding and installation practices. (Note that RP12.6 requires earth ground resistance to be one ohm or less).

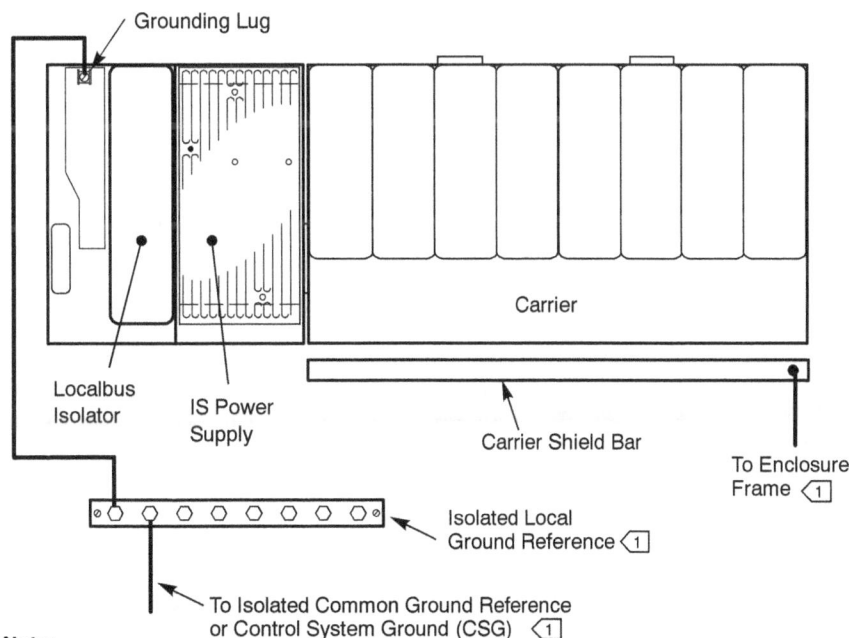

Note:

1 Refer to Figure 4-1 and Figure 4-2 for ground connection details.

Figure 4-14 Typical Grounding Requirements for a Control System Using an Intrinsically Safe (IS) I/O System

Chapter 5   Signal Wiring

This chapter describes methods to minimize electrical interference in signal and ground wiring and properly install and terminate signal wiring.

Signal Wiring Practices

The following recommendations can help you install reliable signal wiring:

■   Field termination connectors on I/O subsystems generally accept up to AWG 12 (3 mm^2) stranded or solid wire. Wire sizes between AWG 14 and AWG 18 (2 mm^2 and 0.8 mm^2) are typically used. AWG 18 is recommended. To select wire, determine the maximum current expected or the maximum voltage drop permissible, and the wire length. Then, use local wiring codes to determine the appropriate wire size.

■   To terminate field wiring cable shields, it is recommended that shield bars be used with I/O carriers.

■   Multi-pair cables should contain at least one spare set twisted-pair wires. It is recommended that each cable contain at least 20 percent spares.

■   For a digital I/O system (for example, FOUNDATION Fieldbus, DeviceNet, Profibus, AS-Interface), use only cables specified by the system.

■   On external terminal blocks, leave a test point for connecting test equipment. Leaving a little extra exposed wire at a terminal or using terminal lugs are ways to provide test points.

■   Use ring tongue or flanged spade tongue connectors on terminal blocks to hold wires in place.

■   Avoid stray strands at stripped-back multi-strand wire that can short signals to other terminals.

■   Inspect all solder joints for proper solder connections.

■   Inspect wires at terminals to ensure that wire insulation is stripped back cleanly and that no insulation is under a wire connection at a terminal block or terminal strip.

■   Ensure that cables are run so that sharp metal edges cannot cut through cable insulation.

■ Shield ends that are not grounded should be tape-wrapped so there is no possibility that a shield can touch any wires or metal objects. Heat-shrink tubing may also be used to isolate the shield from wires or metal objects.

Electromagnetic Interference (EMI) Reduction

Electromagnetic interference (EMI), sometimes called radio frequency interference (RFI), can be caused by such devices as radios, intercoms, AC power lines, motors, and high capacity switches. Most modern control system products are fully compliant with electromagnetic standards for European Union countries. The equipment is shielded to reduce EMI to acceptable levels in normal operating conditions.

To minimize EMI introduced by sources external to your control system, you should take the following precautions:

■ Use fiber-optic cable as much as possible in network communications systems, especially between buildings. Fiber-optic cable provides optimum protection from EMI and nearby lightning strikes.

■ Route network cables as far away from EMI sources (large motors, generators, transformers, and so on) as practical.

■ To minimize electrically induced noise on millivolt signals from thermocouples, resistance temperature detectors (RTDs), strain gauges, and pH electrodes, use individually shielded, twisted-pair cables specified for these devices.

■ Use multiple-pair, shielded, twisted-pair cables for 4–20 mA and 1–5 V field signals.

■ Because pulse count signals contain fast rise-time components that make them both noisy and susceptible to external noise, route them in individually shielded, twisted-pair cables.

■ Run a digital control network through grounded conduit to add mechanical protection and electrical noise attenuation. For best attenuation with copper cable, use shielded Category 5 twisted-pair cable. For maximum protection in high EMC noise areas, use fiber-optic cable. Avoid cable splices in conduit. If splices are necessary, ensure that the connector is completely insulated from the conduit.

■ Electrostatic coupling usually occurs when long signal cable runs are very close together. It is especially severe when wires that carry different types of signals and have different energy levels run closely together. Shield these

wires from one another with metal conduit, covered tray, metal trough, or shielded cable. An overlapping, multiple-folded, foil-shielded cable with a continuous drain wire in contact with the shield is recommended.

■ Individually connect carrier shield bars to isolated ground references.

■ Terminate unused conductors and shields to ground on one end only.

Note ...   *To avoid ground loops, ensure that all shield wires are connected to ground at one end only. It is recommended that measurements be made and documented to ensure single-ended grounding.*

■ Always ground field device cases. Field devices can be grounded to metal conduits, pipes, cable trays, and so forth, if these items are electrically grounded. (Refer to Figure 5-1 and Figure 5-2 for examples.) Use case-ground studs with star washers to cut through protective coatings.

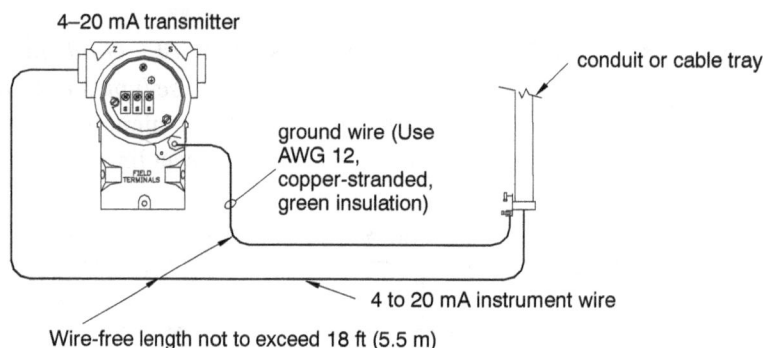

Figure 5-1   Instrument Ground Bond Connections for 4–20 mA Transmitters

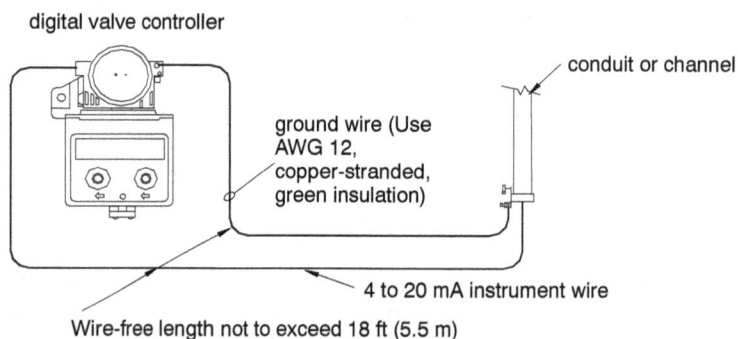

Figure 5-2   Instrument Ground Bond Connections for Digital Valve Controller

Protection from AC Power Line Noise

Inductive coupling interference can occur when signal wiring leads are routed close to power lines or other high current leads. It also can occur when other equipment is not sufficiently separated from power lines, transformers, rotating equipment, or other high power machinery.

Long runs of adjacent cables have a greater potential for noise coupling than short ones. Therefore, the longer the adjacent run, the greater the amount of spacing and shielding required. Redundant shielding can be used to provide noise protection in situations where the recommended distances cannot be met. However, these situations must be evaluated on a case-by-case basis.

In some cases, different cable constructions can provide added shielding. For example, the cable recommended in Appendix A for variable speed drive (VSD) applications provides excellent shielding for power cables (verified by testing). Instrument cables can be provided with a metallic outer armor, but adds cost.

■ Table 5-1, based on IEEE 518-1982, *IEEE Guide for the Installation of Electrical Equipment to Minimize Noise Inputs to Controllers from External Sources*, provides the recommended minimum separation distances between twisted-pair signal cables and AC power lines.

Table 5-1. **Recommended Separation Distances**

Voltage	Current	Minimum Distance Between Signal Cable and AC Power Cable
0 to 125 volts	0 to 10 amps	12 inches (30 cm)
125 to 250 volts	1 to 50 amps	15 inches (38 cm)
250 to 440 volts	0 to 200 amps	18 inches (46 cm)
Note: Signal wiring and AC power cable separated by the recommended distances should not be run parallel for more than 20 ft (6 m). If longer parallel runs are needed, increase the separation distance by 12 in. (30 cm) for each additional 30 ft (10 m) in length.		

■ Signal cable should be separated from devices that generate magnetic fields by a minimum of 5 ft (1.5 m). Ten to fifteen ft (3 to 4.5 m) of separation is recommended. Shielded, twisted-pair cable with approximately eight crossovers per ft (26 crossovers per meter) is five to six times more effective in reducing noise coupling than shielded cable alone.

■ Separate cable trays should be used for signal cables and power cables. Where separate cable trays cannot be used, cable tray dividers must be installed. Dividers must be solid, fixed barriers, and minimum separation distances must be maintained. Signal wiring needs to be completely shielded when AC fields are present.

■ Assuming the installation does not use intrinsic safe barriers, low-level signal lines should be continuous from signal source to receiver. The signal lines should not run parallel to high-current or high-voltage lines.

■ If a magnetic field from AC power leads are a source of interference, the leads can be twisted to reduce field strength. When signal leads and power leads must cross, they should do so at right angles. Both signal wiring and power wiring should be twisted on both sides of the crossing for, at least, the recommended distances given in Table 5-1.

■ Metal conduit is mainly used for physical support, and can add some magnetic field protection if the conduit is not carrying significant ground currents. Metallic conduit should not be buried beneath high-voltage power transmission lines or in known ground currents because it can become a conductor of induced voltages.

■ Generally, aluminum conduit, cable trays, and enclosures provide better shielding at higher frequencies than steel. Steel provides better magnetic shielding at AC power frequencies.

■ If signal cable is individually shielded (and grounded at one end per cable-shield guidelines), armored cable grounded at both ends may be used in place of conduit.

Direct Cable Runs

Figure 5-3 illustrates typical wiring and ground connections for directly connecting signal wiring between field devices and an I/O carrier.

Multi-Pair Cable Runs In Junction Boxes

Figure 5-4 illustrates typical wiring and ground connections when junction boxes or marshalling panels are used. The figure shows both overall cable shields and metal conduit. Overall cable shields are used when multiple-pair cables are installed in cable trays. Conduit is used for support when cable trays or other physical support means are not available.

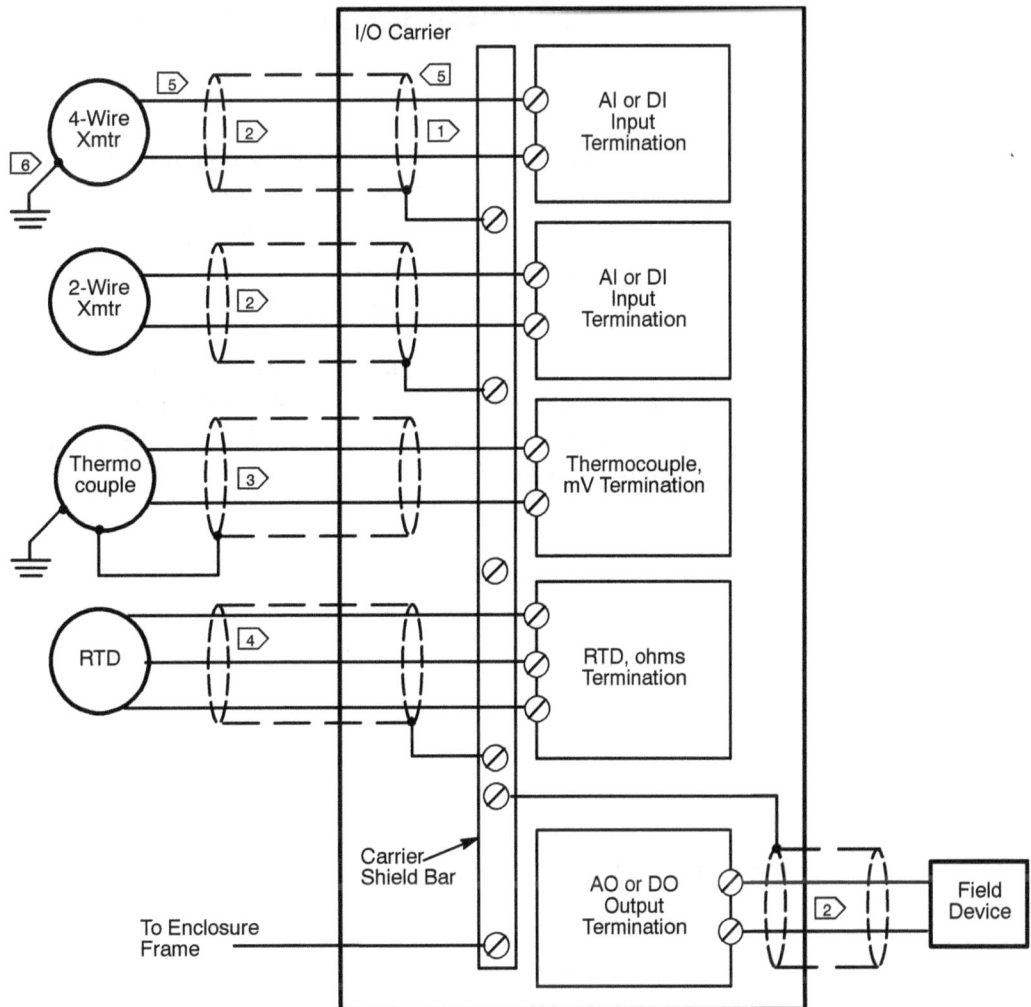

Notes:

1. Consult the manufacturer's installation manual for actual terminal connections.

2. Use 18 to 14 AWG (0.8 to 2.1 mm^2) stranded, twisted-pair with overall shield, drain wire, and outer PVC jacket.

3. Use cable specified for thermocouple signals with overall shield and drain wire and outer PVC jacket. Shield for grounded thermocouple shown. If ungrounded thermocouple is used, install shield in same manner as for a two-wire transmitter.

4. Use triple-wire cable specified for RTD signals with overall shield and drain wire and outer PVC jacket.

5. Shields should extend to within 1 inch of terminations.

6. Normally, the signal is isolated from the case in four-wire transmitters. Therefore, grounding the case does not cause a ground loop situation.

Figure 5-3   Typical Wiring and Ground Connections for Direct Cable Runs

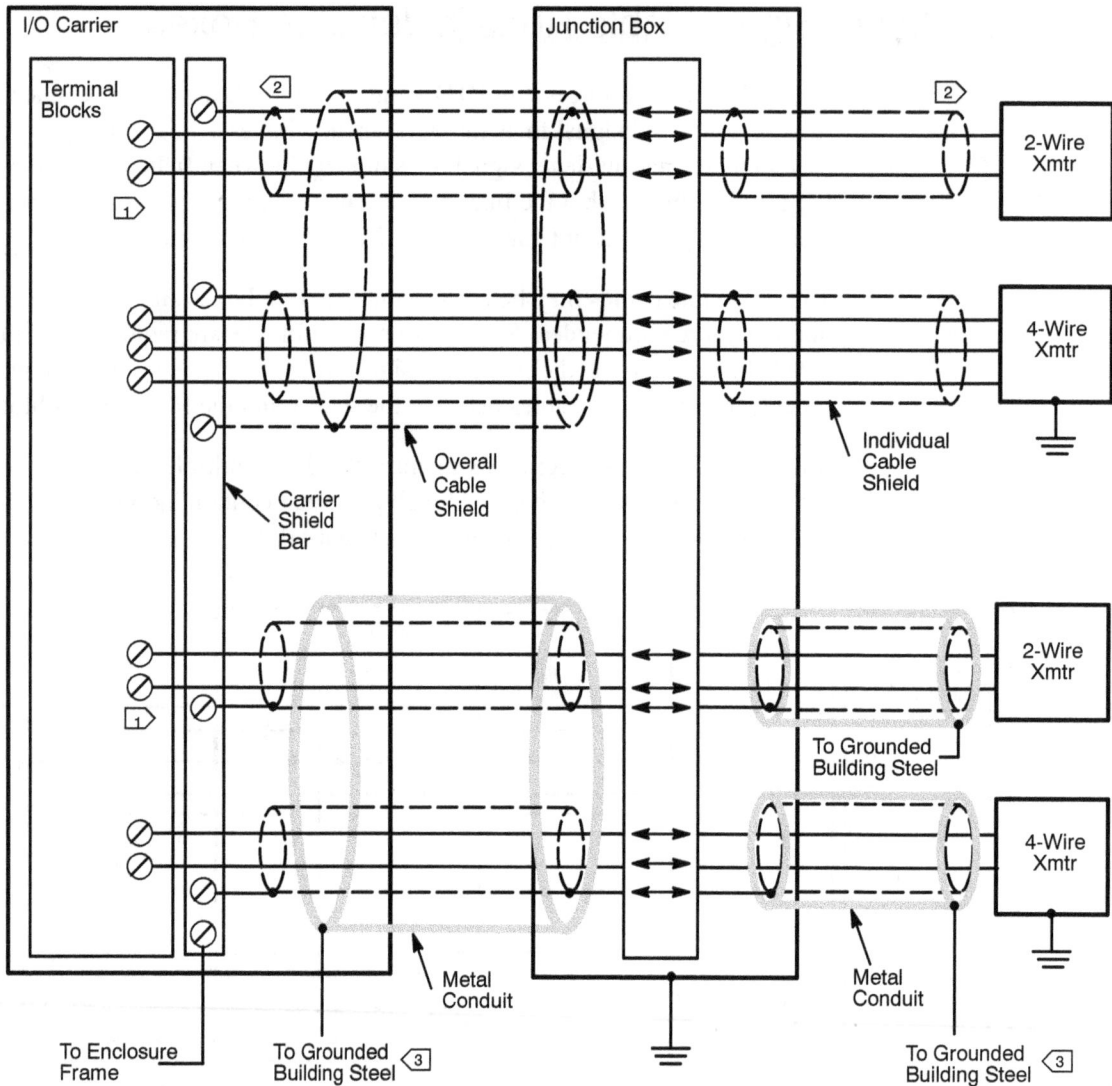

Notes:

1> Consult the manufacturer's installation manual for terminal connections.

2> Shields should extend to within 1 inch of terminations.

3> If the junction block is grounded, conduit can be grounded to the junction block instead of building steel.

Figure 5-4   Typical Wiring and Ground Connections for Multiple-Pair Cable Runs Using Junction Boxes or Marshalling Panels

Low-Level Signal Cable Runs in Junction Boxes

Figure 5-5, using a three-wire RTD as an example, shows typical connections for running low-level signal lines through junction boxes or marshalling panels. It is preferable to run low-level signal lines directly between field sensing devices and I/O terminal blocks, but the lines may be run through junction boxes and marshalling panels as long as the integrity of the shield is not broken.

Thermocouple cables may be run through junction boxes and marshalling panels, but it is not recommended. Special connection blocks are required in the boxes or panels to maintain the thermocouple signal. It is always better to run continuous (uncut) thermocouple cables between the sensor and the I/O terminal block.

Multiple-pair cables specifically manufactured for multiple runs of the same sensing device type may be used. Each pair must contain individual shielding. For adequate cable support, conduit may be used.

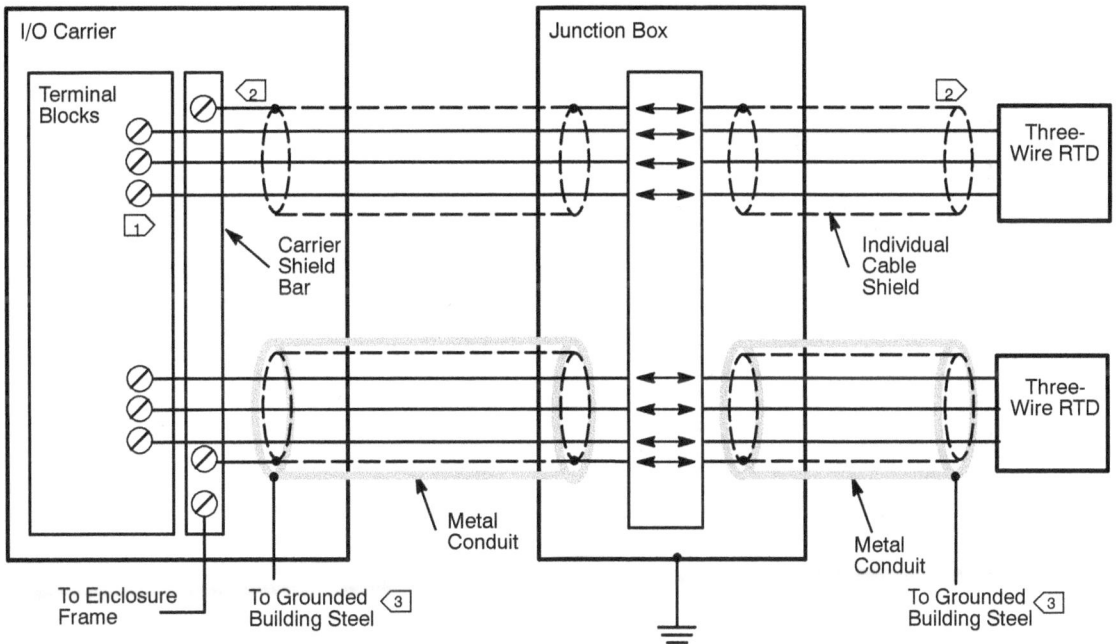

Notes:

1. Consult the manufacturer's installation manual for terminal connections.

2. Shields should extend to within 1 inch of terminations.

3. If the junction block is grounded, conduit can be grounded to the junction block instead of building steel.

Figure 5-5   Typical Wiring and Ground Connections for Low-Level Signal Cable Runs Using Junction Boxes and Marshalling Panels

Class 1 Division 2 Recommendations

The following recommendations are applicable to signal wiring in Class 1, Division 2 locations, including classic, HART, and fieldbus installations:

- If your site has Class 1, Division 2 locations that are based on NEC 500, power must be removed from the wiring to a transmitter prior to disconnecting or opening up the field instrument.

  Quick-disconnect type cable may be used to provide proper protection for personnel in the field. If it is used, you may have to remove power to the affected field instrument. To facilitate unpowering, you can provide an inline switch with the instrument so that, when the switch is open, you are ensured that no energy is present at the instrument.

- If unpowering the system is not feasible, you may use alternatives, such as intrinsically safe or non-incendive systems.

- It is recommended to use 600 VAC-rated cable in applications where the cable might be physically exposed, such as in cable trays. If cable trays are used, ensure that other signal cable types of lesser ratings are not physically exposed.

Refer to applicable NEC, CSA, or local government codes for additional information.

AS-Interface Considerations

The Actuator Sensor Interface (AS-Interface) is a digital, serial, bi-directional communications protocol and bus system that interconnects simple binary on/off devices, such as actuators, sensors, and discrete devices, in the field. The AS-Interface standard is defined by CENELEC standard EN 50295.

The two-conductor, AS-Interface cable supplies both power and data for field devices. An AS-Interface network can include branches. Refer to AS-Interface standard (EN 50295) for design and engineering details on AS-Interface cable.

For more information on the AS-Interface and installation of AS-Interface devices, refer to the AS-Interface web site:

http://www.as-interface.com.

Note ... *It is recommended that you do **not** connect AS-Interface devices directly to AS-Interface card terminals. Use one AS-Interface cable to connect the AS-Interface card to the power supply and use another AS-Interface cable to connect the devices to the power supply. If using extenders and repeaters, refer to the device data sheet for additional cabling recommendations.*

The AS-Interface system is a "floating system." Therefore, an AS-Interface system is never grounded. Only device cases should be grounded to pipes and building steel. These items must provide a good electrical ground to the plant ground grid. The power supply is grounded through the AC ground in the AC power distribution system.

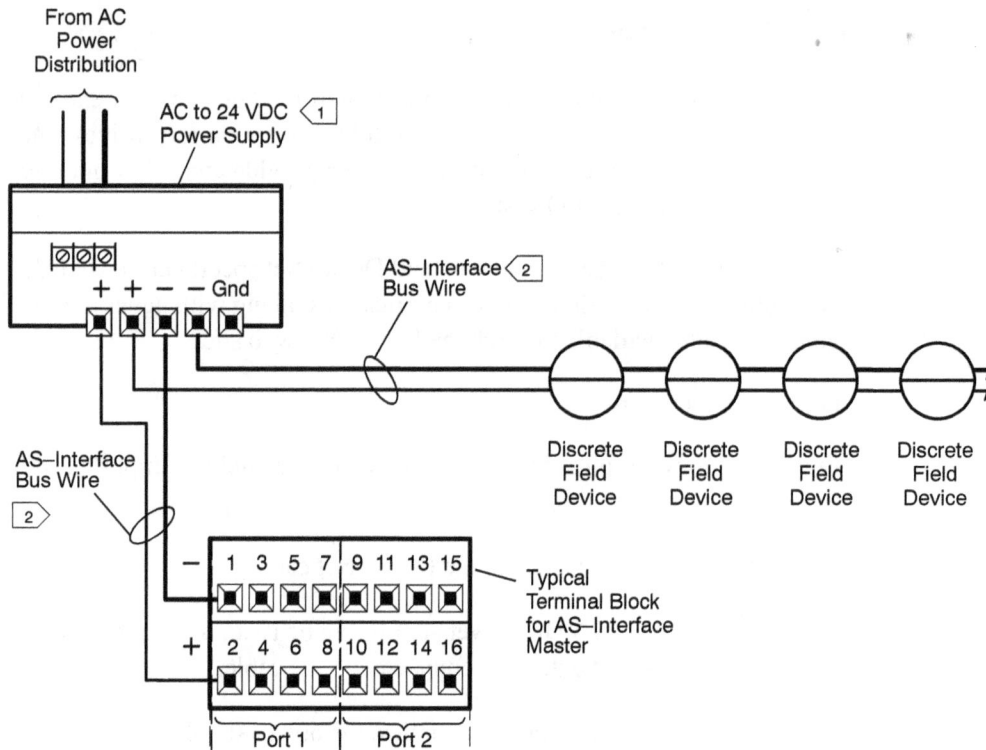

Notes:

1. AC power supply is shown. Plant 24 VDC can be used if the voltage and ground requirements meet AS–Interface Specifications.

2. Brown wire is + voltage; Blue wire is – voltage.

Figure 5-6   Typical Power and Ground Connections for an AS-Interface Network

DeviceNet Considerations

DeviceNet provides single-cable connection of field devices (for example, limit switches, photoelectric cells, valve manifolds, motor starters, drives, and operator displays) to a controller. Many manufacturers provide controller interface specially designed for I/O systems.

It is very important that you understand DeviceNet specifications before attempting an installation. These specifications, along with general user guidelines, are available through the DeviceNet web site:

http://www.odva.org

Figure 5-7 shows typical DeviceNet power and grounding. Important power and ground points are:

- 24 VDC power connects to the DeviceNet cable.

- More than one 24 VDC power supply can be used, depending on the load requirements of the devices connected to the cable.

- When using multiple power supplies, the power supply common voltage cannot vary more than 5 VDC between any two points on the network.

- The current should never exceed the cable and connector ratings.

- The voltage drop in the cable should not exceed 5 VDC between a power supply and its associated stations or nodes.

- To avoid ground loops, ensure that there is only one ground point for the entire network. That single ground point should be as near as possible to the physical center of the network.

- To maintain proper signal and ground isolation, DeviceNet specifications describe certain isolating methods to be designed into field devices. Only devices designed according to the specifications should be used.

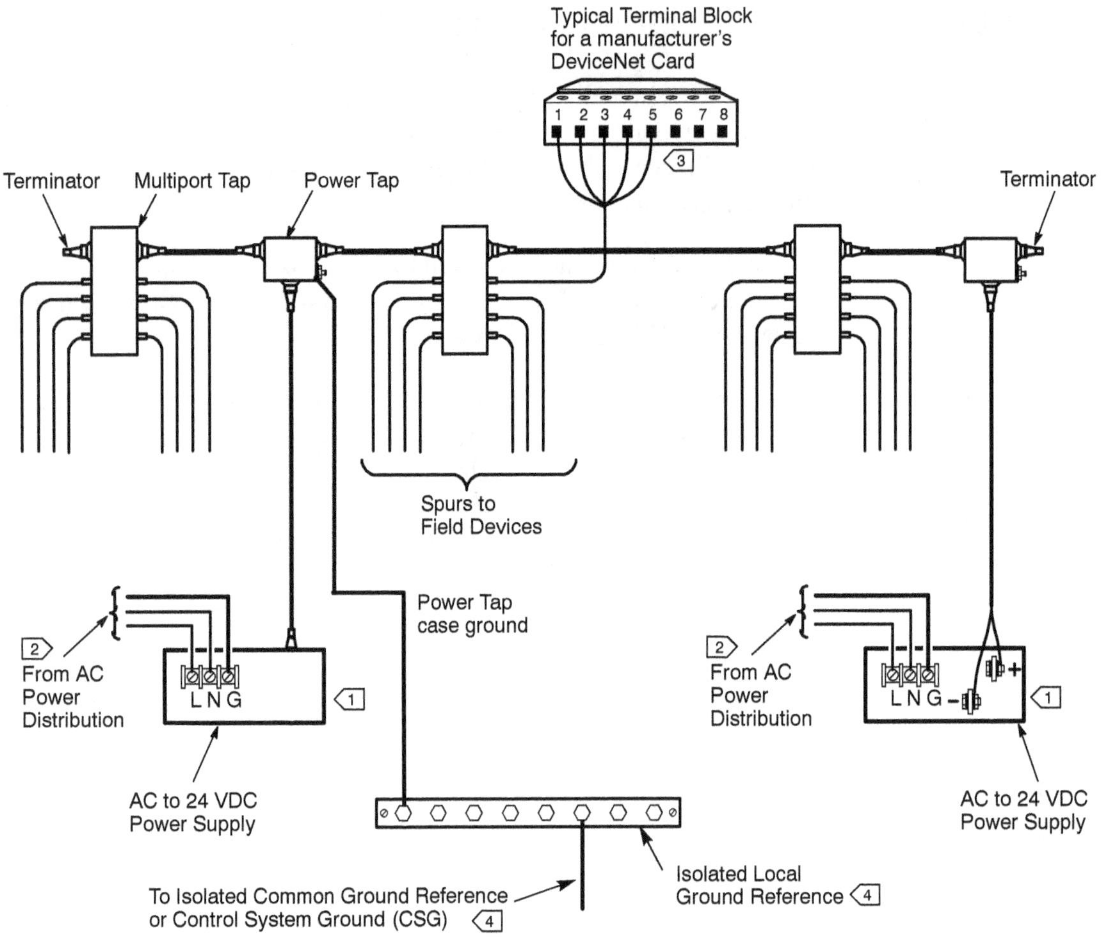

Typical Terminal Block
for a manufacturer's
DeviceNet Card

Terminator   Multiport Tap   Power Tap   Terminator

Spurs to
Field Devices

Power Tap
case ground

From AC
Power
Distribution

From AC
Power
Distribution

L N G   L N G

AC to 24 VDC
Power Supply   AC to 24 VDC
                                                            Power Supply

To Isolated Common Ground Reference
or Control System Ground (CSG)   Isolated Local
                                              Ground Reference

Notes:

1> DC return must be isolated from chassis.

2> AC power supplies are shown. Plant 24 VDC can be used if the
   voltage and ground requirements meet DeviceNet specifications.

3> Terminal Designations on the DeviceNet Card Terminal Block:
   Terminal 1 — –24 VDC (black insulation)
   Terminal 2 — CAN LO (blue insulation)
   Terminal 3 — Shield (drain) (clear insulation)
   Terminal 4 — CAN HI (white insulation)
   Terminal 5 — +24 VDC (red insulation)
   Terminal 6 — Not used
   Terminal 7 — Not used
   Terminal 8 — Shield (drain) (clear insulation)

4> Refer to Figure 4-1 and Figure 4-2 for ground connection details.

Figure 5-7   Typical Power and Ground System for DeviceNet

FOUNDATION Fieldbus Considerations

FOUNDATION Fieldbus H1 is a digital, serial, bi-directional communications bus that interconnects field devices, such as actuators, sensors, and discrete devices, with controllers. It includes both a communications protocol and hardware specifications.

Certain consideration must be given to FOUNDATION Fieldbus H1 installations to ensure proper operation. Figure 5-8 provides a typical wiring overview for the FOUNDATION Fieldbus H1 system.

Pay special attention to the power and ground aspects in these figures. For detailed installation information and checkout procedures, refer to the manufacturers appropriate FOUNDATION Fieldbus H1 documentation.

For more information on FOUNDATION Fieldbus specifications, refer to the FOUNDATION Fieldbus web site:

http://www.fieldbus.org/

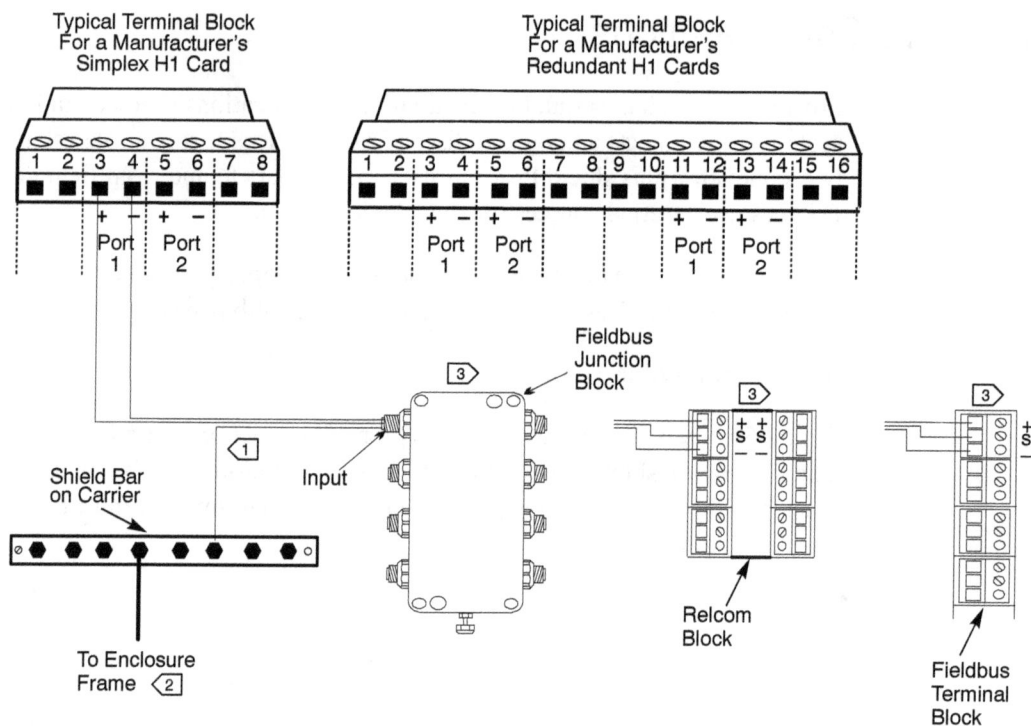

Notes:

1. Fieldbus cabling must be shielded. The shield is grounded only at the H1 Card end on the carrier shield bar.

2. Refer to Figure 4-1 and Figure 4-2 for ground connection details.

3. Various termination blocks are available for FOUNDATION Fieldbus systems. These blocks are examples of some of the types you may use.

Figure 5-8   Typical Ground Wiring for FOUNDATION Fieldbus

Profibus Considerations

Profibus is a digital, serial, bi-directional communications protocol and bus system that connects analog and discrete field devices to a control system. The two-conductor plus shield Profibus cable passes data to and from field devices. Devices use external sources for field device power.

Complete Profibus specifications are needed to properly install a Profibus system. The specifications can be found on the Profibus web site:

http://www.profibus.com

Figure 5-9 shows example power and ground connections for a Profibus system. The illustration shows one terminator. The second terminator is contained in the master. If the master is between cable end points, a terminator must be placed on each end point.

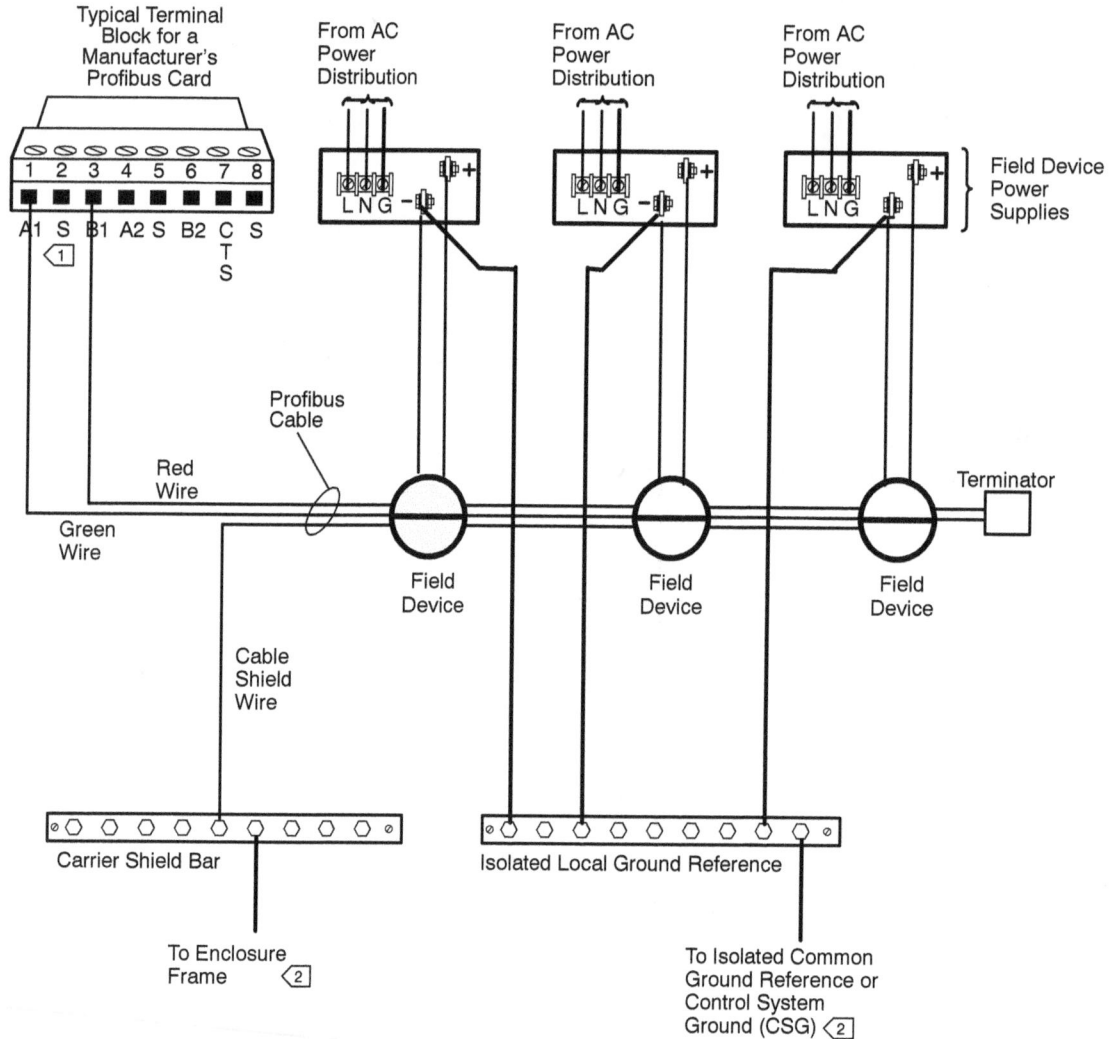

Figure 5-9   Typical Power and Ground Connections for Profibus System

Notes:

1. When the master is at one end of the Profibus cable, the green wire connects to A1 and the red wire connects to B1. If the master is between the cable end points, use the A2 and B2 terminals. Connect the cable shield wire to the carrier shield bar.

2. Refer to Figure 4-1 and Figure 4-2 for ground connection details.

Chapter 6　Alarm Wiring

Most controller and I/O power supplies include an alarm relay connector for output to alarm systems. The contacts may be either daisy chained so that any controller and I/O power supply in a group triggers an alarm or individually connected to an alarm system so that you know exactly which supply generated an alarm condition. Usually, contacts are normally-open when the supply is not powered (sometimes known as shelf condition).

Figure 6-1 shows a typical wiring arrangement for power supplies in a group. Figure 6-2 shows contacts connected from a single power supply to the input terminals for a typical discrete input card.

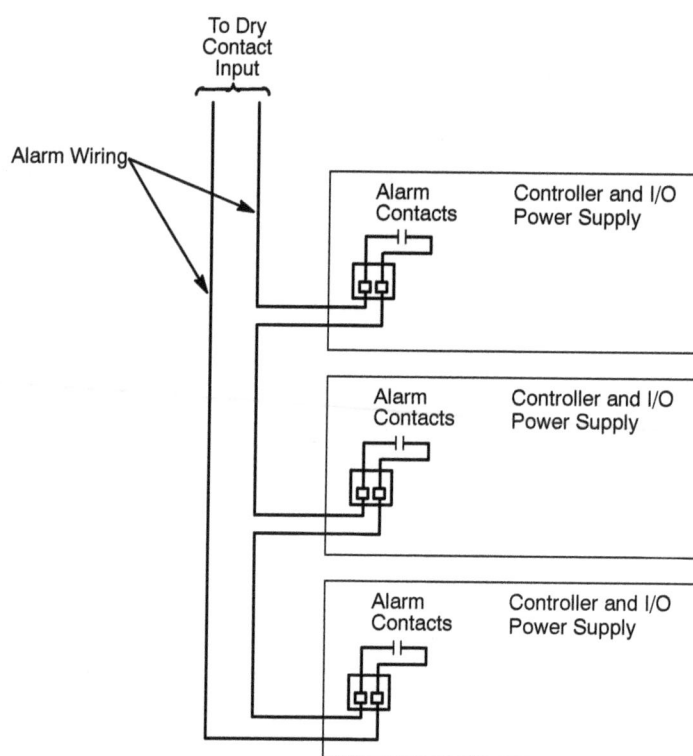

Figure 6-1　Example of Grouped Controller and I/O Power Supplies with One Alarm Output

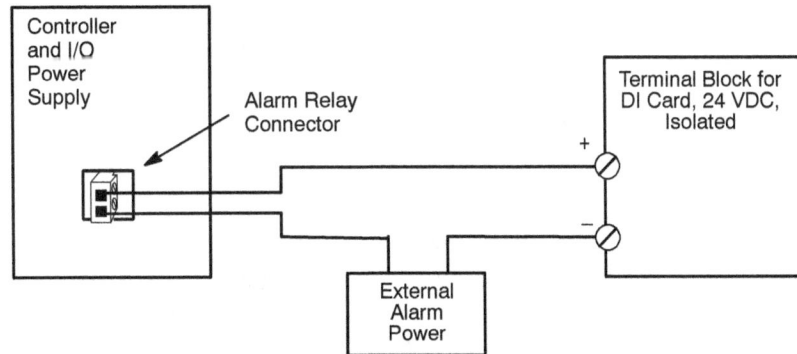

Figure 6-2   Example of Single Controller and I/O Power Supply Alarm through a DI Card

Chapter 7   Earth Ground

An adequate earth ground system is extremely important to both user safety and correct operation of a control system. An earth ground, in parallel with the equipment ground system, safely conducts stray or induced electrical currents to earth. In addition, it considerably reduces transient signals as well as signals induced by continuous electromagnetic fields, both of which can cause erroneous control signals in the control system.

The information in this section provides guidelines for constructing a good earth ground. Building steel should never be considered an adequate earth ground for a control system. However, control room grid systems, if they meet the requirements for good earth grounds, can be used. In all cases, construction of and connection to earth grounds must be in accordance with local, state, and federal codes.

Designing an Earth Ground

For digital switching circuits, several control system industry sources recommend a ground system that has a resistance of one ohm or less between the control system ground and true earth with a maximum resistance of three ohms. A resistance of one ohm or less minimizes the possibility of phantom errors caused by voltage drops in the ground system.

The control system ground must be at least as good as a ground associated with any other system. If a ground used with a radio communication system has a one-ohm resistance to true earth, the control system ground must have a one-ohm (or less) resistance to true earth. Both ground systems should be referenced to the plant grid.

For a plant grid, multiple ground rods provide the most effective ground system for the following reasons:

- The individual rod-to-earth contact resistances are effectively placed in parallel. Adding rods to the system reduces the ground-system-to-earth resistance.

- An element of safety is provided over the single rod system. All ground contact does not depend on a single rod.

The distance between rods in a multiple rod system must be a minimum of twice the immersion depths of the rods. For more information on installing and testing

of ground systems, refer to the publication *Getting Down to Earth* from Megger Group Limited (www.megger.com, formerly AVO International and Biddle Instruments).

Figure 7-1 shows an example of a plant grid system. If an existing plant grid is accessible and the ground-grid-to-true-earth resistance meets the requirements, the existing grid can be used for the control system ground.

A dedicated point close to the control system (preferably a ground rod location) is used for the system ground point. The ground rod is connected to one of the plant ground grid rods with AWG 4/0 stranded-copper wire. The ends of the wires are thermally welded to the rods. If either the existing grid is not accessible or the resistance is not within specifications, a new grid is required.

Note ...   *For platforms used at sea (for example, an oil platform or a gas platform) where process automation equipment is located, one of the legs of the platform is considered to be the connection to the earth ground. Even with floating platforms, the leg is still used. The sea is considered to be earth ground.*

Figure 7-2 through Figure 7-7 illustrate various examples of grounding that may be used.

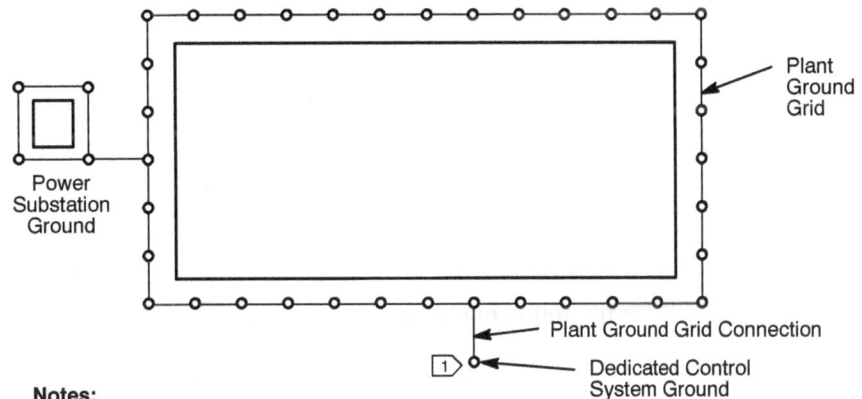

Notes:

o = Ground Rod

All connecting cables should be at least No. 4/0 AWG (107.2 mm^2) copper wire thermal welded to the rods.

1▷ Ground connection can be a single rod or one of the configurations shown in the following figures.

Figure 7-1   Example of Plant Ground Grid System

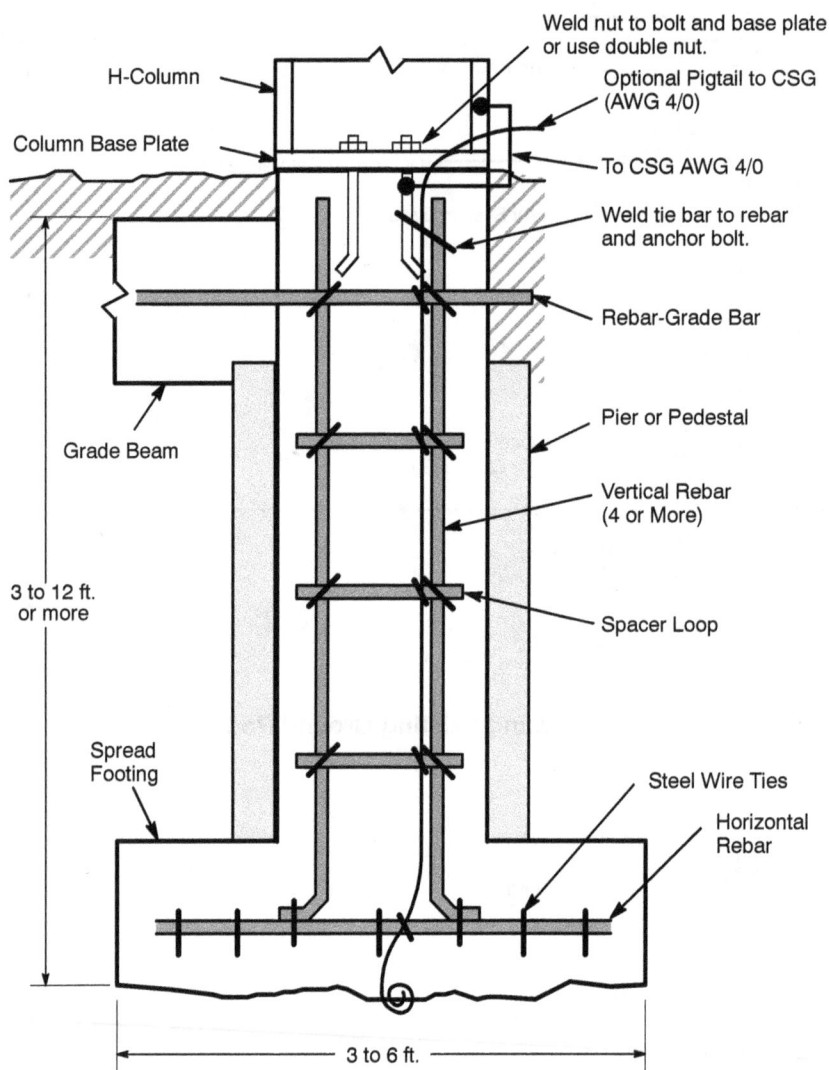

Note:

Ufer ground systems (developed by Mr. Herbert G. Ufer during World War II) should be applied to every column. Normally, they are found only at newer facilities. Beware that some rebar is Teflon coated. If it is, this type ground system will not provide a proper control system earth round.

Figure 7-2   Ufer Ground System Example

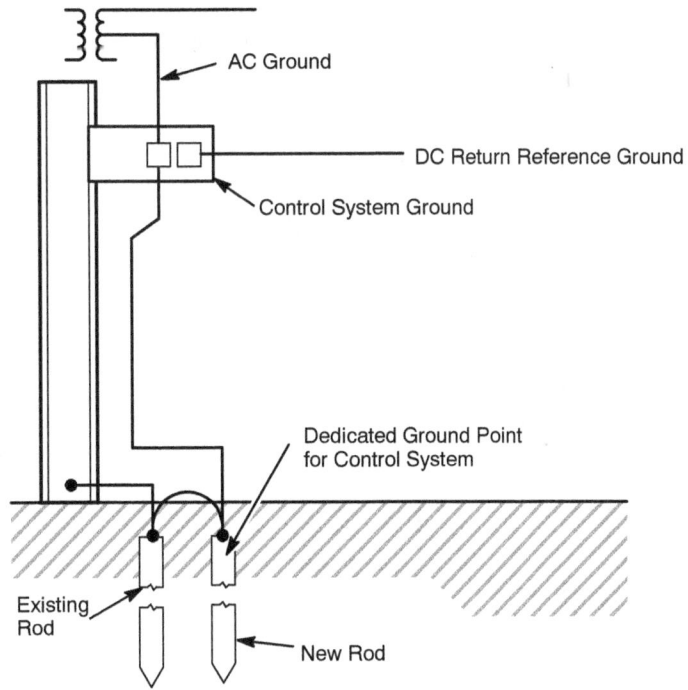

Figure 7-3   Example Using Existing Ground Rods

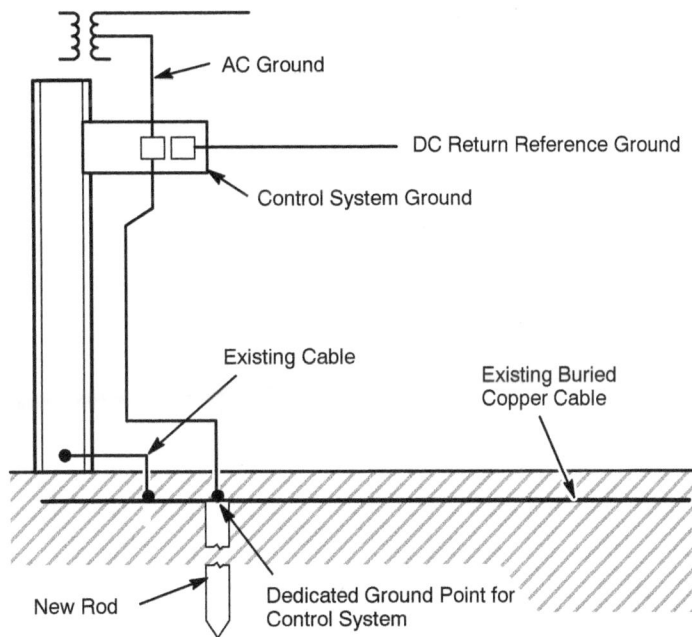

Figure 7-4   Example Using Existing Counterpoise System

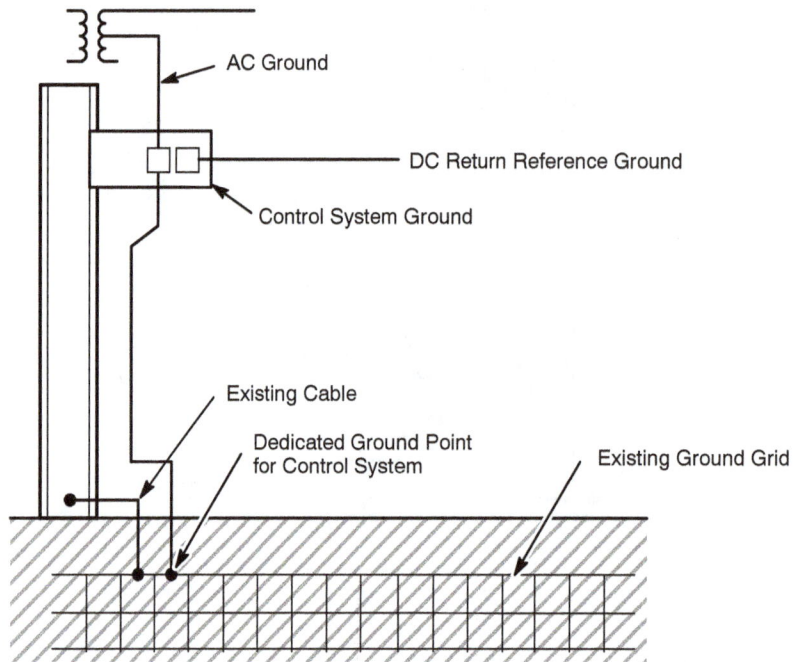

Figure 7-5   Example Using Existing Ground Grid System

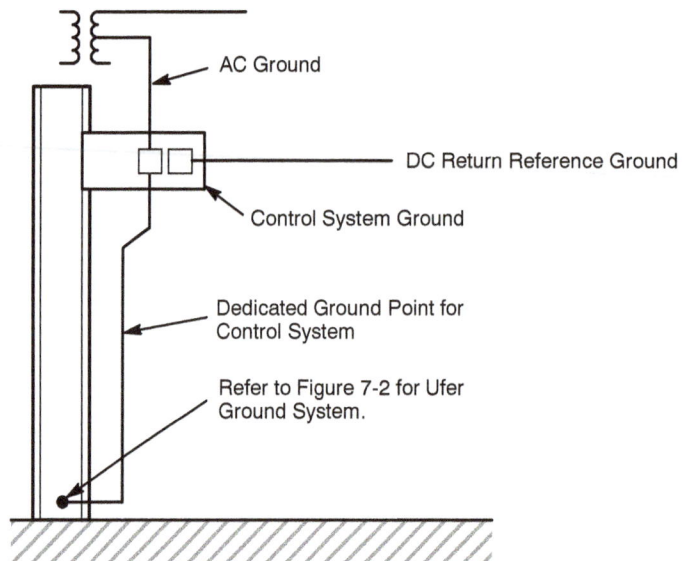

Figure 7-6   Example Using Ufer Ground Connections

AC Ground

DC Return Reference Ground

Control System Ground

Dedicated Ground Point
for Control System

Buried Steel Plate

Existing
Cable

Minimum
5' x 5'

Existing
Ground
Plate

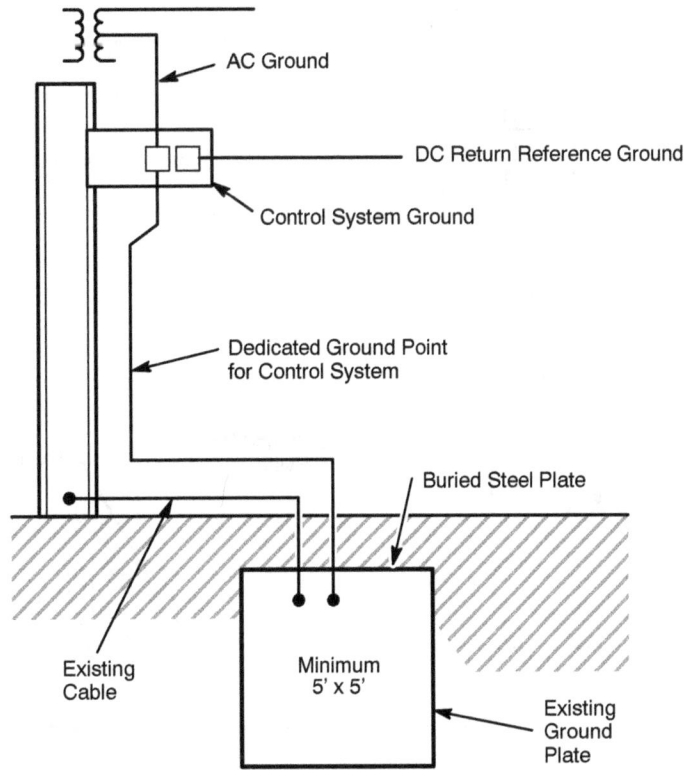

Figure 7-7   Example Using Existing Buried Steel Plate Ground

Testing an Earth Ground

The electrical resistance of an earth ground system must be tested to ensure that the ground meets the requirements for control systems. The total resistance of the ground system is comprised of the resistance of the metal used in the plant earth ground grid, the resistance of ground cable connection to the grid, the resistance of the contact between the grid and earth, and the resistance of the earth itself. For a control system, the recommended earth ground resistance is one ohm or less with a maximum resistance of three ohms.

There are several methods for testing earth ground resistance. The three-point (or Fall of Potential) method, the two-point (or Direct) method, and the Slope method are frequently used. Various manuals are available from tester manufacturers and consulting firms that detail earth ground resistance testing. Also, IEEE Std 81, *IEEE Guide for Measuring Earth Resistivity,Ground Impedance, and Earth Surface Potentials of a Ground System*, contains descriptions of these methods.

Several internet sites also contain helpful information about earth ground resistance measurements and methods in the forms of white papers and technical articles. It is recommended that you access such sites for further information. Some sites are:

■ www.megger.com

■ www.aec–us.com

■ www.hoodpd.com

An earth ground tester is used to determine the total resistance of an earth ground system. (Refer to Figure 7-8). The tester contains an electric current source and connections for probes.

Figure 7-8 shows connections for the three-point ground method. In this method, an electrical current is passed through earth between a current probe (C) and the ground grid being tested (E). At a point between these locations, a third probe (P) is inserted in the ground. Standardized tables are available that give recommended distances between C, E, and P locations to obtain accurate measurements.

A typical point for P is 62% of the distance between C and E, as measured from E. (Refer to Figure 7-8.) The distance between C and E depends on the depth of the ground rods. Table 7-1 lists typical distances.

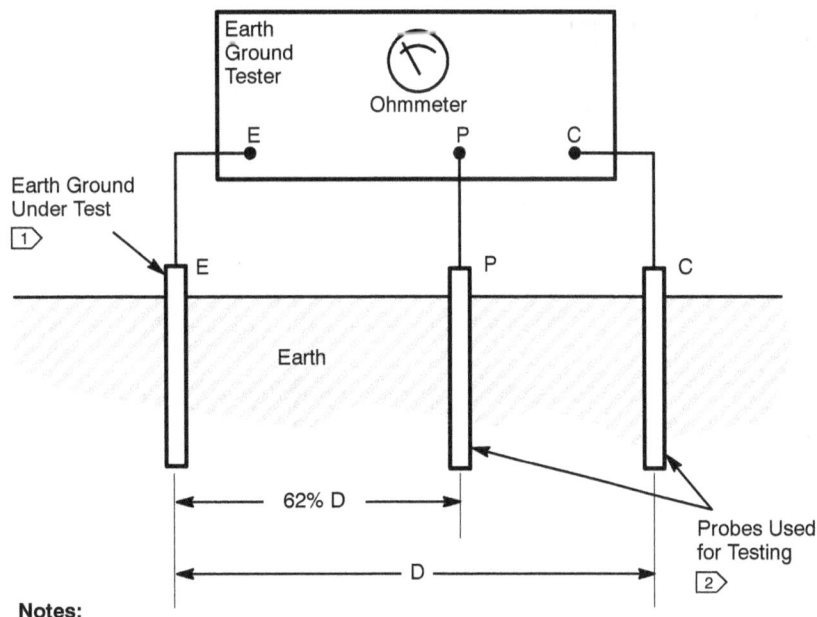

Notes:

1 Disconnect and ground cables from the ground system while testing is in progress.

2 Probes used for testing are placed in a straight line from the earth ground under test.

Figure 7-8   Three-Point Method Setup for Testing an Earth Ground System

Table 7-1.   Typical Distances for Three-Point Method

Depth of Ground Rods at E (ft)	Distance Between E and P (ft)	Distance Between E and C (ft)
6	45	72
8	50	80
10	55	88
12	60	96
18	71	115
20	74	120
30	86	140

Resistance value is obtained by measuring the current between C and E and the voltage between P and E. Many earth ground testers convert these measurements to resistance and show the resistance on an ohmmeter. Ohm's Law can be used to determine the resistance if an ammeter and voltmeter are used on the tester.

It is recommended that measurements be made in several directions from the ground system and the values entered into a table to determine possible high resistance areas and to use for future reference. Since earth ground resistance

varies with changes in ground moisture and temperature, it is also recommended that resistance measurements be made at different times during a year.

The three point method works well if you have enough room to lay out the probe locations. Room can be limited both by obstacles, such as buildings, highways, and railroads, and by other ground systems where the electrical currents in those systems could influence the measurements of the system under test. If measurements are attempted inside the perimeter of a ground grid system, such as that shown in Figure 7-1, the measurements can be influenced, and inaccurate readings can result.

In these cases, the two-point method can be used, as illustrated in Figure 7-9. It is not as accurate as the three-point method. In the two point method, terminals P and C are tied together and connected to a known adequate ground system, such as that of a power utility neutral ground. Then, earth ground resistance can be measured between the known system and the system under test. This setup reduces the distance necessary to make a measurement. If a voltmeter is being used in conjunction with an ammeter, the voltmeter is connected between P/C and E.

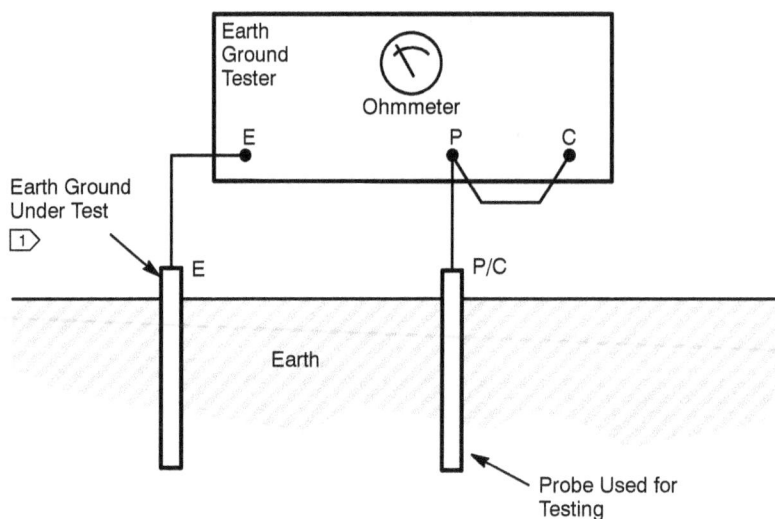

Note:

1️⃣ Disconnect and ground cables from the ground system while testing is on progress.

Figure 7-9   Two-Point Method Setup for Testing an Earth Ground System

The Slope method can also be used when room for inserting probes is limited. This method uses a setup similar to the two-point method, except that the probe P/C (in Figure 7-9) is moved to several locations in a straight line, and a

measurement is taken at each location. Locations are typically at 20%, 40%, and 60% of the distance from P/C. From these readings, a slope coefficient is calculated and the result compared to a standard table of values. From the table, the best location for P/C can be found. Then, an earth ground resistance measurement is made. Because this method can be subject to many extenuating influences, several measurements should be made in all directions from the earth ground system to ensure the accuracy of the results.

When earth ground systems must provide three ohms or less resistance, the resistance of the test leads can no longer be assumed to be negligible, especially since the leads can be many feet long. Therefore, lead resistance should be measured and subtracted from the earth ground reading.

Annual Inspection

As a preventive maintenance action, each connection on an earth ground system needs to be checked annually. This check ensures that connections are tight, ground wires are in good condition, and no contamination or corrosion exists that could otherwise compromise the ground system integrity.

Chapter 8 Lightning Protection

This chapter provides guidelines for protecting control systems from lightning damage. A direct strike can disrupt critical processes, start fires, damage buildings and equipment, and injure personnel. Near strikes can disrupt critical processes and damage electronic circuitry, by inducing voltage in unprotected wiring. Therefore, adequate lightning protection is essential in a production plant.

Reference Documents

The documents listed below provide further information for understanding lightning protection systems:

- Hart, William C. and Malone, Edgar W., Lightning and Lightning Protection, 1st ed., Don White Consultants, Inc., Gainesville, Virginia, 1979, ISBN No. 0932263143

- NFPA-78 Lightning Protection Code 1986, Quincy, Massachusetts: National Fire Protection Association, 1986.

Lightning Risk Determination

Factors that determine the level of protection are:

- Geographic location

- Process criticality

Lightning strikes occur more often in some geographic areas than in others. Elevation, humidity, geographical latitude, and normal weather patterns influence the frequency of lightning strikes in a particular area. Therefore, the typical lightning storm patterns in an area will influence the extent of protection needed.

Process criticality also dictates the sophistication of required lightning protection. The more critical a process, the more important lightning protection is, even though the system may be in an area of low lightning occurrence. If any strike or near strike can cause loss of control of a critical process, severe financial loss, major equipment damage, or danger to personnel, a complete lightning protection system is appropriate.

Lightning Protection Systems

Lightning protection systems provide safe conduction paths to earth ground to minimize equipment damage and personal injury. A complete lightning protection system includes:

- Lightning rods

- Conductor system

- Ground system

- Lightning arrestors and surge protectors

Lightning rods (also referred to as air terminals) intercept lightning discharges above a building or facility. The conductor system is a safe discharge path from the lightning rods to the ground system. The ground system lets the lightning discharge or dissipate safely. Lightning arrestors and surge protectors protect power lines, network cables, instrumentation wiring, and other such equipment from induced voltages. Together, these elements minimize lightning discharge damage.

Although a lightning protection system intercepts, conducts, and dissipates the main electrical discharge, it does not prevent possible secondary effects, such as spark-over, in nearby large metal structures.

Note ... *To minimize secondary effects of lightning strikes, ensure all adjacent metal structures interconnect with and tie to the main conductor system. This construction design maintains the same electrical potential throughout structures in the vicinity.*

Lightning Rods

Lightning rods intercept a discharge above a structure and direct the discharge to a safe path. In particular, lightning rods minimize the possibility of fire. A large plant site area needs a complete system of properly located lightning rods.

Lightning rods should be located on the structures most likely to be struck. Chimneys, ventilators, towers, and other high parts of buildings should have lightning rods installed. Roof edges are the parts of flat-roofed building most likely to be struck.

Conductor System

Once intercepted, a lightning discharge follows a low-impedance path to the earth (path of least resistance). Normally, the least resistant path is metal. A conductor system consists of one or more such metal paths. Each path must be continuous from the lightning rod to the ground. Paths must not have any sharp bends or loops. This ensures that the system provides the most direct path to earth for lightning discharge.

As illustrated in Figure 8-1, no bend in a lightning conductor should form an angle of less than 90 degrees, and no bend should have a radius less than 10X the diameter of the cable used for the path. A non-ferrous metal such as copper or aluminum is the preferred material.

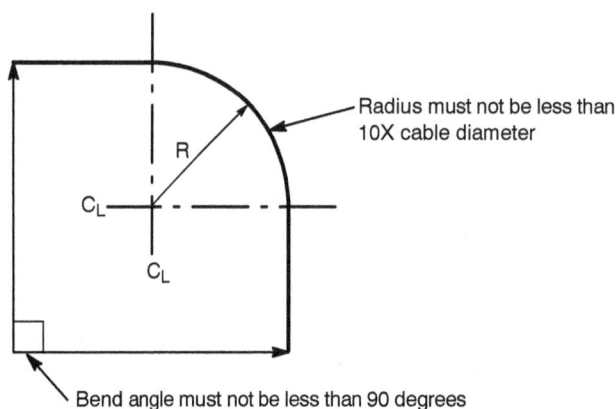

Figure 8-1   Minimum Conductor Bend Radius Requirements

The impedance of a conductor system is inversely proportional to the number of separate discharge paths. Therefore, increasing the number of paths decreases the impedance. In a multi-path conductor system, the paths (wires) should form a cage around the structure. The steel framework of a structure can substitute for separate conductors, but smooth connection straps must span any sharp bends or other hindrance. Figure 8-2 shows a typical protection system with a conductor, grounded steel framework, and connection straps.

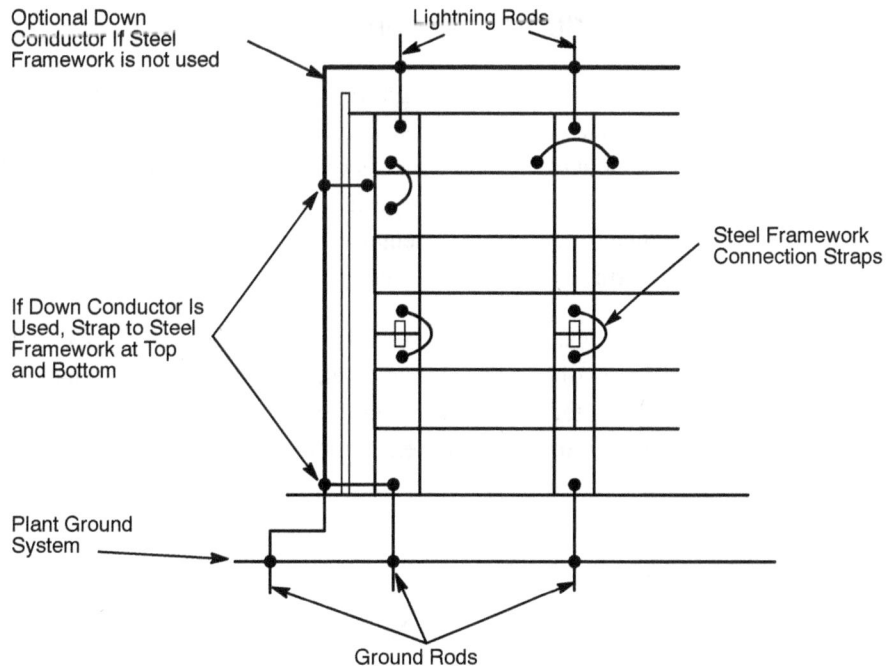

Figure 8-2   Typical Lightning Protection System

In metal structures, the conductor system can use the structure framing instead of separate conductor cables. In such cases, lightning rods should be electrically bonded to the top part of the framework, and ground terminals should be bonded to the bottom. Structures with electrically continuous metal exteriors may not require separate lightning rod and conductor systems, if the metal is at least 3/16 inch (4.8 mm) thick. The metal exterior itself can intercept lightning and conduct it to ground.

Ground System

Proper grounds are essential for effective lightning protection. Each ground connection, and each branch of each ground connection, should extend below, and at least 2 feet (0.61 m) away from, a building's foundation walls. This construction minimizes wall damage in the event of a lightning strike.

Lightning Arrestors and Surge Protectors

Note ... *Lightning arrestors assist in the isolation of coaxial type highways, but strikes can jump those systems. For the fullest protection, fiber optic links are recommended.*

Lightning arrestors and surge protectors minimize current induced in the wiring of a control system. Induction can occur in two ways:

■ A lightning discharge passing through the conductor system generates a transient magnetic field, which induces current in nearby wiring

■ As the ground system dissipates a discharge in the earth, a step difference in potential develops in the earth itself. This difference induces current in underground control system wiring

Protection devices use three main types of circuits:

■ Varistors

■ Semiconductors (avalanche diodes)

■ Gas discharge tubes

Varistors and semiconductors provide protection from lower current levels. Gas discharge tubes protect the system from high current and voltages levels. Most protection devices use a combination of these circuits.

Implementing Lightning Protection

When planning a lightning protection system, carefully consider environmental conditions and plant requirements. Guidelines for implementing a lightning protection system include:

■ Ground the building and plant site to a single ground system. If there is a remote or separated building at the plant site, isolate the signals, power, and communications systems of the remote building;

■ Use overhead cables when possible. The air around overhead cables acts as an insulator, making them less susceptible to lightning induced voltage than underground cables. Also, overhead cables are less susceptible to step-potential induction in the earth than underground cables. Ground all cables, including the conduits, pipe racking, and cable trays;

- Install surge protection on all metallic cables running outside of buildings by:

  - ☐ Installing a protection device for power wiring at either the substation entrance (preferred), or at each piece of powered equipment

  - ☐ Installing a protection device on each phone line entering the building

  - ☐ Installing a protection device on instrumentation wiring, if the structures are not all grounded to a plant-wide system, or the equipment is not designed to handle an induced voltage surge

- For Control Network cables, install a surge protector in each building at the cable entry site

- Install protected cables in conduit or cable trays

- When building a lightning protection system, use mechanically strong materials with physical properties that resist rust and corrosion

Inspection of Lightning Protection Systems

Building additions or structural repairs performed without consideration for a lightning protection system can reduce the system's effectiveness. Deterioration of or mechanical damage to the system may reduce its effectiveness.

To prevent loss of protection, evaluate all proposed structural changes for effects on the protection system, and ensure that no structural repairs inhibit system protection. Inspect the structure periodically, at least annually, for deterioration and mechanical damage. Thoroughly inspect and test the lightning protection system every five years.

Appendix A Variable Speed Drive Considerations

A significant amount of electrical interference is caused by AC Variable Speed Drives (VSD), sometimes called Variable Frequency Drives (VFD). Most VSDs use Pulse Width Modulation (PWM) to control motor speed. The operating characteristics of PWM can create electrical interference because of the fast rise and fall times of the signals used by the PWM control circuits.

This interference can affect the reliable operation of control systems, VSDs, and electric motor components. VSD and motor issues, well documented in technical papers, include failures of motor bearings and stator windings, high frequency electrical interference on control systems, and power cable failures. In addition, common mode electrical noise and reflective wave issues have been experienced with many brands of VSDs. According to VSD manufacturers, electrical interference related issues can be avoided by following the manufacturer's installation recommendations.

Recommendations

Actions to minimize interference from VSDs include:

- Minimize Cable Lengths between the VSD and Motor

  Cable length and type are critical variables in determining the amount of noise generated in a system. The fast rise time of voltage pulses generated by insulated gate bipolar transistor (IGBT) AC drives react with the transmission line. The longer the line, the greater the potential for reflected voltages. The drive manufacturer needs to specify maximum cable length for the given installation. Generally, the cable length should be no more than 200 ft.

- Use Lowest VSD Carrier Frequency

  Maximum permitted cable length varies with the switching (carrier) frequency used by the VSD. The lower the frequency, the greater the length of cable that can be used. Drive electronics operate at the lowest switching frequency available for the drive. The frequency of the electrical noise in the motor power wiring is higher than the VSD switching frequencies because of transmission line effect. The electrical noise can reach frequencies of 30 megahertz.

■ Use Armored Power Cable

The power cables used between the VSD and motor should have a metallic outer armor (copper or aluminum) to shield the system components from the high frequency electric fields. (The armor also provides a low impedance path back to the AC drive power supply.) Copper or aluminum should be used because steel does not provide effective shielding at higher frequencies.

Extensive testing of various types of cables produced the following results (Table A-1):

Table A-1.   Cable Shield Properties

Cable Type	Per Unit Cross Talk	Per Unit Common Mode Current
Tray Cable — No shield or Armor (3 grounds)	56 db	11.9 db
MC Type Cable — Aluminum Armor (3 Grounds)	5.0 db	1.0 db
MC Type Cable — Steel Armor (3 Grounds)	44 db	3.9 db

MC Type cable with a continuous corrugated aluminum armor and three symmetrically placed ground wires offers the best shielding for motors that are connected to variable speed drives. (See subsection *VSD Cable Specifications*). Aluminum conduit is not a recommended solution because the threaded couplings corrode and jeopardize the electrical continuity of the shield.

TC cable is available with an overall foil shield. Although the foil shield does not provide the same protection as armored cable, it may offer adequate protection. In some installations, the foil shield has provided the necessary protection and significantly lowered the voltage levels of the electrical interference created by the VSD.

■ Use Isolation Transformers for VSD Power

An isolation transformer (separate from the one dedicated to control system power) provides isolated power and a dedicated ground system for the VSD. This ground system creates a path to eliminate unwanted signals. Without the ground system, unwanted signals are routed through the main power system grounds, causing the signals to become common mode noise. The isolation transformer should include electrostatic shields between the windings.

■ Use Other Noise Suppression Components

AC drive installations should use components that are designed and tested to meet electromagnetic compatibility (EMC) directives and European Union (CE Mark) approval. Input power filters, output power filters, and common mode chokes often help suppress electrical noise in VSD applications. These noise suppression devices may be required if electrical noise levels prevent normal control system operation.

■ Use Control-Matched Inverter-Rated Motors

Inverter-rated motors should be used in VSD applications. Inverter-rated motors are designed to withstand the added stresses encountered when controlled by VSDs. Motor and VSD compatibility should be added to the criteria when selecting motors for VSD applications. Motor failures can be reduced if motors are compatible with both the application and the VSD.

Electrical Noise Examples

Figure A-1 and Figure A-2 show oscilloscope traces of common mode noise on traditional analog and HART signals caused by a low power, variable speed drive. The noise is seen as the large, high frequency spikes superimposed on what would otherwise be fairly smooth trace lines.

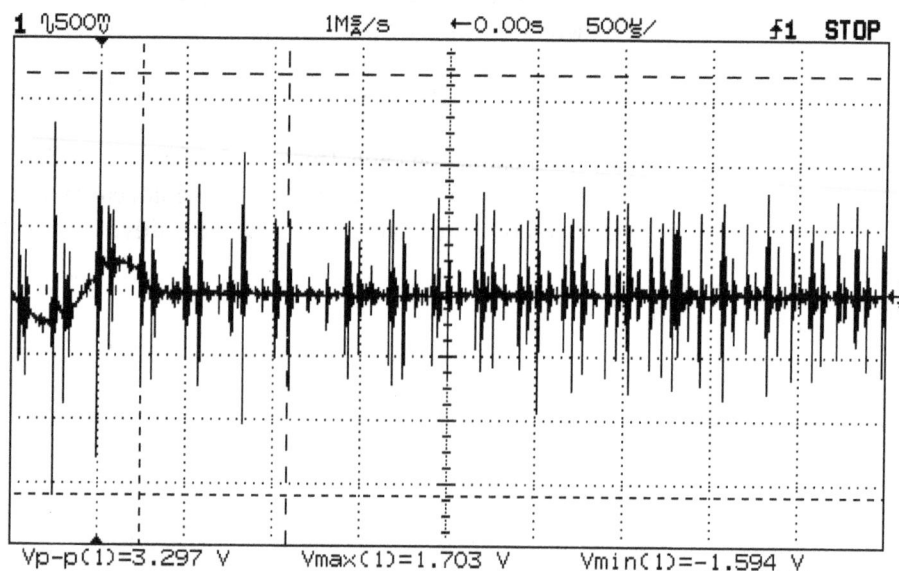

Figure A-1   Common Mode Noise on an Analog Signal (Display A)

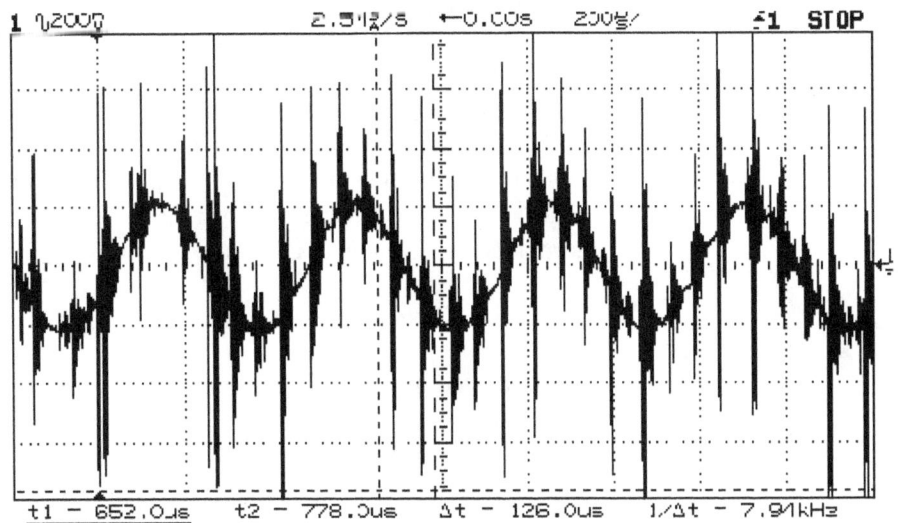

Figure A-2   Common Mode Noise on an Analog Signal (Display B)

Ground Considerations

By their nature, VSD drive systems using IGBT inverters produce high electrical noise levels. The noise is high frequency signals (up to 30 mHz) that are generated by the fast rise times (square-wave type signals) in the systems. The high frequency signals are capacitively coupled from the power cable and the motor to surrounding metal and are magnetically coupled from ground conductors into nearby field instrumentation and its wiring.

Figure A-3 illustrates a typical voltage output of the inverter with superimposed noise. The level of the noise along the power cable and at the motor terminals varies, depending on cable length and other factors. The waveform shown in the figure may vary from manufacturer to manufacturer, but the fast rise times and attendant noise are characteristic of IGBT drives.

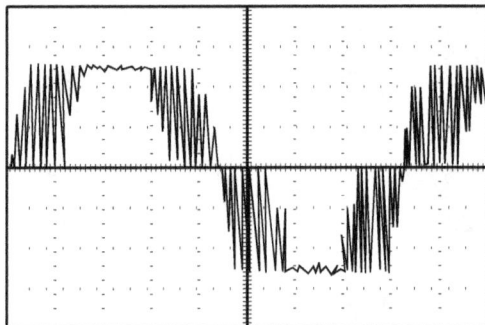

**Figure A-3   Voltage Output Inverter Signal
with Superimposed Noise**

If not adequately controlled, the high frequency noise can cause motor stator winding failure from arcing, motor bearing failure from metal transfer, and impeded control system operation from direct coupled noise and stray ground currents. To minimize the effects of the noise, the following is essential:

■ Use armor power cable connected at *both* ends. Usually, shielding is connected at one end only, but for VSD drives, all internal cable shields and the armor shield must be grounded at both ends.

■ Use the shortest power cable length possible.

■ Use motors with shielded stator winding.

■ Keep VSD ground cables on or as near as possible to building steel to ground out high frequency signal radiation.

■ Be sure that the building ground system provides a low impedance return path to true earth ground (normally, your plant ground grid).

Figure A-4 illustrates power and ground cabling for VSD systems which, if properly installed, normally reduces electrical noise to low enough levels that motor damage is minimized and control system operation is not impeded.

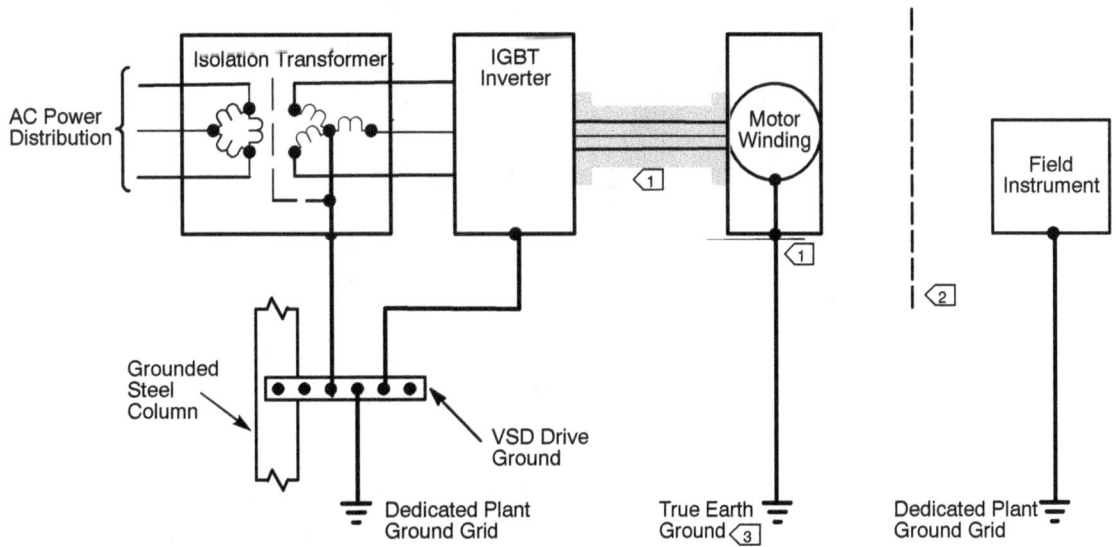

Figure A-4   Ground Wiring for Adequate Noise Protection with IGBT Inverters

Notes:

1> See Figure A-5 for ground connection details.

2> Keep field instrumentation and VSD equipment separated as far as possible and use armored power cable, shielded motor stator windings, and proper grounding to minimize affects of stray high frequence signals.

3> Motor frame may be connected to building steel if the steel provides a good, low-impedance path to true earth ground (dedicated plant ground grid in most cases).

4> Waveform is typical of high frequency stray signals possible on VSD ground wiring and radiated like radio signals.

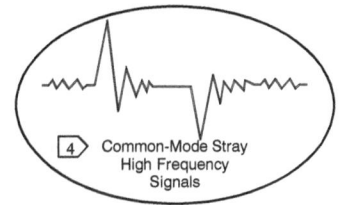

Motor Ground Connections

Figure A-5 shows details for ground connections of the motor power cable. As shown, the motor frame is grounded to true earth ground. For new buildings, the frame can be is grounded to a UFER ground or to effectively grounded building steel. For existing buildings, the frame can be grounded to the nearest available effectively grounded building steel or grid. Effectively grounded means that building steel provides a good low impedance path to true earth ground. Figure A-6 shows the drop out path from a cable tray to the motor.

Inverter End of Cable

**Motor End of Cable
(Conduit/Terminal Box)**

Bonding
GND Bushing

Phase
Conductors

Phase
Ground
Conductors

Main Drive
Ground Bus

Cable Armor
Ground

Bonding
GND Bushing

Phase
Conductors

Stator
Window

Phase
Ground
Conductors

Ground
Terminal

Cable Armor
Ground

Motor
Frame

Note:

1. Cable grounds are grounded at both ends as shown.

2. Motor frame may be connected to building steel if the steel provides a good, low-impedance path to true earth ground (dedicated plant ground grid in most cases).

To True
Earth
Ground 2

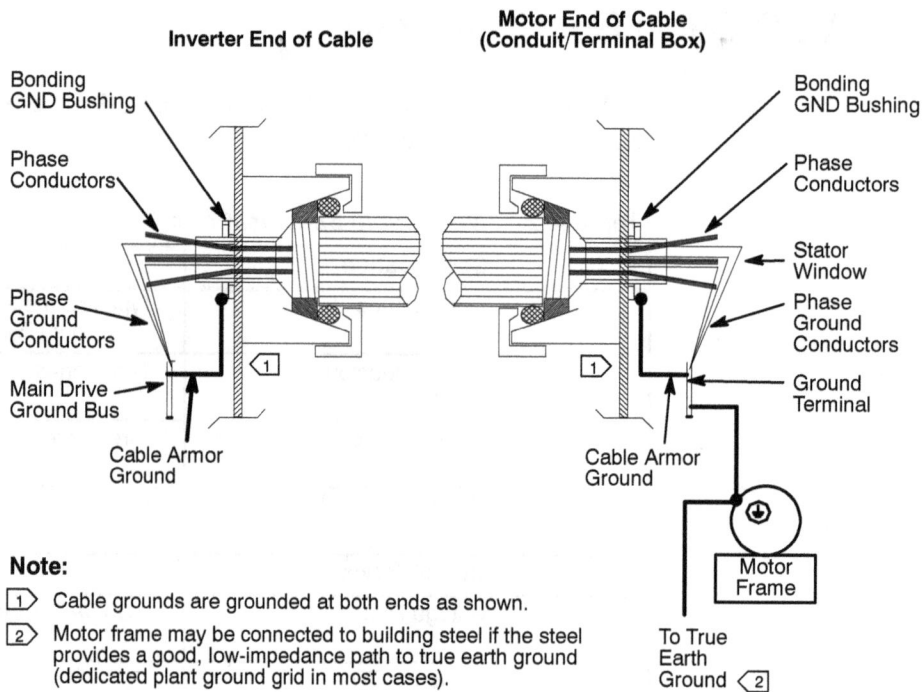

Figure A-5   Ground Connections at Motor Power Cable Ends

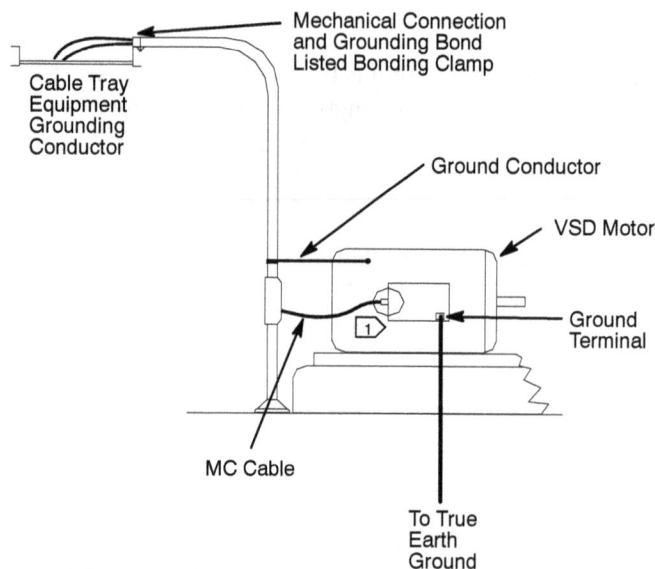

Mechanical Connection
and Grounding Bond
Listed Bonding Clamp

Cable Tray
Equipment
Grounding
Conductor

Ground Conductor

VSD Motor

Ground
Terminal

MC Cable

To True
Earth
Ground

Notes:

1. See Figure A-5 for ground connection details.

Figure A-6   Dropouts from Cable Tray to VSD Motor

VSD Cable Specifications

Table A-2 lists specifications for recommended VSD cables. Several cable manufacturers produce VSD cable.

Table A-2. **VSD Cable Specifications**

Specification	Description
Cable Type	Type MC metal-Clad three-phase cable per NEC 334-1, UL listed
Phase Conductors	Three phase-conductors with ampacity per NEC
Ground Conductors	Three bare ground conductors per NEC article 250
Metallic Sheath [1]	Continuous corrugated aluminum (welded or seamless)
Overall Jacket	PVC
Voltage Rating	$\leq$ 480 VAC, Cable insulation rating of 600 VAC
Connector Type	See Figure A-5
1. Alternate metallic shields may be copper tape spiral shield with galvanized steel interlocked armor or no shield with aluminum continuous armor.	

Critical Cable Lengths

The critical cable length is based on the pulse rise and fall time. The critical length is the minimum length that does not produce reflected voltage waves at the motor terminals. Table A-3, Table A-4, and Table A-5 show various factors for determining maximum cable length.

Table A-3. **Critical Cable Lengths by Pulse Rise Time**

Pulse Rise Time	Maximum Cable length
50 nsec	13 feet
100 nsec	25 feet
200 nsec	50 feet
400 nsec	100 feet
600 nsec	148 feet
1 μsec	246 feet
2 μsec	592 feet
4 μsec	984 feet
Note: Cable lengths greater than 100 feet are not recommended because of the potential for causing unwanted signals.	

Table A-4. *Critical Cable Lengths by Switching Type*

Switching Device	Maximum Cable length
IGBT 0.1 µsec	100 feet
Bipolar	250 feet
GTO 1.0 µsec	600 feet
SCR 4.0 µsec	600 feet

Table A-5. *Critical Cable Lengths by Switching Frequency for an IGBT Drive*

Switching Frequency	Maximum Cable length
1 kHz	200 feet
3 kHz	175 feet
12 kHz	100 feet
Note: It is recommended to use the lowest possible frequency to minimize the potential for unwanted signals.	

The above tables illustrate the importance of cable length, switching device, and switching frequency in minimizing reflected voltage waves at motor terminals. If the cable is long enough, the voltage at the terminals can be twice the inverter output voltage.

Appendix B Testing with an Oscilloscope

An oscilloscope can be used to locate electrical interference on control signals and ground systems. A 500 mHz oscilloscope with digital storage capability is preferred. There are several brands available, and most use PC interface software to perform trace data captures.

Warning... *Improper use of an oscilloscope can lead to electrical shock, and it may affect the normal operation of the manufacturing process. Be extremely careful when connecting to high voltage power cables and process signals. Read and follow the directions provided by the oscilloscope manufacturer prior to its use.*

Note ... *If you are using an AC-powered oscilloscope, it must use a floating input signal ground to obtain correct readings. However, as noted above, a floating input signal ground may be hazardous.*

General Monitoring

The presence of electrical noise may mean that system operation is jeopardized. Monitor the following signals to determine the presence of electrical interference:

- **Analog Input and Output Signals**: Electrical noise on analog-only signals usually does not cause faults because the unwanted electrical noise is filtered out.

- **Digital Communications**: Electrical noise can cause faults in digital communication signals by adding spikes and other noise that the communication system interprets as part of the digital signal.

- **Power Supply Ground to +24 VDC**: Electrical noise will be present. However, the noise spikes should be less than one volt peak to peak. If the noise spikes are greater than 1-volt peak to peak, further investigation is required. The noise source should be located and remedial actions taken.

- **Power Supply Ground to Cabinet Ground**: Electrical noise will be present. However, the noise spikes should be less than one volt peak to peak. If the noise spikes are greater than one volt peak to peak, further investigation will be required. The noise source should be located and remedial actions taken.

Fieldbus Segment Monitoring

When a Fluke 123 oscilloscope is used to monitor signals on a fieldbus segment, electrical noise caused by the oscilloscope connections (not by noise on the segment) may show on the trace. To minimize noise caused by the oscilloscope connections, use the ground-clip lead that is part of the probe. Do not use the common ground lead connection, which can be susceptible to electromagnetic noise in the plant and, thus, introduce electrical noise on the trace.

Appendix C Wire Size and Color Codes

Table C-1 provides the recommended size and color coding for power and ground wiring used in control systems with classic I/O and HART I/O subsystems. In most cases, you must also meet applicable government wiring codes.

Wiring for fieldbus systems, such as AS-Interface, DeviceNet, FOUNDATION Fieldbus, and Profibus, may use color codes other than those shown in the table. When available, fieldbus color codes should take precedence over those shown in the table.

Your site may use other color codes. In either case, it is vital that there be consistent color coding for each type of wire used. Also, all ground wiring should be identified and tagged. Inconsistent color application and untagged ground wiring can make subsequent troubleshooting very difficult.

Table C-1.   Wiring Specifications

Use	Size	North American Insulation Color	European Insulation Color
AC Line	Per Code for distance	Black	Brown
AC Neutral	Per Code for distance	White	Blue
AC Ground	Per Code for distance	Green	Green with Yellow Stripe
+12 VDC	8 AWG (8.3 mm^2)	Dark Blue	Dark Blue
+12 VDC Redundant	With Bus Bar: 8 AWG (8.3 mm^2) between supply and bus bar, and 14 to 12 AWG (2.1 to 3.3 mm^2) between bus bar and devices.	Dark Blue with Yellow Stripe	Dark Blue with Yellow Stripe
−12 VDC		Black with Yellow Stripe	Black with Yellow Stripe
+24 VDC Primary		Red	Red
+24 VDC Redundant	Without Bus Bar: 14 to 12 AWG (2.1 to 3.3 mm^2)	Red with Yellow Stripe	Red with Yellow Stripe
−24 VDC		Black	Black
DC Ground to Local Ground Reference	8 AWG (8.3 mm^2)	Green [1]	Green [1]
Cable Shield Ground	12 AWG (3.3 mm^2)	Light Green	Light Green
Ground to DIG	1/0 to 4/0 AWG (50 to 120 mm^2) [2]	Green	Green

1. Green always means that the wire is a ground reference with no current flowing through it under normal operating conditions.

2. The wire size is chosen depending on the distance to the DeltaV Instrument Ground. See Table C-2 below.

Table C-2 lists the wire sizes to use for wiring between an isolated local ground reference and an isolated common ground reference as well as between an isolated common reference point and the control system ground (CSG), depending on distance. Use multiple-strand wire – not solid wire.

Table C-2. Ground Wire Sizing Chart

Wire Length	Wire Size
Up to 25 feet	1/0 (50 mm^2)
Up to 50 feet	2/0 (70 mm^2)
Up to 200 feet	4/0 (120 mm^2)

Index

Notes

www.ingramcontent.com/pod-product-compliance
Lightning Source LLC
Chambersburg PA
CBHW081548220326
41598CB00036B/6600

www.ingramcontent.com/pod-product-compliance
Lightning Source LLC
Chambersburg PA
CBHW041706210326
41598CB00007B/553